SHAPING AMERICAN TELECOMMUNICATIONS

A History of Technology, Policy, and Economics

TELECOMMUNICATIONS
**A Series of Volumes Edited
by Christopher H. Sterling**

Selected titles include:

Borden/Harvey • The Electronic Grapevine: Rumor, Reputation, and Reporting in the New On-Line Environment

Cherry/Wildman/Hammond • Making Universal Service Policy: Enhancing the Process Through Multidisciplinary Evaluation

Gattiker • The Internet as a Diverse Community: Cultural, Organizational, and Political Issues

Gillett/Vogelsang • Competition, Regulation, and Convergence: Current Trends in Telecommunication Policy Research

Hudson • From Rural Village to Global Village: Telecommunications for Development in the Information Age

Pelton/Oslund/Marshall • Communications Satellites: Global Change Agents

Sterling/Bernt/Weiss • Shaping American Telecommunications: A History of Technology, Policy, and Economics

SHAPING AMERICAN TELECOMMUNICATIONS

A History of Technology, Policy, and Economics

Christopher H. Sterling
George Washington University

Phyllis W. Bernt
Ohio University

Martin B. H. Weiss
University of Pittsburgh

LEA Lawrence Erlbaum Associates
Taylor & Francis Group

New York   London

Lawrence Erlbaum Associates, Inc., Publishers
10 Industrial Avenue
Mahwah, New Jersey 07430
www.erlbaum.com

Cover image by Jen Sterling

Cover design by Tomai Maridou

Library of Congress Cataloging-in-Publication Data

Sterling, Christopher H., 1943–
    Shaping American telecommunications : a history of technology, policy, and
  economics / Christopher H. Sterling, Phyllis Bernt, Martin Weiss.
        p. cm. — (Telecommunications)
    Includes bibliographical references and index.
    ISBN 0-8058-2236-4 (cloth : alk. paper)
    ISBN 0-8058-2237-2 (pbk. : alk. paper)
    1. Telecommunication—United States. 2. Telecommunication policy—United States.
  I. Bernt, Phyllis. II. Weiss, Martin B. H. III. Title. IV. Telecommunications
  (Mahwah, N.J.).

HE7775S.S76 2006
384.0973—dc22   2005045616
                                                                      CIP

Books published by Lawrence Erlbaum Associates are printed on acid-free paper,
and their bindings are chosen for strength and durability.

Printed in the United States of America
10 9 8 7 6 5 4 3 2

For Ellen (as always, and for every reason),
Jennifer and Robin (our best products from four decades),
and Rachel (our first grandchild)
—Chris Sterling

For Joe, who brings me joy and laughter;
and for my colleagues at the former Lincoln Tel & Tel,
who faced challenging times with grace and teamwork
—Phyllis Bernt

To Karen for her constant love and support,
to Evan and Maddy, in lieu of time I would
have otherwise spent with them,
and to Eric and Carena, just because
—Martin Weiss

Contents

Preface   xv

About the Authors   xvii

1   Introducing Telecommunications   1

    1.1 Technology—The Means   *2*
        Basics   2
        Telephone Systems   4
        Local Loop   6
        Digital Transmission Systems   7
        High-Capacity Channels   8
        Data Networks   10
    1.2 Economics—Paying for It   *12*
        Supply and Demand   12
        Market Structure   13
        Cost–Benefit Analysis   18
    1.3 Policy—Building Boundaries   *20*
        Communication Common Carriers   21
        Public Utility Concept   22
        Legal Bases of Regulation   23
        Constitutional Bases of Regulation   26
        Regulatory Theories   27
        Regulatory Structure   30
        Jurisdictional Issues   32

2 Telegraph to Telephone (to 1893)   36

    2.1 Technology—Limits and Opportunities   37
        Electrical Industry   37
        Role of Patents   38
    2.2 Telegraphy—Creating New Patterns (to 1860)   38
        Morse   38
        Extending the Telegraph   40
        Railway Synergy   41
        Western Union   42
    2.3 Building a Telegraph Industry (1860–76)   43
        Supportive Role of Government   44
        Monopoly   45
        Network Economics   47
        Impact of Instant Communication   48
        Submarine Telegraphy—Going International   49
    2.4 Telephony—Adding Voices (1876–85)   50
        Bell's Device—"Nothing Much New"   50
        Search for Support   51
        Building a Telephone System   53
        Patent Fight   55
    2.5 Developing the Telephone Industry (1885–93)   56
        Fractious Local Exchanges   56
        Western Electric   57
        Developing AT&T   58

3 Era of Competition (1893–1921)   64

    3.1 Improving Technologies   65
        Switching—Connecting Everyone   65
        Automatic Switching   66
        Long Distance Transmission   67
        Wireless Telegraphy and Telephony   68
        Patent Pool   70
    3.2 Competition Appears   70
        Rise of the "Independents"   71
        Vail and AT&T Centralization   72
        Bell's Response to Competition   74
    3.3 Twisting Path of Government Regulation   77
        Interconnection   78
        Mann–Elkins Act   78
        1913 Antitrust Suit   79
        Kingsbury Commitment   80
        Wartime Government Control   81
        Willis–Graham Act   82

4 Regulated Monopoly (1921–56)   86

    4.1 Further Technological Advance   *86*
        Transmission—Multiplex   86
        Transmission—Microwave   87
        Switching   89
        Radio Broadcasting   90
        Transistors   91
    4.2 Economics   *93*
        Concepts of "Natural Monopoly"   93
        Rate-of-Return Regulation   95
        Separations   99
    4.3 Regulation—Setting a Pattern   *104*
        Radio Act   104
        Splawn Report   104
        Communications Act   105
        FCC Telephone Investigation   106
        Walker Report   107
    4.4 Antitrust: The 1949 Suit   *108*
        1949 Complaint   109
        Derailing the Suit   110
        1956 Final Judgment   112
        Aftermath   113

5 Competition Reappears (1956–74)   118

    5.1 Technology—Adding Competitive Options   *118*
        Solid-State Electronics   119
        Computer Revolution   120
        Communication Satellites   121
        Coaxial Cable   123
    5.2 Terminal Equipment Competition   *124*
        Hush-a-Phone   124
        Carterfone   125
        Bell System PCAs   126
        FCC Certification   126
    5.3 Opening Up Transmission   *127*
        Above 890 Decision   127
        TELPAK   129
        Seven-Way Cost Study   130
    5.4 Creating Specialized Carriers   *131*
        Formation of MCI   131
        Approving MCI   133
        Interconnection Negotiations   133
        More Applicants   134
        SCC Decision   134
        Domsats   135

5.5 AT&T Responds to Competition   136
   Changing Image   137
   NARUC Speech   137
   Defeating Datran   138
   FX Service Decision   139
   Going Public   140

6 Breaking Up Bell (1974–84)   145

6.1 Filing Suits   147
   MCI Brings Suit   147
   Justice Files Its Case   149
   AT&T Responds   150
6.2 A Defining Interlude—Execunet   151
   A New Service   152
   FCC Versus the Court   153
   ENFIA   153
6.3 Trial and Settlement   154
   Paper Chase   154
   Attempts to Settle   155
   Trial   156
   Settlement   157
6.4 Breakup: Defining the MFJ   158
   Agreement   158
   Defining "Local"   159
   Reorganization   160
   Line of Business Restrictions   162
6.5 Baby Bells   164
   Concept—Limited Regional Power   164
   People   167
6.6 Changing Economics   168
   Rising Costs   168
   Bypass Fear   170

7 Operating Under the MFJ (1984–96)   178

7.1 "Free the RBOC Seven"   178
   Petitions for Change   179
   Triennial Review   182
   Decision   184
7.2 Long Distance Marketplace   185
   AT&T Trivestiture   186
   Lucent   187
   Bell Labs   188
   Long Distance Competition   188
7.3 Changing Economic Policies   189
   Equal Access   190
   Access Charges and NECA   193

Switched and Special Access 194
Impact on Universal Service 197
Accommodating Competition 199
Implementing Price Caps 201
7.4 To Regulate or Not to Regulate 204
FCC Computer Inquiries 204
Computer Inquiry III and ONA 208
Jurisdiction—Congress, the Courts, or the FCC? 210
7.5 Exporting Deregulation 212
Privatization and Liberalization 212
Market Flexibility 214
Opening Markets 217
Global Investments, Mergers, and Joint Ventures 218
Trade Issues and the WTO 219

8 Innovating New Services (1980s/1990s)   228

8.1 Digital Age 228
Basics 228
Switches 229.
8.2 Wireless Telephony 231
Emerging Role of Wireless 231
Cellular Systems 232
Changing Cellular Standards 233
8.3 Managing Spectrum 234
Implementing Lotteries 235
Auctioning Spectrum 236
Auction Troubles 237
Spectrum Task Force 238
Finding 3G Spectrum 239
8.4 Cable Convergence 240
Rise of Cable Television 241
Cable/Telephone Ownership 241
First Merger Moves 243
Broadband Gambles 244
8.5 Internet and Information 246
Early Information Services 246
ARPANET 248
Commercialization 249
DSL Versus Cable Modems 249
Voice Over IP 253

9 1996 Act and Aftermath (1996–2000)   258

9.1 Why Change Took So Long 259
Earlier Attempts 260
1986 Dole Bill 260

Industry Complexity 261
Political Factors 262
9.2 Legislating Competition and Convergence 263
(Re)Introducing Local Competition 264
The "War of All Against All" 266
Convergence? 267
9.3 Streamlining Government 268
Easing FCC Rules and Regulations 268
Redefining State and Local Roles 269
9.4 Redefining Universal Service 270
Creating a National Policy 271
. . . and Paying for It 273
From Universal Service to Universal Access? 278
9.5 The FCC Implements New Regulations 279
Interconnection 279
Pricing Guidelines 282
Triennial Review 284
Section 271 Proceedings 286
9.6 Merger Fever 288
RBOC Mergers 289
Bellcore 291
Rise of WorldCom 292
9.7 Appeals, Delay, Decisions, and Impact 294
Appeals 294
The Courts Rule 296
Winners and Losers 299
Prospects 301

10 Meltdown . . . and the Future (Since 2000)   313

10.1 Overcapacity *314*
Fiber Links 314
Wireless Networks 316
Broadband Services 318
Poor Predictions 319
10.2 Economic Pressure *320*
Investment: Frenzy to Freeze 320
Stock Price Declines 322
Layoffs 323
Fraud 325
Bankruptcy 327
10.3 Waves of Disaster *328*
First: Dot.coms 329
Second: Manufacturers 330
Third: CLECs 331
Fourth: Interexchange Carriers 334
Fifth: Wireless Providers 335
Smaller Waves? 337

10.4 Policy Confusion   *338*
  "Promise" of the 1996 Act   339
  Tauzin–Dingell Debate   339
  FCC Deregulatory Mantra   340
  Auction Debacle   342
  Antitrust as Policy   343
10.5 Looking to the Future   345
  Clearing the Glut   345
  Consolidating Further   346
  Reforming Management   348
  Continuing Research   349
  Reviving Policymaking   350
  Summing Up   351

Appendices

A. Regulatory Concepts in Telecommunications
   Economics, Finance, and Accounting   360

   A.1 Regulatory Regimes   *360*
   A.2 Cost Issues   *367*
   A.3 Pricing Issues   *372*

B. Glossary   376

C. Chronology   384

D. Historical Statistics   394

References   398

Author Index   411

Subject Index   415

Preface

This book—like so many others—has been a long time coming.

Phyllis Bernt and Martin Weiss (who earlier authored a book about international telecommunication) began an effort a decade ago to author a comprehensive introductory history of American telecommunications. But then both got pulled into administrative and other duties on their respective campuses. A bit later, Christopher Sterling had launched a separate effort to co-author a book about the dramatic changes wrought on AT&T. Both projects were under contract to Lawrence Erlbaum Associates for LEA's Telecommunication Series. About five years ago, the two projects were merged as Sterling joined with Bernt and Weiss to revive and revamp their project. But academic administration (each of us has held an associate deanship at some point in this process) kept intruding, the field—and thus our story—kept changing, and the book's manuscript eluded completion. Over the past two years, Bernt and Sterling, each finally freed of such "administrivia," focused on getting the book completed.

Our intent throughout this long writing process (during which the industry has taken an almost 180-degree turn from dramatic growth to drastic decline and, more recently, elements of consolidation) has been to provide a historical analysis of the technical, economic, and policy basics of American telecommunications over about 160 years. We seek to provide historical context for what is happening today. This is a history filled with fascinating people, often complex companies, and changing government policies. As with any industrial sector story that has a complex background and is loaded with jargon, those seeking to learn more may find the process difficult going. It is an effort worth pursuing, however, because telecommunications is central to our daily lives in ways many of us never realize. We hope that this book makes that effort a bit less onerous.

ACKNOWLEDGMENTS

We express our sincere thanks to Jill F. Kasle of George Washington University. Jill and Chris Sterling embarked back in 1995 on a book about AT&T's story amidst all the dramatic changes we trace on the following pages. As will sometimes happen with team projects, however, we concluded that we had different concepts of what such a book should be and contain. She very kindly let Sterling take their shared though incomplete work product to become a part of the book you now hold.

Linda Bathgate, our editor at Lawrence Erlbaum, has once again defined patience as she first shepherded two books, and then one, through to the publication stage. She demonstrates just the right balance of push and patience (and a sense of humor, which always helps) and we are all grateful to her for staying the course.

Trevor Roycroft, then at Ohio University, was the first "outsider" subjected to reading the whole manuscript, and we are grateful for his many suggestions, nearly all of which are reflected in changes made here.

On a more personal level, Chris Sterling thanks Ellen (in our 40th anniversary year!) for her continuing support and interest in this latest project. While the material may not fascinate her as it does her husband, she has provided invaluable back-up to help get it written. Phyllis Bernt thanks the students in her introductory regulation courses for their feedback and encouragement when asked to use various drafts of the manuscript as required reading material.

C. H. S.
P. W. B.
M. B. W.
July 2005

About the Authors

Christopher H. Sterling has taught media and telecommunications courses at George Washington University for nearly a quarter century. He's authored or edited 20 books, including a four-volume anthology of excerpted documents from the break-up of AT&T. With research interests focusing on media and telecommunications history and policy, Sterling also edits *Communication Booknotes Quarterly* and the "Telecommunication Series," both published by LEA, and serves on the editorial boards of seven journals. He was a senior staffer at the Federal Communications Commission in the early 1980s, and holds a PhD from the University of Wisconsin–Madison.

Phyllis W. Bernt is a professor of communication systems management at Ohio University. Her research interests include the viability of universal service in a competitive telecommunications marketplace, trends in international telecommunications, and the social effects of information technologies, including issues of privacy and gender. For most of the 1980s she worked for Lincoln (Nebraska) Telephone, now part of Alltel, where she was responsible for rate development, cost analysis, and tariff preparation. She has been an institute associate at the National Regulatory Research Institute, and is currently the lead Principal Investigator for an NSF grant investigating gender and racial representation of information technology in middle school students' media environment. She holds a PhD from the University of Nebraska–Lincoln.

Martin B. H. Weiss serves as Associate Dean for Academic Affairs and Research at the School of Information Science and chairs the Information

Science and Telecommunications department at the University of Pittsburgh. His research interests include the analysis of situations where
competing firms must cooperate, cost modeling of new technologies in
telecommunications systems, and the evolution of telecommunications
industry. For a decade beginning in 1978, he worked at Bell Labs, then
Mitre, and finally with Deloitte, Haskins, and Sells. With Phyllis Bernt, he
co-authored *International Telecommunications* (Howard Sams, 1993). He
holds a PhD from Carnegie Mellon University.

Chapter 1

Introducing Telecommunications

More than other sectors of a nation's economy, telecommunications operates through a unique melding of technology, economics, and policy. At the most basic level, telephone, wireless or radio communication, and Internet connectivity all rely on technology. All forms of telecommunication involve a message, a sender, and a receiver that utilize telecommunications transmission. There are many variations in this transmission process—telephone networks and data networks may move, or switch, a message differently; and messages may be sent over narrow copper wires or over larger capacity fiber optic cables or through the air. Each mode operates under fundamental principles, and a basic knowledge of these is needed to understand how a telephone call takes place, or a byte of data moves through a network, or an Internet connection is made.

An understanding of telecommunications also requires knowledge of economic and policy factors. U.S. telecommunications has developed from a long history of monopoly and regulation. While until recently telecommunication systems in virtually every other country in the world were government owned, in the United States telecommunications has operated as a commercial enterprise almost from its mid-19th-century inception. Government involvement in deployment and development of telecommunications took place here more through regulation than ownership. For decades the U.S. telephone system was largely controlled by the Bell System and a host of small local telephone companies dependent on Bell.

The potential for substantial economic gain can, of course, lead to abuses when an essential service (like local telephony) is provided by a monopolist free from competitive pressures when making pricing and production decisions. Regulation at both the federal and state levels emerged early in the 20th century as the best way to control these possible abuses of monopoly power. In regulating telecommunications, the Federal Communications Commission (FCC), the 50 state public utility commissions (PUCs), and the D.C. commission do more than just control the price of service. Through regulatory decisions they have also structured the industry, determined how services would be provided and by whom, and set service quality standards. Understanding the sometimes convoluted development of telecommunication thus requires some knowledge of these regulators' modes of operation, sources of authority, and motivating principles.

This chapter introduces the fundamentals of telecommunications technology, economics, and policy, including basic principles that will be helpful in understanding the history, issues, and concepts discussed in the chapters that follow.

1.1 TECHNOLOGY—THE MEANS

As the historical core of telecommunications for more than 130 years, *telephony* is the process of transmitting sound—usually the human voice—through electrical signals. For this transmission to occur, sound waves must be converted to electrical signals at the originating point of a call and then the electrical signals must be reconverted to sound waves at the terminating point of the call. A microphone, or *transducer*, converts the human voice to electrical signals and then a speaker converts the electrical signals back to sound waves. In a modern telephone, the microphone and speaker are combined into one telephone handset, so that the telephone set can both generate and receive calls. A *transmission medium*, such as a copper wire, a fiber optic cable, or the air waves, carries the electrical signals between telephone handsets.

Basics

Human speech occupies a variable range of frequencies that depend on the speaker and the moment (as when people raise and lower the pitch of their voices to convey meaning). Generally human speech—and hearing—can range from a low of about 300 Hz (Hertz) to a high of about 10,000 Hz (or 10 kHz). Experiments have found that language can be understood and speakers readily identified using a narrower frequency range of 300 to 3400 Hz. The International Telecommunication Union (ITU) recommends that telephone systems be designed to carry at least this range, referred to as a signal's *bandwidth* (for this and other terms, *see Appendix B*).

While telephone systems have long been designed to transmit human speech, increasingly systems are used for other purposes as well, such as connecting computers (via modems) and transmitting documents (via fax). The traditional telephone network was largely analog; in other words, the electrical signals transmitted were analogous to, or replicated, the sound waves formed by human speech. Data emanating from computers, on the other hand, consisted, not of electrical waves, but rather of bits and bytes of data, or combinations of zeros and ones. Modems (modulator/demodulators) were needed to convert the digital data stream of bits and bytes into an analog speech-like signal (unintelligible to most people) that the largely analog telephone network could carry.

The transmission medium traditionally consisted of copper wires which were connected to the transducer via a circuit called a *hybrid*. The function of the hybrid was (and is) to separate signals being sent from those being received so that the received signal can be directed to the speaker, as illustrated in Fig. 1.1.

FIG. 1.1.   Basic telecommunications link.

By the early 20th century, the most widely used transmission medium was a *twisted pair* of copper wires. Earlier transmission most commonly utilized a single strand of iron wire, which had also been the standard for telegraph systems. Such "open wire" systems were relatively cheap but were more prone to interference, especially from the electric power systems that were proliferating in the late 19th and early 20th centuries. Today, *fiber optic* cable is beginning to replace twisted pair copper.

All transmission media can be evaluated according to their attenuation and their susceptibility to noise. *Attenuation* is the reduction of signal strength, and this increases with both distance covered and the frequency used. A weaker signal is, of course, more difficult to hear. The *bandwidth* of a transmission medium really is the range of frequencies that a transmission medium can carry with acceptable attenuation for a specified distance. The history of transmission media can be seen as a quest for increases in acceptable bandwidth. Twisted pair copper was able to transmit a wider range of frequencies over longer distances than iron wire; today, fiber optic cable is able to transmit much higher bandwidth than twisted pair copper. As is discussed in later chapters, local telephone companies (providers of traditional telephone network service based on twisted pair copper) are under increasing pressure to replace copper with more fiber optic cable, also called broadband facilities. The questions of how the local telephone companies will pay for these broadband facilities, and how they will recover the cost of the twisted pair copper already in use, are major issues of concern to telecommunications policymakers (as discussed in chaps. 8–10, of this volume).

Noise refers to unwanted signal energy (sometimes called *interference*) from any source. In other words, the intended signal is the electrical energy that has been converted from the sound waves; noise is any other electrical energy that manages to invade the transmission path of the intended signal. Although intended signals suffer attenuation, noise normally does not because it can enter the transmission medium at any point, not just at the beginning point of the signal. Thus the effect of attenuation is not just that the signal (the desired information) becomes weaker, but also that the signal becomes more difficult to distinguish from noise. Engineers capture this notion in the *signal-to-noise ratio*, in which the signal power is divided by the noise power. If this ratio is small, then the signal will be more difficult to discern than when the ratio is large. It is the combination of the bandwidth of a transmission medium and the signal-to-noise ratio that governs the capacity of a channel to carry information.

Telephone Systems

Telephone systems involve key components that must fit and work together to enable telephone service. To place a telephone call, a subscriber must have equipment, usually a telephone, through which he or she is able to connect with the telephone network in order to send a message to a desired telephone number. The subscriber establishes a connection to the network by going "off hook," or lifting the telephone receiver. By doing so, the subscriber establishes a connection with a switch. The subscriber then "tells" the switch the number he or she wishes to reach by dialing that number. The switch recognizes the number and routes the call—or rather establishes a channel through other switches and transmission media—between the calling subscriber and the called number.

For policy reasons that will be discussed further (*see sections 6.4, 7.1, and 7.2*), telephone network service is provided by two different types of companies: local telephone companies or local exchange carriers (LECs) and long distance companies or interexchange carriers (IXCs). LECs provide local service and own the facilities closest to the subscriber; IXCs provide long distance service and use the LECs' facilities, along with their own facilities, in order to provide long distance calls. This relationship is illustrated in Fig. 1.2.

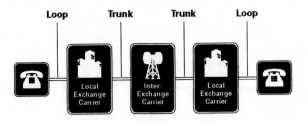

FIG. 1.2. Industry structure.

A LEC's network is structured as shown in Fig. 1.3, with the LEC owning the local connection to the subscriber (the loop); as well as Central Office (CO), or local, switches; tandem switches through which LECs accumulate traffic from several central office switches; and transmission systems connecting switches. IXC networks do not include the local connection to the subscriber or local switches. Instead IXC networks consist of large capacity long distance switches and transmission systems that haul telephone calls from city to city and state to state. (As is discussed in section 9.2, this neat distinction is changing because of Congress's passage of the Telecommunication Act of 1996. As a result of that act, such LECs as Verizon are now offering long distance service once restricted to IXCs, and some IXCs, such as AT&T, are offering local service once reserved for LECs.)

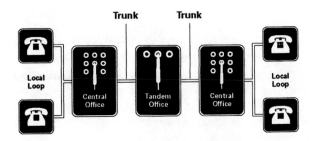

FIG. 1.3. Local exchange network.

The components of a telephone network consist of customer premise equipment (CPE), such as telephones, modems, and fax machines; the local loop; switches (CO and tandem); and transmission systems. Since the late 1970s, subscribers have been able to purchase the CPE of their choice (as long as it meets specific FCC technical standards); earlier CPE could only be leased from the local telephone company.

The contracts that long governed the interconnection of the LEC and IXC networks were termed the "separations and settlements" procedures discussed later (*see section 4.2*), since replaced by "access charge" arrangements (*see section 7.3*). IXCs pay LECs access charges for the use of the LEC network. As a result of Congress's desire to see the development of competition in all parts of the telephone system, competitive local exchange carriers (CLECs) developed in parallel to the LECs. IXCs negotiate arrangements with the CLECs to use their facilities, just as they have used the LECs' facilities. For reasons explored later, however, few CLECS have survived into the 21st century.

Local Loop

The *local loop* refers to the set of technologies used to connect end user locations with the LEC's CO. From the late 19th century well into the 20th, this was done via a direct connection between these users over twisted pair wire. This type of wire is inexpensive and easy to manage, though its transmission characteristics are far from ideal. The local loop has been one of the costliest components of a network, in large part because a separate connection is required for virtually every customer. Shared telephone infrastructure used by multiple consumers did not appear until development of the CO switch.

In the last 40 years, local loop technology has undergone significant changes. The consequence of these changes has been to move the point at which the infrastructure is shared gradually closer to the customer, so that the local loop could become less costly. Today's local loop consists of a combination of loop carrier systems and remote switching partitions/ concentrators.

Loop carrier systems are designed to connect clusters of customers with the CO over transmission facilities shared by many end users. That is, individual customers no longer have dedicated connections to the CO. Under most circumstances, this is transparent to customers because most people use their telephone for only a small percentage of the time. Unlike loop carrier systems, *host-remote switches* consist of a portion of the CO switch that has been placed among a cluster of customers. Host-remote switches handle some of the switching functions, which can further reduce the capacity needed to connect the remote to the main CO switch. In effect, the switch is partially *distributed* throughout the service area using such an architecture.

Local loop facilities will appear in this book as a point of controversy. While local telephone companies have sought ways to reduce loop costs through carrier systems and host-remote switching configurations, existing loop investment remains a major portion of LEC costs. The issue of how this cost should be recovered, whether through local rates or through long distance charges, has been a major point of contention since the 1930s and continues to be controversial today.

Digital Transmission Systems

Transmission systems are facilities intended to connect switches with other switches. Since the 1920s, they have often been high capacity systems. As twisted pair cables offer relatively poor performance over longer distances, engineers developed such other media as coaxial cable, waveguides, microwave, satellite, and, most recently, optical fiber, all of which are superior to twisted pair in capacity and noise performance.

To overcome the attenuation effects associated with distance, engineers developed electronic (vacuum tube-based) amplifiers early in the 20th century. These systems boosted the power of the incoming signal (as well as the incoming noise, unfortunately), which meant that the signal-to-noise ratio deteriorated over distance despite the amplifiers, limiting the information-carrying capacity of telephone channels. The only solution to this problem was to discover ways of regenerating only the desired signal and not the noise—the solution being the development of *digital* transmission beginning in the 1950s.

Unlike traditional analog transmission techniques, in which every incremental signal level had meaning, digital systems depended only on two signal levels. Analog amplifiers had no choice but to amplify noise because there was no way to distinguish signal from noise. Digital systems, however, would detect only high or low voltage levels, so any additive noise was not meaningful and could be discarded. Thus digital repeaters could be engineered to regenerate signals that were nearly indistinguishable from the original signal, thus avoiding the deterioration of the desired signal-to-noise ratio over distance. With digital transmission systems, however, the end-to-end signal can still deteriorate if the signal-to-noise ratio at the input of any repeater is too low for it to correctly distinguish a high or a low voltage; this inability would result in bit errors, which, though correctable, introduce noise on an end-to-end basis.

Since voice or music signals originate at a microphone in analog format, it is necessary to convert them to a digital format if the full advantage of digital transmission systems is to be gained. The location of this conversion has evolved over time, starting at the transmission system terminal, then moving to the local loop side of the CO switch. There has been a slow attempt to finally shift this process to the telephone itself, though this has not happened yet on a widespread basis. A *codec* device converts an analog signal into the digital pulse code modulation (PCM) format. To accomplish this, the codec samples the analog telephone signal 8000 times per second, and then converts each sample into digitized samples of zeros and ones which are converted back to analog format at the other end without loss of fidelity.

High-Capacity Channels

Since the bandwidth of a voice signal remains constant, higher capacity media like optic cable would be of little value without some additional equipment to take advantage of this higher capacity by finding ways to transmit multiple voice channels across a single medium. In other words, instead of handling one voice channel, these media would handle hundreds of voice channels at once. The collection of technologies that achieve this are referred to as multiplexers, and are of two types: frequency division multiplexing (FDM) and time division multiplexing (TDM).

FDM is the technique used initially and consisted of subdividing the high bandwidth media into many subchannels, each of which matched the bandwidth of a telephone channel. This is completely analogous to the broadcast spectrum, in which AM and FM bands are separated into many different channels, each of which is occupied by a single broadcaster per location. Figure 1.4 illustrates the operation of FDM.

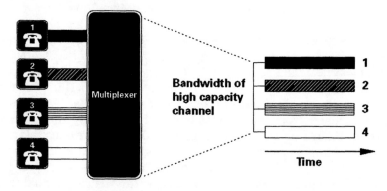

FIG. 1.4. Frequency division multiplexing.

In order to flexibly provide resources to the network as needed, and because different media had different capabilities, *multiplex hierarchies* were developed. These allowed managers to use a building block approach to provisioning capacity and allowed for manufacturing efficiencies. The basic level of multiplexing in the FDM hierarchy was a *group* of just 12 telephone channels. The next level was a *supergroup* (five groups, or 60 telephone channels), or a *mastergroup* (10 supergroups, or 600 telephone channels), which could be multiplexed onto a transmission system. Higher capacity systems were also available, up to a *jumbogroup*, which consisted of 10,800 telephone channels over a single transmission system.

Carrier systems based on this multiplex hierarchy were the mainstay of the telephone network until the 1960s, when they were gradually supplanted by digital transmission systems that used the time division multiplex hierarchy. Just as FDM was well suited to analog transmission, so too is TDM suited to digital transmission. Unlike FDM, in which the bandwidth of the transmission system is subdivided into subbands, TDM allocates the entire bandwidth to a single channel for a short period in regular intervals. This is illustrated in Fig. 1.5.

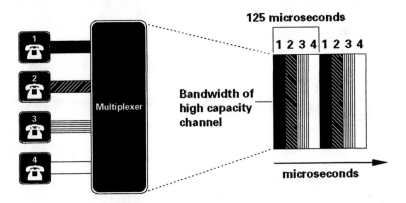

FIG. 1.5.   Time division multiplexing.

With TDM, each time the bandwidth of the transmission system is devoted to a particular telephone channel, the entire digitized sample (i.e., 8 bits) must be transmitted. Since the telephone channel is sampled 8000 times per second, the transmission system must devote a slot to each voice channel every 125 microseconds (μsec) (=1/8000). As more channels are carried, the amount of time in which the 8 bits must be transmitted decreases, so that the bit rate of the transmission system must increase. It can be demonstrated relatively easily that an increase in bit rate is directly proportional to an increase in the bandwidth required by the system. Thus, using a different technique, TDM systems perform essentially the same functions with the similar trade-offs as do FDM systems.

As with FDM systems, TDM employs a multiplex hierarchy. The original digital transmission system was the T1 carrier system and represented the first level of multiplexing in the TDM hierarchy. The T1 system consists of 24 telephone PCM telephone channels (and is sometimes called a *digroup* because it is equivalent to two FDM groups). The next level of TDM that is in common use is the T3 carrier system, which consists of 28 T1 channels or 672 telephone channels. As was the case with FDM, the TDM hierarchy defined by the ITU is not consistent with this one. In the 1980s, an international standard-setting effort sought development of a very high speed multiplex hierarchy called SONET (for Synchronous Optical NETwork) or SDH (Synchronous Digital Hierarchy). SONET is now the standard for most heavy-duty (carrier grade) networks, although T1 and T3 services are still used as well.

Development of multiplexing changed the relationship between the telephone companies and their users, especially larger customers. As the telephone industry developed an ability to handle 24 calls over one digital T1 line, large customers demanded T1 service at a discount, rather than paying for 24 single voice channels. In effect, the large customers wanted to profit from the cost savings that multiplexing provided, rather than letting the telephone companies reap all of the benefits. To retain their customers, telephone carriers (still heavily regulated at the time) asked state utility commissions and the FCC for the authority to provide T1, T3, and higher capacity services at discounted rates.

Data Networks

Telephone networks operate using a technique called *circuit switching*. This process dedicates bandwidth exclusively to a specific call. In other words, a transmission path is created through the network for the call and that path is maintained until the call is ended. Circuit switching made a lot of sense historically and is generally well suited to voice communications which involves a continuous stream of communication.

As digital computers emerged in the 1950s and 1960s, users soon sought to interconnect them. The ubiquitous (though analog) telephone network was the obvious medium for such interconnection. To allow the network to carry such digital signals, engineers had to develop *modems* that would convert the data stream from computers to tones that could be transmitted over the telephone network. As users gained experience with data transmission, it soon became clear that using circuit switching for data networks was not efficient because computers tend to send bursts of data followed by pauses. A sporadic rather than a dedicated transmission path would be more efficient.

A better technique for data is to collect the transmitted data from many computers, carry them over a common network, and then distribute them to their respective destinations. To accomplish this, it was necessary to encapsulate the data in *packets*, which would provide instructions on where to send the enclosed data and how to treat it. This technique, called *packet switching*, makes more efficient use of the network because it is highly likely that each computer will send its burst of data at different times than other computers.

Since devices made in different nations need to interconnect, data communications was an area ripe for standardization. Initially, standardization was provided within a given manufacturer's set of products. As the need for more and faster data communications capabilities grew and demand for public data networks emerged, manufacturer standards were no longer sufficient. To address these emerging needs, the Swiss-based International Standards Organization (ISO) and International Telecommunication Union (ITU) began to develop standards for data communications. While they coordinated their efforts, each had different objectives. The ITU sought standards for public data networks, while ISO was developing architectures and protocols for general computer interconnection.

The eventual product of the ITU's work was recommendation X.25, which specified the interconnection between a public data network and a user. By contrast, ISO produced its Open Systems Interconnection Reference Model (OSI-RM) and eventually many of the protocols within the reference model. The OSI-RM is a standard way to organize the required functionality of data communications systems. It is organized around seven layers, each of which can be implemented by one or more distinct network protocols. Another similar, though unrelated, set of protocols were also developed under the auspices of the ARPANET project of the U.S. Department of Defense. These TCP/IP and associated protocols became the *de facto* worldwide standards, despite the work of ISO and ITU. These are routinely used to interconnect computers worldwide and are the underlying protocol for the Internet.

While traditional telephone networks form the industry's underlying infrastructure, the dynamic growth of today's data networks (and especially the Internet) strongly suggests that data networks will replace traditional voice links. Until recently, voice and data networks were separate, with voice traffic traversing circuit switched networks and data being carried across packet networks. As technology developed, it became increasingly possible to "packetize," combine, and transmit all video, audio, data, and voice signals across one network. For large businesses, this presented the opportunity to save money by combining voice and data services. Telephone company owners of the traditional voice network have also recognized, as they continue to install digital transmission media and digital switches, that there are efficiencies inherent in moving from a network based on circuit switching to a packetized network.

Even more significant perhaps, newer telephone company competitors recognize huge business opportunities inherent in offering *Voice over Internet Protocol* (VoIP). By packetizing voice and sending it over the Internet, competitors including cable television and Internet Service Providers (ISPs) can offer customers an alternative to the public circuit switched telephone network, often at substantial savings. Because the Internet is not regulated, these VoIP providers do not face the regulatory costs and restrictions borne by the telephone companies. As the quality of VoIP improves, and as CPE capable of handling VoIP is developed, VoIP is fast emerging as a viable competitor to the traditional voice network—and, as we shall see, raises new policy concerns.

1.2 ECONOMICS—PAYING FOR IT

In most sectors of the U.S. economy, firms are free to set the prices they will charge and how much they will produce. Their only guidance in making these decisions are the realities of the marketplace. In a few sectors, however, regulatory bodies, rather than market conditions, constrain what firms can charge and produce—and telecommunications is one such sector (others include power and transport). Because telecommunications was largely a monopoly for most of the 20th century, policymakers determined that consumers should be protected from potential monopoly abuses through economic regulation. Instead of the workings of a competitive market, regulators have controlled the price and provision of telecommunications services at the federal and state levels.

Supply and Demand

The marketplace is the usual determinant for what and how much is produced and for the price that is paid for the results of production. The relationship between supply, demand, and price can be quite clearly stated:

> The price in a free market . . . is determined by the interaction of supply and demand. Suppliers (firms) attempt to maximize profits and compete with one another to obtain sales. Demanders (consumers) attempt to allocate their limited incomes over the available goods and services so as to satisfy their preferences. As these two groups interact under the given market conditions, an equilibrium price is established. It acts as a balancing mechanism to ration or allocate scarce economic resources.[1]

The function of price as a balancing mechanism between supply and demand can be represented graphically by the movements of the supply and demand curves (*see Fig. 1.6*). Because buyers are more likely to purchase more units of a product as price declines, the demand curve is downward sloping. On the other hand, suppliers are more likely to supply more products the higher the price; this suggests that the supply curve is upward sloping. The price of a product is the point at which the supply and demand curves intersect, as shown in Fig. 1.6. Price, then, is a

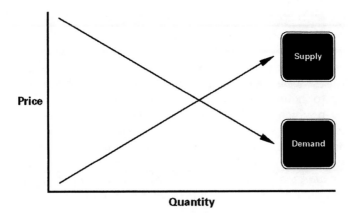

FIG. 1.6. Supply and demand.

measure of the value placed on the product or service by both buyer and seller.

Whether the price that results from the intersection of supply and demand allocates economic resources in the most efficient manner is largely dependent on market conditions under which buyers and sellers make decisions. If they face few constraints, then market forces are operating efficiently and no intervention is necessary to protect the interests of either party. If, on the other hand, either is constrained in their choices, regulators may find it necessary to intervene in order to protect consumers or to control the effects of market imperfections. The level of constraint facing buyers and sellers is affected by market structure.

Market Structure

Economists examine the interrelationships of price, supply, and demand through the use of several theoretical market structures. At one extreme is perfect competition; at the other is total monopoly. In between are oligopoly and monopolist competition. Each structure has a different effect on the interplay between supply and demand. The underlying assumptions of the theoretical market structure of perfect competition are that:

- The products offered are homogeneous and therefore perfectly substitutable for one another;
- There are a large number of buyers and sellers; none of the buyers or sellers is large enough to affect total supply or total demand;
- Buyers and sellers have complete information about market conditions;

- There are no constraints on the market or on any buyer or seller; and
- There is complete freedom to enter and exit the market.[2]

The perfect competition model is obviously only a theory as no markets meet all these ideal conditions. However, the model does present an opportunity to examine what would happen in a perfectly free marketplace, one in which no seller or buyer could control price or quantity.

In a perfect competition model, maximum efficiency is achieved for the seller, for the buyer, and for society in general. To understand the seller's situation, it is important to understand his cost function, especially the concept of *marginal cost*, which is defined as the cost of producing the last unit of production. The formula for marginal cost is:

$$MC = \frac{\Delta Cost}{\Delta Quantity}$$

Put into words, for small changes in quantity, marginal cost reflects the increase in cost divided by the increase in quantity produced. A firm's marginal cost curve reflects the concept of *diminishing returns*: Beyond a certain level of production, the per-unit cost of producing more units will increase because the firm will have to add additional variable expenses like labor. This suggests that the marginal cost curve slopes upward (*see Fig. 1.7*). At the same time, if quantity increases and price decreases as a

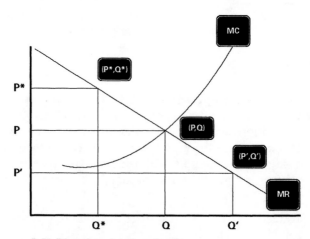

P, P', P*= price at various levels
Q, Q', Q*= quantity at various levels
MC= marginal cost
MR= marginal revenue

FIG. 1.7.   Marginal cost and marginal revenue curves.

result, a firm's marginal revenue curve will tend to slope downward as shown in Fig. 1.7. The formula for marginal revenue is:

$$MR = \frac{\Delta Revenue}{\Delta Quantity}$$

It is intuitively obvious that, if price decreases, marginal revenue will decrease. The firm will get less revenue from the last units sold because of declining prices.

Because the firm's ultimate goal is to maximize profit, it is not economically efficient for the firm to sell products when the marginal cost is greater than the marginal revenue. In Fig. 1.7, if a firm sold a product for less than price P and at more than quantity Q, it would be incurring more in cost than it would be recovering in revenue. For example, if the firm sold at (P',Q'), it would lose money on each unit because at (P',Q') marginal cost is higher than marginal revenue. On the other hand, from the buyer's perspective, a price higher than P and production at quantities less than Q would not be economically efficient. If the firm sold at (P*,Q*), a significant number of buyers who were willing to pay up to price P, a point at which the firm would still be making a profit, would be left out of the market. It is only at (P,Q) that the price and quantity are at their most efficient level. At point (P,Q), marginal cost equals marginal revenue, suggesting that the point at which MC = MR is the most efficient price level.

In the perfect competition model, the firm's demand curve is horizontal rather than downward sloping because no seller or buyer can affect the market. Sellers are all price takers because total market demand has determined the price. If a firm wishes to participate in the market, it must do so at the market price. To be efficient, the firm must produce at the quantity that will maximize profit. As shown in Fig. 1.7, that quantity is at a point at which marginal cost is above or equal to marginal revenue. Since there is only one price in the market, marginal revenue is equal to price. In order to maximize profit, the firm will stop production when marginal cost equals price; otherwise, the firm will lose money. The firm will continue to produce until the point at which marginal cost equals marginal revenue because the firm will make a profit on each of the units sold until that point is reached. The graphic representation of this perfect competition model can be seen in Fig. 1.8.

No markets replicate the conditions presented in the perfect competition model. Indeed Shepherd refers to the perfect competition model as "a cartoon version of effective competition."[3] The perfect competition model, however, does provide an opportunity to examine optimal conditions. In a perfectly competitive market, price represents the perfect equi-

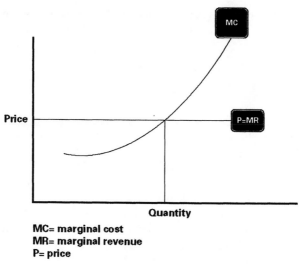

MC= marginal cost
MR= marginal revenue
P= price

FIG. 1.8. Perfect competition model.

librium, with supply at its most efficient level and with consumer demand met. At this level, marginal revenue equals marginal cost, and, of even greater significance, marginal cost equals price. This perfect situation can be described as follows:

> Producing more of one good involves giving the consumer additional units that are valued at less than the cost of production to society; producing less involves foregoing units that the consumer values more than the cost of production. Only when price is equal to marginal cost is consumer satisfaction maximized.[4]

Competition can work to assure that price, if not equal to marginal cost, is at least not substantially higher than marginal cost. But if there is no competition, price can diverge greatly from marginal cost, allowing sellers to profit at the expense of consumer satisfaction. The problems that can arise in a market structure lacking competition are illustrated in the theoretical model of pure monopoly.

In a pure monopoly, there is only one provider of a good that has no substitutes. Far from being a price taker, as is the case in a perfectly competitive market, the monopolistic seller is a price setter. With no other options available to the buyer, the monopolist controls the market. The monopolistic seller has the incentive to produce less than the socially optimal level of production and to charge more than the socially optimal price. Rather than pricing at a level equal to marginal cost, the monopolist will price as far above marginal cost as possible without pricing all buyers out of the market. In Fig. 1.7, the monopolist would price at P* and produce at Q*.

As a specific example, assume that a monopolist faces the following situation:

- Thirty consumers are willing to pay $12 for a product—say, a telephone.
- At 30 units of production, the monopolist's costs are $6 per unit.
- Forty consumers are willing to pay $8 for the same telephone.
- At 40 units of production, the monopolist's costs are $5 per unit. (Because of economies of scale, across a certain level of production, the per-unit cost declines with the more units that are produced.)

In this situation, the monopolist will face the following profit calculations:

- At 30 consumers and a $12 price:
  - 30 × $12 = $360 in revenue
  - 30 × $6 = $180 in cost
  - $360 – $180 = $180 in profit
- At 40 consumers, and an $8 price:
  - 40 × $8 = $320 in revenue
  - 40 × $5 = $200 in cost
  - $320 – $200 = $120 profit

Here the monopolist can make more profit by selling fewer telephones at a higher price. If the monopolist does so, however, economic efficiencies will not be realized. The monopolist can produce 40 telephones at a more efficient cost than it can produce 30, and can still sell those phones at a profit. All 40 buyers in this example are harmed. The 30 buyers who pay an inflated price of $12 are harmed, as are the 10 potential buyers who are denied the opportunity to make a purchase at a price that would still provide the monopolist with a profit.

Perfect competition and pure monopoly represent two extremes. In between these two extremes are more realistic models of market structure, including the dominant firm, tight oligopoly, loose oligopoly, and monopolistic competition.[5] In a dominant firm structure, one firm controls over 40 percent of the market and has no close rivals. This was the case in the early years of long distance competition when AT&T had well over 40 percent of long distance market share. In tight oligopoly, a small number of firms control more than 60 percent of the market. Both the dominant firm structure and tight oligopoly exhibit the effects of a monopoly market structure, with price above marginal cost and quantity at less than socially optimal levels. Like the monopolist, the dominant firm and the firms in the tight oligopoly are able to control the market and abuse their market dominance. Loose oligopolies and monopolistic competition, however, exhibit the characteristics of a competitive market structure. Loose oligopolies are made up of many firms, none of which has market power; furthermore, no combination of firms is large enough to control the market or to fix prices. In monopolistic competition, many firms compete, and market entry is reasonably easy.

While perfect competition may be impossible, effective competition is possible as long as there is "parity and strong mutual pressure among many firms."[6] Such parity and mutual pressure is possible as long as (a) there are many firms incapable of colluding over price or production, (b) there is no dominance by any one firm or combination of firms, and (c) it is reasonably easy for new competitors to enter the market.

The motivations of firms in monopolistic market structures are much different than those of firms in competitive market structures. Firms in monopolistic market structures try to maximize their profits by keeping prices high; they face few pressures to lower price toward marginal cost. Firms in effectively competitive markets cannot control price because they face pressure from equal rivals. They do have the incentive to lower price toward marginal cost in order to be competitive in the marketplace.

Cost–Benefit Analysis

While most markets are allowed to function without regulatory interference, there are instances in which markets are affected by regulatory action. The existence of a monopoly represents one instance in which regulators have stepped in to make up for the absence of competitive forces. Without competitors, a monopolist is able to abuse its monopoly power by reaping excess profits through higher than optimal prices and through lower than optimal production levels. By regulating price and requiring socially desirable levels of production, regulators seek to control the imperfections of a monopoly market structure. Because, as will be further discussed in section 4.2, telecommunications was, until relatively recently, considered a natural monopoly, regulation of price and service provision has been a standard feature of telecommunications policy.[7] For much of the 20th century, state utility commissions controlled the price local telephone companies charged for local service and intrastate long distance, while the FCC regulated the price of interstate services. The major motivation for this regulatory oversight was to protect consumers from potential abuses of monopoly power.

As telecommunications has moved from monopoly to increasing competition, regulators have faced increasing challenges in determining the appropriate regulatory response to changing market structures. These challenges have been especially severe because telecommunications companies provide a range of services and products in a range of different markets and market structures. Indeed, telecommunications companies have for some time provided a mixture of monopoly and competitive services. At the onset of long distance competition, AT&T faced vigorous competition for its business services, while maintaining a virtual monopoly over residential long distance. As a more current example, local telephone companies provide local service and long distance service to regular subscribers, as well as access services to IXCs. Local service and access services have been (until the changes projected by the Telecommunications Act of 1996) monopolies. Because of the slow development of CLECs, incumbent local telephone companies continue to have significant market dominance over local service and access services. At the same time, the incumbent local telephone companies face significant competition for long distance, as well as competition for emerging broadband services.

This mixture of competitive and monopoly services is problematic because of the divergent motives of the monopolist on the one hand and of the competitive firm on the other. A firm facing competition has an incentive to drive prices, and therefore revenues, down. If that same firm has the ability to make up those lost revenues by raising prices for other products—namely products in monopoly markets—it is safe to assume the firm will act as a monopolist in those markets unless restricted from doing so. A good deal of the history of telecommunications regulation involves regulatory efforts to protect the consumers of those monopoly products. The FCC required the former Bell System to create separate subsidies and the newly formed Regional Bell Operating Companies to formulate special accounting methods so that they would not be able to shift costs out of competitive services to be recovered through monopoly offerings. To prevent AT&T from shifting costs out of business services and into residential long distance, the FCC moved to "price cap" regulation (*see section 7.3*).

Efforts to protect consumers from the potential excesses of the former Bell System monopoly, or to protect the captive consumers of monopoly services from price gouging by companies facing increasing competition for their other services, have no doubt resulted in benefits to consumers, especially residential consumers, through lower prices. These efforts, however, have not been without cost. Indeed, a growing number of economic studies suggest that regulatory intervention may have resulted in substantial economic, as well as social, cost.

One of the earliest such studies analyzed the effects of rate-of-return regulation that limits a firm's price to its cost and an agreed-upon return on its investment. It claimed that this method encouraged the firm to inflate its investment base and therefore its price. In other words, the firm would benefit from "gold plating" rather than from efficiency.[8] Arguing that regulated firms benefit from regulation that controls entry into the market, another economist argued that firms may seek such regulation, but that "the benefit to the industry will fall short of the damage to the rest of the community."[9] In other words, the cost of regulation outweighs its benefits.

Far from replicating the pricing efficiencies of the competitive marketplace, several economists have claimed that regulatory pricing has been highly inefficient, citing a loss in consumer welfare ranging from $2.4 billion to $14.1 billion per year.[10] The introduction of new services is often met with regulatory delay as regulators debate such issues as market structure and pricing in time-consuming proceedings. In measuring the economic cost of regulatory delay in the introduction of voice messaging services and of cellular telephone service, one researcher found the economic cost to be significant in both cases.[11] And finally, in an analysis of the effect of deregulation on several industries, another researcher found that the ending of regulation had produced substantial benefits to society through lower prices and increased efficiencies.[12]

It is not clear whether economic regulation of telecommunications has been a successful effort to protect the consumer from monopoly abuses, or an expensive, unnecessary, and inefficient intrusion into the marketplace. What is clear, however, is that regulation has been, and continues to be, an important aspect of telecommunications service provision in the United States.

1.3 POLICY—BUILDING BOUNDARIES

American telecommunications has operated as a regulated industry for almost a century, as will be detailed later in this book. To fully understand the significance of that history, as well as the current challenges facing a competitive marketplace, it is useful to examine the underpinnings of regulatory theory, as well as the specific regulatory structure that has been crafted to oversee the provision of telecommunications. This section discusses the legal and constitutional provisions that underlie regulation, as well as the arguments and historical precedents that have presented regulation as an attractive policy alternative.

Communication Common Carriers

Providers of telecommunications services—local telephone companies, long distance providers, cellular telephone companies—are treated as *common carriers*, and, as such, face certain restrictions and requirements in providing service. The 1934 Communications Act (the legislation that governs telecommunications) defines common carriage in a rather circular manner:

> The term "common carrier" or "carrier" means any person engaged as a common carrier for hire, in interstate or foreign communication by wire or radio or in interstate or foreign radio transmission of energy, except where reference is made to common carriers not subject to this Act; but a person engaged in radio broadcasting shall not, insofar as such person is so engaged, be deemed a common carrier.[13]

The apparent significance of this definition is that common carriers provide communication services through both wire and wireless means; that they provide these services for a profit; and that common carriage does not apply to broadcasting. The deeper significance of this definition involves the rights and obligations that telecommunications providers incur as a result of their common carrier status.

The Communication Act specifies that common carriers will "provide service on demand, at tariffed rates that are just and reasonable, and without any unreasonable discrimination or undue preference."[14] In other words, carriers are to provide service to all willing customers in their service area; they are to charge publicly filed (tariffed) rates which are to be "just and reasonable"; and they are not to discriminate among customers by charging different rates for the same service.[15] In return for these restrictions, common carriers are granted some limitations on liability. Specifically, communication common carriers are not responsible for the content of the messages they transmit.[16]

As common carriers, the former Bell System, the current Regional Bell Operating Companies and the hundreds of independent telephone companies[17] have long been subject to restrictions on their ability to enter and exit markets and on their ability to change rates without regulatory approval. The basis for these restrictions goes beyond the concept of common carriage. Though the recent history of telecommunications shows an easing of regulatory restrictions and oversight, these companies have traditionally been closely regulated because they have, until recently, been considered a public utility.

Public Utility Concept

As is the case with common carriage, it is difficult to formulate a clear, simple definition of a *public utility*. Public utilities are natural monopolies that provide essential services.[18] Other researchers note that public utilities are capital intensive, sell services rather than goods, and must engineer systems to accommodate peak demand. They further note that public utilities have been divided into two major classes: "(1) those enterprises which supply, directly or indirectly, continuous or repeated services through more or less permanent physical connections between the plant of the supplier and the premises of the consumer, and (2) the public transportation agencies."[19] While there is some disagreement about the concept of a *natural monopoly*,[20] it can certainly be argued that the local telephone companies fit the other criteria listed. They are capital-intensive undertakings; their networks have been engineered to accommodate peak demand; they provide services rather than goods. It can also be argued that they provide essential services. When local telephone companies were the monopoly providers of essential service, their status as a public utility seemed clear.

An important feature of a public utility is that it is a *private* enterprise subject to government regulation. In a country dedicated to the principles of private ownership, it may seem strange for government to have the ability to restrict the freedom of a privately owned enterprise to determine its prices and the terms under which it will provide service. However, the concept of an enterprise somehow invested with a public interest, and so subject to government oversight, has a long history. One of the earliest efforts to regulate private industry was the early Catholic Church doctrine of *justum pretium*, or the just price doctrine.[21] According to this notion, a seller should not be unjustly enriched through a transaction. Rather, the seller should receive a just price for his goods, that is, a price that will cover the seller's costs and provide enough profit so the seller can have enough for his economic support. It is easy to see the basis for "just and reasonable" rates in the just price doctrine. Other precursors of the public utility concept included the medieval guilds, which held monopolies of specific crafts and, in return, were closely regulated to assure that they provided service to all at reasonable rates; the French royal charters that granted monopolies to plantations and trading companies and in turn regulated their activities to assure that government goals were met; and the occupations under English common law that were referred to as "common callings" and, as such, were subject to special rights and duties.[22]

In all of these earlier examples, special rights or privileges were granted in return for specific obligations. In the public utility approach to regulation, a private enterprise was granted an exclusive franchise to provide monopoly service and the right to charge rates sufficient to cover all costs and to generate a reasonable profit. In return, the private enterprise accepted the obligation to charge reasonable, nondiscriminatory rates; to serve all customers; to provide adequate service; and to seek approval before extending, abandoning, or changing its services. Until competition began to make inroads in the 1990s, telephone companies were treated as public utilities. With the entry of new long distance carriers and of competitive local access providers into the marketplace, the public utility approach is, at least in theory, being replaced with the dynamics of a competitive market.

Legal Bases of Regulation

There are many industries that appear to be "affected with a public interest," such as meat packing or drugstores, that are not regulated as public utilities, while other industries, such as grain elevators and electricity, are deemed to be of such a high degree of public interest that they warrant close regulation. According to one observer, there is no "definite way of foretelling the necessary characteristics for distinguishing an industry affected with a public interest from one that is not sufficiently affected to require detailed public regulation." Industries are treated as public utilities because (a) the public has demanded that they be regulated, (b) a legislature has found it necessary to regulate the industry in the interests of the public, and (c) the courts have recognized a need for regulation.[23]

Efforts in the United States to regulate private enterprises in the public interest have fallen in and out of favor.[24] The Colonial period witnessed efforts to control the price of food and tobacco, and the early 19th century saw the granting of exclusive charters to corporations building canals, bridges, turnpikes, and railroads. The price controls of the Colonial period were soon rescinded, and, as the 19th century progressed, the courts refused to recognize the exclusivity of the charters. The emphasis during the middle of the 19th century was on private rights and on the benefits of competition, not on regulation. During the later part of the century, the courts and the public changed their views about the need for government regulation. Economic recessions, the development of large monopolies, the growing number of mergers, the decline in business competition, and pricing discrimination by the railroads made regulation an attractive prospect. Perhaps the loudest voices advocating government action were those of the Grangers (the Patrons of Husbandry), who called for the regulation of the railroads.

Congress and state legislatures responded to public sentiment by creating regulatory bodies and passing laws with the purpose of protecting the public interest. Congress created the Interstate Commerce Commission; state legislatures established railroad advisory boards; and, as will be discussed more fully, state legislatures passed laws to regulate the prices charged by grain elevator operators, insurance companies, and even grocers. These legislative actions did not go unchallenged, however. Companies, facing regulation, challenged the government's right to interfere with the actions of private businesses. The courts, through a series of landmark cases, established a web of legal precedent that supported the government's right to regulate private enterprises in order to protect the public interest.

In the first landmark case, *Munn v. Illinois* (1877),[25] the Supreme Court found that the presence of a monopoly justified government regulation. Acting on the authority granted to it by the 1870 revision of the Illinois State Constitution, the Illinois legislature passed a law requiring grain elevator operators in Chicago to obtain a license, file their rates, and charge no more than a legally established maximum rate. The operators of one grain elevator, Munn and Scott, were sued for failure to comply with the law. Munn and Scott contested the Illinois law, claiming that it violated Article I of the Constitution because it interfered with the U.S. Congress's right to regulate interstate commerce and violated the Fourteenth Amendment to the Constitution by denying Munn and Scott due process of law.[26] The majority opinion, expressed by Chief Justice Waite, found against Munn and Scott. In his opinion, Justice Waite relied heavily on the precedent of English common law, stating that "when the people of the United Colonies separated from Great Britain, they changed the form, but not the substance, of their government."[27] Quoting heavily from the words of a 17th-century English chief justice, Justice Waite articulated the basis for government regulation of certain types of businesses:

> This brings us to inquire as to the principles upon which this power of regulation rests, in order that we may determine what is within and what without its operative effect. . . . We find that when private property is "affected with a public interest, it ceases to be *juris privati* only." This was said by Lord Chief Justice Hale more than two hundred years ago . . . and has been accepted without objection as an essential element in the law of property ever since. Property does become clothed with a public interest when used in a manner to make it of public consequence, and affect the community at large. When, therefore, one devotes his property to a use in which the public has an interest, he, in effect, grants to the public an interest in that use, and must submit to be controlled by the public for the common good, to the extent of the interest he has thus created.[28]

Noting that the grain elevators in Chicago were controlled by nine busi-nesses and that these met annually to set their rates, Justice Waite found these elevators to be a "virtual" monopoly standing in the "very 'gateway of commerce,' and take[ing] toll from all who pass."[29] As such, the grain elevators were invested with a public interest and so subject to regulation.

The *Munn v. Illinois* case established the precedent that the existence of a monopoly could justify regulation; future court cases expanded the ba-sis for regulation beyond the existence of a monopoly. In *German Alliance Insurance Company v. Lewis* (1914), the courts found that the necessity of a service could trigger the need to regulate its provider. Under Kansas law, the superintendent of insurance could require the posting of insur-ance schedules and could mandate changes in insurance premiums; in 1909 the superintendent ordered a 12 percent reduction in fire insurance premiums. German Alliance Insurance Company, though complying with the order, asked the Supreme Court to declare the Kansas law un-constitutional. Arguing that the fire insurance business was not a mo-nopoly and that fire insurance companies received no special privileges or immunities from the State of Kansas, the German Alliance Insurance Company argued there was no basis for state regulation of fire insurance. The court disagreed with German Alliance Insurance. Writing for the majority, Justice McKenna found that

> the business of insurance has very definite characteristics, with a reach of influence and consequence beyond and different from that of the ordinary business of the commercial world. . . . To the insured, insurance is an asset, a basis of credit. It is practically a necessity to business activity and enter-prise. It is, therefore, essentially different from ordinary commercial trans-actions, and, . . . is of the greatest public concern.[30]

The *German Alliance v. Lewis* case expanded the reach of regulation be-yond monopoly. *Nebbia v. New York* (1934) established the precedent that virtually *any* enterprise could be regulated. The State of New York had es-tablished a milk control board to regulate the price of milk in order to "remedy conditions of oversupply and destructive competition or curtail-ment of the dairy industry, a paramount industry of the state."[31] A gro-cer sold milk for less than the minimum price and was convicted of vio-lating the Milk Control Board's order. The grocer (Nebbia) appealed his conviction to the Supreme Court. In a sweeping opinion, the court found against him, stating:

> If the law-making body within its sphere of government concludes that the conditions or practices in an industry make unrestricted competition an in-adequate safeguard of the consumer's interests, produce waste harmful to

the public, threaten ultimately to cut off the supply of a commodity needed by the public, or portend the destruction of the industry itself, appropriate statutes passed in an honest effort to correct the threatened consequences may not be set aside because the regulation adopted fixes prices reasonably deemed by the legislature to be fair to those engaged in the industry and to the consuming public. And this is especially so where, as here, the economic maladjustment is one of price, which threatens harm to the producer at one end of the series and the consumer at the other. The Constitution does not secure to anyone liberty to conduct his business in such fashion as to inflict injury upon the public at large, or upon any substantial group of the people. Price control, like any other form of regulation, is unconstitutional only if arbitrary, discriminatory, or demonstrably irrelevant to the policy the legislature is free to adopt.[32]

According to *Nebbia v. New York*, any enterprise, not just a monopoly or a virtual necessity, could be regulated if the legislature determined a need for regulation in order to protect the public.

Constitutional Bases of Regulation

While the courts may have established legal precedent to support the government's right to regulate the activities of private businesses, the government's ultimate authority to regulate private enterprises rests in the Constitution of the United States. The federal government's power to regulate businesses arises from Article I, Section 8. This article, known as the "interstate commerce clause," states that the Congress shall have the power "to regulate commerce with foreign nations, and among the several states." Moreover, Section 8 of Article I also gives Congress the power to "make all laws which shall be necessary and proper for carrying into execution the foregoing powers, and all other powers vested by this Constitution in the government of the United States."

The Constitution thus gives the government the broad power to make all laws necessary to regulate interstate business. This power is not absolute, however. Limitations on governmental power are found in the Fifth Amendment which requires that private property not be taken without due process of law and just compensation. Specifically, the language of the Fifth Amendment states that no person shall be "deprived of life, liberty, or property, without due process of law; nor shall private property be taken for public use, without just compensation." This amendment is also referred to as the "takings clause." Telecommunications companies have invoked this amendment when contesting FCC decisions the companies believed to be confiscatory, that is, depriving them of their property. In particular, telephone companies have used the takings clause when arguing against FCC limitations on the amount of earnings companies are allowed to realize, or against rates a telephone company believes to be inadequate to cover costs and provide an adequate profit.

States derive their power to regulate private entities from the Tenth Amendment to the Constitution, which simply says that "the powers not delegated to the United States by the Constitution, nor prohibited by it to the states, are reserved to the states respectively, or to the people." The Tenth Amendment is the source of what are called the "police powers" of the states; that is, the "broad powers of the states to protect the health, safety, morals, and general welfare of their citizenry."[33] As with the federal power to regulate business, the states' powers are also limited by requirements for the due process of law. According to the Fourteenth Amendment, no state shall "deprive any person of life, liberty, or property, without due process of law; nor deny to any person within its jurisdiction the equal protection of the laws."

In considering the effects of both the Fifth and the Tenth Amendments, it is important to remember that corporations have the legal status of persons. The due process, just compensation, and equal protection provisions outlined in both of these Constitutional amendments apply, therefore, to corporations as well as to individuals.

Regulatory Theories

Many theories have been proposed to explain both why regulatory bodies have developed and why they function as they do.[34] While no one theory seems to explain all of the actions taken by regulatory bodies, such theories do provide possible ways to understand regulatory behavior and possible measures by which to gauge the effectiveness of state and federal regulatory commissions. Although the various theories of regulation are presented here in four seemingly neat categories, there is actually much overlap among the various theories. The concern in analyzing the theories of regulation is not into which category the theory seems to fit, but rather the underlying assumptions about the role of regulation that each theory presents.

Public Interest Theory. This is the oldest, and perhaps the most obvious, theory of regulation. Its basic premise is that the primary goal of the regulator, and the primary motive underlying regulatory activity is, or should be, the protection of the public from potential abuses in the marketplace. Public interest theory developed in the early years of regulation, during which some industries (notably utilities) were regarded as natural monopolies. The assumption underlying the concept of a natural monopoly is that, in some industries, because of the high costs of market entry or because of economies of scale and scope, one company can provide service more cheaply and efficiently than can two or more. However, even though a monopoly may be efficient, the monopoly provider can, if unchecked, abuse its power by refusing to provide service, pricing unfairly, or providing substandard service quality. The role of the regulator, according to the public interest theory, is to prevent these abuses of monopoly power by regulating price, service provision, and service quality. Some authors discuss several variants of the public interest theory, which they also call the "consumer protection theory." They list several goals or rationales for regulation in the public interest, including correcting economic, social and political failures.[35] They also note that the public interest theory is normative; that is, it is a theory that explains what regulators *should do*, rather than explaining why regulators actually act as they do. Another researcher agrees with this assessment, calling this theory the "yardstick by which regulation is measured."[36]

Private Interest or Interest Group Theory. Students of regulation became disenchanted with the public interest theory, regarding it as naive and incapable of explaining the actual behavior of regulatory bodies. Several variants of private interest or interest group theories have been developed that claim to be more realistic and better able to explain why regulatory bodies do not always seem to regulate purely in the interest of the public. At the core of these theories is the idea that individuals and groups use the regulatory process to serve their own private needs and interests, rather than to advance the public good. In this view, everyone from stockholders, to managers of companies, to interveners in regulatory proceedings, to employees of the regulated companies, to the regulators themselves seek to use the regulatory process to further their own ends. One observer presents an interesting variation called "Coalition-Building Theory."[37] According to this notion, regulators regulate by building political coalitions; in other words, for regulation to be possible at all, consensus must be reached. It is the role of the regulator to build this consensus; in the process, the regulator amasses political power, thus using regulation to serve his or her own interests.

Regulatory Capture Theory. While there are many parties with an interest in the regulatory process, Regulatory Capture Theory focuses on the concerns of the regulated industry itself. This theory claims that the regulatory body is captured by the industry it regulates. Arguments in support of this theory often point to the revolving door that exists between the regulatory body and the industry it regulates, noting the frequency with which former regulators move into jobs with the regulated firms; regulators hoping to gain more financially rewarding positions with the regulated firm will be less likely to act in a way that will harm the firm's interests. Other arguments offered in support of this theory cite the much stronger lobbying presence that the regulated industry can maintain compared to the lobbying power of consumer interests. For whatever specific reason, according to the capture theory, the regulatory body comes to serve, and even identify with, the interests of the regulated industry rather than the public interest. An interesting variant of the capture theory is the "Life Cycle Theory," which sees regulatory bodies passing through various life stages.[38] As the regulatory body is formed and in its very first years, it is filled with a sense of mission and vigor; the emphasis during this period is very much on serving the public interest. In its youth, the regulatory body is still effective in fulfilling its role as protector of the public good; however, as the regulatory body progresses into maturity and then old age, it comes to identify more and more with the industry it regulates. Finally, in its old age, the regulatory body serves the interests of the regulated firm. The life cycle theory suggests that the regulatory body is initially formed to serve the public interest, and that, over time, it becomes the captive of the regulated industry. A more sinister variation, the "Conspiracy Theory" claims that, from its very inception, the regulatory body was created to serve industry interests rather than the public interest.[39]

Equity-Stability Theory. This theory claims that regulation exists because of legislators' desire to further social equity and fairness. Legislators create administrative bodies like regulatory commissions in order to "protect society from the unimpeded operation of the market forces."[40] If left to the marketplace, society's resources would not be distributed in a socially equitable manner; the role of regulation is to act as a corrective and to assure equity and fairness. Whereas this theory emphasizes social equity for its own sake, one variation, the Capitalist State Theory, suggests a less altruistic rationale for regulation. This theory holds that regulation exists "due to the inability of the market to regulate capitalist behavior."[41] In effect, it is the role of regulation to assure the continuation of the capitalist system by controlling its excesses.

There are other theories of regulation in addition to the four major categories listed above. For example, there are theories that focus on the regulatory body as an organization, and explain that regulators operate out of a sort of organizational imperative. They are regulators, so given the opportunity, they will regulate. All of these theories have shortcomings. Focusing on the life cycle of a regulatory body does not explain why some regulatory bodies have gone through periods of resurgence and re-dedication to serving the public interest. Capture theories do not adequately explain why regulatory bodies often require firms to act in ways that are counter to the firms' interests. Though these theories may not be totally accurate in explaining regulatory behavior, they do point to the varying interests and stakeholders affected by the regulatory process, and they do provide a framework for analyzing regulatory activity.

Regulatory Structure

There are no references to regulatory bodies like the FCC or state utility commission in the U.S. Constitution, as such bodies were later created by legislation to fill a perceived void in the governing structure. Regulatory bodies are administrative in function; their role is to fulfill specific goals set out for them by the legislatures that have created them. While the legislatures specify the goals they are to attain, regulatory commissions have a good deal of latitude in deciding how those goals are to be met.

Prior to the creation of regulatory commissions, several earlier forms of regulation were found to be ineffective in overseeing the behavior of firms in industries deemed to be "invested with a public interest" and so in need of governmental oversight.[42] Relying on the courts to regulate the behavior of firms (judicial regulation) proved to be inefficient and ineffective. The courts do not start proceedings but rather wait for those seeking redress to initiate court action. As a result, the courts are reactive rather than proactive. They cannot set policy in advance; they can only act after the fact. The reactive and ad hoc nature of court proceedings make them a poor vehicle for formulating coherent regulatory policies. Relying on legislation to regulate an industry (direct legislative regulation) also proved ineffective. The legislative process is notoriously slow and inflexible. Laws passed to regulate firms are difficult to change when the need arises. The fact that it took Congress years to substantially amend the Communication Act of 1934 is proof of the slowness of the legislative process. Regulation by local franchise also proved inefficient. Once a franchise agreement is in place, it is difficult, if not impossible, to change for the duration of the agreement. Like direct legislative regulation, local franchise regulation proved to be ineffective in regulating industries that changed quickly as a result of technological developments or increasing numbers of new entrants.

The railroads proved to be the catalyst that triggered the creation of a more effective means of regulation. Before 1870, state legislatures called on fact-finding and advisory boards to assist them in crafting railroad legislation. As the years progressed, these boards were given the authority to regulate railroad rates and behavior. At the federal level, Congress determined a need for a federal commission to oversee the railroads and created the Interstate Commerce Commission (ICC) in 1887. The scope of activity accorded to the state railroad commissions was expanded in the early part of the 1900s to include electricity, gas, and communications (telephone and telegraph). In 1907, both New York and Wisconsin passed legislation to create such broad-based regulatory commissions; by 1920, two-thirds of the states had followed suit. At the federal level, the ICC's duties were expanded to include oversight of telephone and telegraph with the Mann–Elkins Act in 1910. Discontent with the ICC's efforts to regulate communications, Congress passed the Communication Act of 1934 and thereby created the Federal Communications Commission to regulate communication through the air waves and by wire.

The Communication Act of 1934. This gives the FCC its authority and establishes its mission as "regulating interstate and foreign commerce in communication by wire and radio so as to make available . . . a rapid, efficient, Nation-wide and world-wide wire and radio communication service with adequate facilities at reasonable charges."[43] The Act is organized in several parts, or "titles." The FCC's authority to regulate communication common carriers is presented in Title II.[44] Part I of Title II specifies that communication carriers are to provide nondiscriminatory service, charge just and reasonable rates, and file schedules (or tariffs) of their services and charges.[45] To assure compliance with these requirements, the FCC is given the power to investigate, suspend, and prescribe rates; and the authority to hear complaints and impose fines.[46] There have been several attempts to substantially rewrite the Communication Act since its passage in 1934; however, most of these attempts failed (*see section 9.1*). Finally, Congress, in the Telecommunications Act of 1996, rather than rescinding the 1934 Act, augmented it by adding the requirement that competition be introduced into all aspects of telecommunications. The 1996 Act left intact in Title II the common carrier provisions, along with the associated powers granted to the FCC to oversee common carriage.

Just as federal legislation created and empowered the FCC, legislation in the various states created and empowered state utility commissions. As a result, communication is regulated by public utility commissions in all 50 states and in the District of Columbia, as well as by the FCC. While these commissions do not fit neatly neatly into any of the three branches of government (executive, legislative, judicial), they have strong connections to each of them, especially the first.

The Executive Branch. FCC commissioners, and the members of state utility commissions in 37 states and the District of Columbia, are nominated by the executive branch (the President at the federal level; the governor at the state level; the mayor of the District of Columbia). The executive branch oversees the regulatory commission's budget requests; also, the Department of Justice at the federal level, and the states' attorneys general at the state level, enforce and defend the commissions' orders in court.

The Legislative Branch. The legislative branch creates regulatory commissions. Congress passed the Communication Act in 1934 and thereby created the FCC. At the state level, state legislatures passed laws creating and empowering state utility commission. The legislative branch establishes the commission's powers and also identifies the commission's goals. For example, in 1996, Congress, in passing the Telecommunications Act of 1996, amended the 1934 Act and gave the FCC a mandate to introduce competition into telecommunications. The legislative branch approves nominations to the commissions and controls the commission's purse strings by approving, or disapproving, the commission's budget requests.

The Judicial Branch. All commission decisions are subject to judicial review. It is the duty of the courts to assure that the commissions have observed due process in their deliberations and decisions, and have not acted in an arbitrary manner.

The Federal Communications Commission. The FCC was established by the Communication Act of 1934 which gave the commission authority to regulate interstate and international wire and radio communication. As part of that authority, the FCC was given the power necessary to assure that communication common carriers provide nondiscriminatory service at just and reasonable rates. The FCC has both investigative and enforcement powers. It can investigate rates to assure they are reasonable and can investigate complaints against carriers; it can enforce its own rates and can require the payment of damages.

State Commissions. There are 51 state utility commissions, including the District of Columbia. There are differences in makeup and function among the various commissions. In 11 states, the commissioners are elected; in 2 states, they are chosen by the legislature; in 37 states they are nominated by the governor; in the District of Columbia, they are nominated by the mayor.[47] State commissions regulate a range of industries, including electric, gas, water, railroads, trucking, buses and cabs, grain elevators, and telephone. Not all commissions regulate all industries. Decisions of state commissions are also subject to judicial review; those disagreeing with the decisions can appeal to the state courts.

Jurisdictional Issues

There is a system of dual jurisdiction in telecommunications, with regulation at both federal and state levels. A complicating factor in this arrangement is the fact that the same facilities are used to provide federal and state services. For example, a call from Chicago to Denver uses the same local loop and the same local telephone switch as does a call from Chicago to Peoria. The call to Denver, however, falls under the jurisdiction of the FCC, while the call to Peoria falls under the jurisdiction of the Illinois utilities commission. As long as the FCC and the Illinois commission are in agreement about how these facilities should be regulated, there are no major points of contention. If, however, the FCC and the state commission disagree on the specifics of regulation, the result can be contention, confusion, and court action.

Regulation of telecommunications emerged and evolved in the United States in response to developments in technology and in the marketplace. As early attempts at competition in the telephone industry failed and were replaced by monopoly, policymakers accepted monopoly as the market structure best able to create a viable telecommunications infrastructure and created regulatory bodies to serve the public interest by controlling the monopolist. As new technologies eroded the Bell system's monopoly position, policymakers came to regard competition as the market structure best able to serve the public interest and redirected regulatory efforts toward facilitating the development of a competitive telecommunications marketplace.

The development of the telecommunications industry, with its technological innovations and growing governmental and regulatory involvement, began with the introduction of the telegraph. Allowed to develop without regulation and with little governmental involvement, the history of the telegraph nevertheless is instructive about the economic forces and technological challenges inherent in a networked industry.

NOTES

1. Keith M. Howe and Eugene F. Rasmussen. 1982. *Public Utility Economics and Finance* (Englewood Cliffs, NJ: Prentice Hall), p. 181.
2. Charles F. Phillips, Jr. 1993. *The Regulation of Public Utilities: Theory and Practice*, Third Edition (Arlington, VA: Public Utilities Reports), p. 61.
3. William G. Shepherd. 1997. *The Economics of Industrial Organization: Analysis, Markets, Policies*, Fourth Edition (Upper Saddle River, NJ: Prentice Hall), p. 38.
4. Phillips, pp. 61–62.
5. Shepherd, p. 16.
6. Ibid., p. 18.
7. Price is only one aspect of regulation to control monopoly power. Other monopoly abuses include poor service and refusal to provide service. Regulators seek to control these abuses of monopoly power as well.
8. Harvey Averch and Leland L. Johnson. 1962. "The Behavior of the Firm under Regulatory Constraint," *The American Economic Review*, 52:1052–1069 (December).

9. George J. Stigler. 1971. "The Theory of Economic Regulation," *The Bell Journal of Regulation and Management Science*, 2:3–21 (Spring).

10. Robert W. Hahn and John A. Hird. 1991. "The Costs and Benefits of Regulation: Review and Synthesis," *The Yale Journal on Regulation*, 8:261–262.

11. Jerry A. Hausman. 1997. "Valuing the Effects of Regulation on New Services in Telecommunications," Brookings Papers on Economic Activity. *Microeconomics*, 1–38.

12. Clifford Winston. 1993. "Economic Deregulation: Days of Reckoning for Microeconomists," *Journal of Economics Literature*, XXXI:1263–1289 (September).

13. 47 U.S.C. 153 (10).

14. Peter W. Huber, Michael K. Kellogg, and John Thorne. 1999. *Federal Telecommunications Law*, Second Edition (Gaithersburg, MD: Aspen Law & Business), p. 279.

15. The opposite of "common" carriage is "private" carriage. Private carriers do not hold themselves out as providing service to all potential customers. As private carriers, they are not held to such requirements as publicly filing rates or charging nondiscriminatory rates.

16. This lack of control over content is significant in defining a communication common carrier and identifying the major distinction between common carriers and broadcasters or cable providers. Both broadcasters and cable providers control content as well as transmission. Communication common carriers control only transmission.

17. There are over 1,300 independent telephone companies. They are traditionally called "independent" because they were not part of the old unified Bell System. They range in size from the large GTE (which has now merged with Bell Atlantic to form Verizon) to mid-size companies like Alltel to small "ma-and-pa" operations with a few hundred telephone lines.

18. Howe and Rasmussen, p. 2.

19. James C. Bonbright, Albert L. Danielsen, and David R. Kamerschen. 1988. *Principles of Public Utility Rates*, Second Edition (Arlington, VA: Public Utilities Reports), pp. 8–11.

20. This is discussed further in section 4.2.

21. Martin G. Glaeser. 1957. *Public Utilities in American Capitalism* (New York: Macmillan), p. 196.

22. Phillips, pp. 90–91.

23. Ibid., pp. 93–94.

24. For detailed discussions of the following historical developments, see Robert Britt Horwitz, 1989. *The Irony of Regulatory Reform* (New York: Oxford University Press), pp. 48–65; Glaeser, pp. 14–78; and Phillips, pp. 91–93.

25. According to Glaeser, this case has "by common consent, been placed at the threshold of our modern treatment of the public utility problem," p. 206.

26. *Munn v. Illinois*, 94 US 113 (1877).

27. Ibid.

28. Ibid.

29. Ibid.

30. *German Alliance Insurance Company v. Lewis*, 233 US 389 (1914).

31. *Nebbia v. New York*, 291 US 502 (1934).

32. Ibid.

33. Howe and Rasmussen, pp. 43–44.

34. This overview of the theories of regulation owes much to the discussions found in Bonbright et al., pp. 33–66; Horwitz, pp. 22–45; Barry M. Mitnick, 1980. *The Political Economy of Regulation: Creating, Designing, and Removing Regulatory Forms* (New York: Columbia University Press), pp. 79–241; Phillips, pp. 182–187; and Harry M. Trebing,

1981. "Equity, Efficiency, and the Viability of Public Utility Regulation," in *Applications of Economic Principles in Public Utility Industries*, edited by Werner Sichel and Thomas G. Gies (Ann Arbor: University of Michigan Press), pp. 17–52.

35. Bonbright et al., p. 33.
36. Horwitz, p. 27.
37. Trebing, pp. 25–28.
38. Howe and Rasmussen refer to these stages as "incipiency, youth, maturity, and old age" (p. 169). Horwitz discusses the life cycle theory as described by Bernstein; according to Bernstein, the stages are gestation, youth, maturity, and old age.
39. Horwitz, pp. 31–38.
40. Trebing, p. 28.
41. Horwitz, p. 41.
42. Much of the discussion in this section is based on Phillips, chapter 4.
43. 47 U.S.C. Section 151.
44. The Communication Act also gives the FCC regulatory authority over radio and other broadcast media, cable television, and new communication technologies. See Peter W. Huber, Michael K. Kellogg, and John Thorne. 1999. *Federal Telecommunications Law*, pp. 220–221.
45. 47 U.S.C. 201, 202, and 203.
46. 47 U.S.C. 204, 205, 208, 209.
47. See Phillips, pp. 133–140; also *Telecommunications Reports International*, http://www.tr.com/online/election/election_chart.html.

Chapter 2

Telegraph to Telephone (to 1893)

Effective means of communicating rapidly over distance have been sought for centuries. Original distance signaling meant using shouts, drums, fire or smoke, or the use of human or animal couriers. All of these were developed to a high degree by the Roman Empire.[1] Only after centuries of using similar methods did a more systemic approach finally appear in preindustrial Europe.

Several mechanical semaphore signaling systems were placed into service in France and Britain beginning in the late 18th and early 19th centuries. Using a standard set of movable wooden paddles (mounted on wooden or stone towers to be more easily seen) that could switch to agreed-upon positions to indicate different letters and numbers, a signal could be sent from the coast to Paris or London in a matter of minutes. But the mechanical systems worked only in daylight hours, and in clear weather, when the towers holding paddles could be seen one from another.[2] They required constant upkeep, were cumbersome to use for long or complex messages, and called for a fairly large trained body of operators. They were thus expensive and only governments could afford to operate them. Something better, and useful at all hours, was needed.

The answer began to develop in the mid-19th century with innovations in both England and the United States that would lead to wired electrical telegraph (code) systems. These in turn would begin to give way to wired telephony late in the century. Within these new modes of communication would come the shape of the broader telecommunication industry to come. Development of the telegraph and telephone took place within, and sometimes against, a developing electrical industry in both

Britain and the United States. That industry first demonstrated the importance of research and development and patents as a means of industrial dominance.

2.1 TECHNOLOGY—LIMITS AND OPPORTUNITIES

Any innovation is defined by the available technical knowledge and capacity of its time. Electricity was not yet well understood or widely applied in mid-19th-century Britain or America, as widespread use of electric power and lighting was still several decades away. The telegraph was an outgrowth of fundamental experiments in electromagnetism conducted by Hans Christian Oersted during the early 19th century. Michael Faraday continued these experiments until he discovered the principle of electromagnetic induction in 1854. While neither Oersted nor Faraday were interested in the commercial implications of their work, the invention of a practical electrical battery around 1800 would help to make the telegraph possible. Experiments in building electric motors proliferated in the 1840s and 1850s, but these applications were short-lived as only battery power was then available at a steep price margin over then-common steam engines.[3]

Electrical Industry

Prompted in part by manufacturing of telegraph apparatus, the electrical industry developed rapidly after the Civil War. The first commercial electrical product was arc lighting, which was first demonstrated in 1808.[4] The Brush Electric Company, formed in 1872, was a pioneering electrical firm and primarily manufactured telegraph equipment and dynamos until it expanded into arc lighting in 1878. Thomas Edison's first electrical company was founded in 1875. Edison's development of a practical incandescent lamp in 1880 sparked dramatic growth. Westinghouse was established in 1884, and General Electric (GE, formed in 1892) was an amalgamation of two older firms, one of them Thomas Edison's.

Both here and abroad the growing electrical industry sparked exhibitions and demonstrations of electricity for the public, helping to build expectations and a future market. The American Institute of Electrical Engineers (AIEE) was formed in 1884, an indicator of the growing number of technical personnel (dubbed "electricians") at work.[5]

It quickly became obvious that the most effective way to stay ahead of one's competition in this new technology-driven industry was to continue undertaking research to find improved ways of making products or providing services. The connection of research to continued company survival meant that the research effort became a vital part of the electrical business from the start, often directed by the key inventors (Morse, Bell) themselves, at least for a time. Not all business managers realized this—a generally cautious lot, they often had to be pushed to support needed research.

Role of Patents

All of the early electrical companies were established on the basis of patents held for specific products. As long as a patent was in force (up to 17 years in the American system), the company controlling it enjoyed a monopoly on any benefits it provided. But as will be seen below, when basic patents expired, others could readily enter a business as competitors. The constant striving for patents on new and improved devices was almost entirely an attempt to prolong a company's dominance of a particular aspect of electrical manufacturing by controlling the key patents which defined that industry.

Westinghouse and GE competed fiercely and had a number of patent arguments until 1896, when they agreed that GE should receive two-thirds of the business growing from their shared patents. This was the first patent "pool," wherein two or more companies agreed to share (pool) their patents and divide up the resulting market. A kind of cartel agreement, this early pool was a precedent for a much larger agreement concerning wireless equipment after World War I.

2.2 TELEGRAPHY—CREATING NEW PATTERNS
(TO 1860)

One of the most important contributors to the developing electrical industry was the electric telegraph. A code-based system of electrical telegraphy was developed by American artist Samuel F. B. Morse (1791–1872) beginning in the 1830s. British telegraph services using different systems had begun operation in conjunction with developing railway services several years earlier.[6]

Morse

Of the many telegraph-like devices being developed at about the same time, Morse's device was the simplest. First conceived in 1838, it consisted of a battery and a key at the transmitting end and an electromagnet at the receiving end. When the key was depressed, an electric circuit was completed causing current to flow through the wire and through

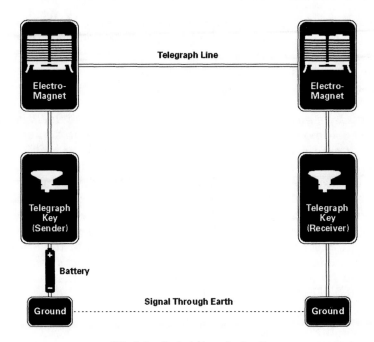

FIG. 2.1.   Basic telegraph circuit.

the electromagnet. The current flow through the electromagnet gener-ated a magnetic field, which caused an iron lever (called the *sounder*) to be attracted to the electromagnet with an audible "click."

By converting an English language character string into a three-element code ("dot," "dash," and "space"), message text could be trans-mitted at a distance virtually instantaneously.[7] As "e" was the most commonly used letter, it was designated by a simple single "dot." Letters used less often were given more complex dot and dash combinations.

At first Morse had considerable difficulty raising financial support. His system posed some significant limitations on the type of service that could be provided. As with the mechanical semaphore systems it would soon replace, Morse's telegraph required intervening trained operators who were capable of generating and understanding the code used. This meant that the actual end users were forced to rely on others to transmit their messages, which created the opportunity for error and also enabled eavesdropping by operators and telegraph company personnel on mes-sages being sent.[8] Furthermore, the capacity of the telegraph line was limited by the speed at which operators could signal and interpret infor-mation, which rarely reached 200 characters per minute.

After several often difficult years of development work, with little pri-vate investment in sight, a grant of $30,000 (or a bit more than

$500,000 in 2005 values) from Congress provided crucial development funds for designing and constructing telegraph wires between Washington and Baltimore. For three years (1844–47) the telegraph operated as a part of services offered by the U.S. Post Office. But few used the service because telegraphy was an expensive luxury, used primarily by government and those businesses that could afford to pay a premium to send messages by a medium faster than overland mails. Given that Washington and Baltimore were only about 30 miles apart, other less expensive means of carrying a message took most of the business.

Faced with continued losses, in 1847 postal officials made a decision their successors regretted for decades afterwards.[9] The Postmaster General sold the initial telegraph line connecting Baltimore and Washington to private interests. He argued that the government had no business operating such a service, especially since commercial operations elsewhere were well under way. Surely the commercial failure of the Baltimore–Washington line was in part due to the fact that the cities were already connected by a regular and efficient rail service, and that not enough other cities could be reached by telegraph.[10]

Extending the Telegraph

In order to exploit their invention, the Morse group began licensing their patent to private companies to build regional telegraph networks under terms that generally required a substantial equity to be turned over to the partnership.[11] Licensing others rather than taking on the whole massive construction process themselves was necessary because a national capital market, as we know it today, did not exist, and neither Morse nor his backers were sufficiently wealthy to self-capitalize the system. Thus, capital for the construction of lines could best be raised by local promoters that sold "subscriptions" to lines to investors. These promoters traveled from town to town along the projected route of a new line selling stock in the system to local investors. Unfortunately this business model led to a fragmented, often uncooperative collection of regional companies, which would make a large-scale, reliable network more difficult to realize.

The first of the licensed commercial lines connected New York and Philadelphia and opened on January 1846.[12] It produced enough revenue to cover operating expenses and provide a modest return on invested capital in its first five months of operation.[13] A New York–Boston line opened six months later and was profitable as was a New York–Buffalo line, which also opened that year. The New York–Buffalo route was the longest and most profitable of these lines, lending credence to the developing realization that telegraph profitability was a function of distance—

the longer the distance, the greater the demand for telegraph services because of the absence of viable speedy alternatives, although this line was also the most reliable of the lines built to that point.

Meanwhile several competing telegraph companies had initiated service and expanded in the late 1840s. Some of them were based on technical systems different than those of Morse. In 1846 Royal E. House filed for a patent on a printing telegraph[14] which was to be refined and later used by the Morse group's rivals.[15] With the success of telegraph service in the East, the Morse backers began planning expansion westward. The Morse group granted a broad license to Henry O'Rielly for service to the Middle West. After contract disputes developed, O'Rielly sought an alternative to the Morse telegraph to hedge his risk as the contract issue worked its way through the courts.

The Morse interests then granted new licenses to other operators to service the Middle Western cities, creating competition. Price competition usually appeared when two or three rivals served the same markets. While the Morse interests eventually prevailed in this and other disputes, the technological development spurred by competition made it difficult to maintain a patent-based monopoly. Soon additional competing systems were constructed in the profitable East Coast routes as well as in the Western arena. As a result of the Morse group's policy of licensing systems to local ownership without maintaining strong operational control, telegraph service was very fragmented. This made it difficult to maintain a monopoly on the service.

Railway Synergy

In 1848 the New York and Mississippi Valley Printing Telegraph Company was formed by Hiram Sibley (1807–88) and associates of Rochester, New York, and purchased the patent rights to the House telegraph. Sibley was one of the first telegraph leaders to realize that a substantial opportunity lay in development of a coordinated system. Financing difficulties, however, left his company with only a single line connecting Buffalo to Louisville by way of Cleveland, Columbus, Dayton, Cincinnati, and Frankfort by the end of 1853.[16] In 1854, the company reorganized. Sibley and his assistant, Anson Stager (1825–85), were able to negotiate very favorable agreements with several railroads in what is now the northern Midwest. These contracts allowed the New York and Mississippi Valley Company to grow without the draining capital investments that retarded or limited the growth of earlier systems. Other companies had negotiated contracts with railroads prior to this, but it was Stager's railroad experience that allowed the synergies to be communicated fully to the railroad managers.

The typical contract had the railway company build the lines (with their capital) along their rights of way along with telegraph offices; in exchange, New York and Mississippi Valley would operate the system and provide free priority transmission of railway messages and give the railroad Mississippi Valley stock for each mile of telegraph they built. This was valuable to the railroads because many of the routes had only single tracks; having a telegraph system allowed them to coordinate the use of these tracks for improved safety and efficiency without incurring the expense of operating and maintaining a telegraph system. By following the railroad rights of way, the telegraph company managed to avoid some of the competition that made competing systems unprofitable, because the railroad rights of way did not follow the routes of existing telegraph companies.

Thus despite the acrimony within the telegraph industry, by the 1850s close relations had developed between telegraph companies and rapidly expanding railway lines.[17] While the telegraph usually *followed* the railroad, the first coast-to-coast telegraph line opened in 1862, seven years *before* the first trans-continental rail line was completed. It immediately made money, again demonstrating telegraphy's value in long-distance communication.

Western Union

Sibley's railroad contracts were the basis for telegraph expansion and consolidation. With additional capital investments, Sibley was able to purchase marginal or failing systems at "fire sale" prices.[18] In addition, the New York and Mississippi Valley Company was able to purchase several of their significant competitors. Sibley chose to focus on takeovers rather than building new lines that would result in unprofitable competition between systems. When he purchased Morse licensees, he also purchased that company's rights to use Morse's patent. In 1856, the Mississippi Valley Company was renamed Western Union, as it now consisted of the (albeit sometimes involuntary) union of the "western" (today's Midwest) telegraph companies.

By 1857 Western Union controlled most of the routes to the west, had use of both the Morse and the House patents, and had exclusive contracts with railroads to protect its routes. Consequently, it was in a very strong position, particularly considering that the routes were substantially the economically valuable long distance ones. The lack of competition allowed the firm to charge higher prices and become profitable, whereas its predecessors had usually been only financially marginal operations. A similar consolidation took place on the eastern seaboard under the auspices of the American Telegraph Company.

In 1857 the six largest telegraph companies entered into a cartel called the "Treaty of Six Nations." This agreement set out four operating principles:

1. Members should interconnect only with each other and discourage competitors;
2. Where members had competing lines, they should divide up the traffic according to the business received from them;
3. Unions of competing lines should be attempted where possible; and
4. A Board of Control should be established to settle disputes.[19]

This set of principles evolved during negotiations, so that by the time the agreement was completed, the signatories divided the country into six sections and assigned monopoly control of each section to one firm. Some of the smaller competitors[20] objected and began to undertake the construction of competitive lines. Negotiations to satisfy these firms were concluded in 1859 (under the auspices of the North American Telegraph Association), which essentially resulted in their buyout by or merger with the six major firms. Thus, only 15 years after the first telegraph line had entered service, the consolidation of the industry from lively competition to a cartel of a few small firms was complete.[21]

In 1860 Western Union was successful at obtaining partial government financing (under provisions of the Pacific Telegraph Act) to extend a line via Salt Lake City out to the Pacific coast, prior to the completion of the transcontinental railroad. This line was completed in four months and was immediately profitable. This action was contrary to the "Treaty of Six Nations," however, and was one of the first important frictions within the cartel. The establishment of this line also nurtured the development of telegraph systems in California and the far West. In the ensuing years the United States Telegraph Company, the dominant firm in this market, would build additional transcontinental lines to compete with Western Union.

2.3   BUILDING A TELEGRAPH INDUSTRY (1860–76)

Expiration of the Morse patents in 1860 and 1861 did not detract from the profitability of the major companies, despite the emergence of new competitors. With the breakout of Civil War hostilities in early 1861, profitability of all telegraph lines was quickly and hugely enhanced due to needs of the military and high demand for news. Both the North and the South recognized the benefit of the telegraph for tactical and strategic purposes and constructed many telegraph lines to support the war effort. More than other telegraph companies, Western Union's lines were located substantially in the North. Companies whose lines bridged north and south found their networks split into two pieces during the war, which reduced their operating efficiencies. With the associated destruction and eventual defeat of the South, these companies sustained significant damage to their systems and business.

Supportive Role of Government

Despite Morse's and, later, Kendall's earnest efforts, the federal government largely remained disinterested in the development of the telegraph industry. Morse's original vision was for a single system, albeit one run by the government, not one run by commercial interests. The only government monetary contributions to help develop the telegraph were the investments of $30,000 in 1844 for construction of the Baltimore–Washington experimental line and another $40,000 (about $800,000 in 2005 values) government contribution toward building a telegraph line out to the Pacific Coast in 1860. But Western Union benefited greatly from the contribution of 15,000 miles of lines constructed by the military as part of its Civil War effort (granted to the company in 1865, in compensation for its losses suffered during the war). Western Union also enjoyed huge profits from its wartime business. That Anson Stager served as head of military telegraphs for the government *and* as general manager of the commercial firm at the same time certainly helped this process along.[22]

The Post Roads Act of 1866 provided another important government-sponsored benefit for telegraph companies: they were given the right to place telegraph wires along the nation's postal roads, and were permitted to cut trees from public land for use as telegraph poles in exchange for priority use of the lines by the government at rates fixed by the Postmaster General. In addition, cooperating telegraph companies were required to provide service to all without discrimination, introducing the notion of a "common carrier" to telecommunications for the first time.[23]

The Post Roads Act gave the government the option (for five years or until 1871) of purchasing all companies that took advantage of provisions of the act, effectively creating "postalization" (government ownership) as a policy option for federal planners. Advocates recommended that the government purchase Western Union and other telegraph operators, and then operate the single merged system as a government service much like the mails. This was the approach then being taken throughout most of the world, and, it was argued, would prevent problems of discrimination (both in price and service) by commercial carriers. Though this effort would lead to temporary control of telephone, telegraph, and wireless by the federal government during World War I, it never became American policy on a permanent basis.

Monopoly

By the end of the Civil War in 1865, Western Union was by far the largest single telegraph company in the United States. As the war drew to a close, the demand for telegraph services dropped off, causing many of the new competitors to go bankrupt or seek acquisition by one of the major firms. By 1866 the telegraph system consisted of three major companies: Western Union, American Telegraph, and United States Telegraph (which emerged from the Pacific coast and expanded eastward). Western Union was by far the largest, with a capitalization of more than $21 million (nearly $235 million in 2005 values).

Relative capitalization (i.e., the ratio of capitalization to gross receipts) provided a clearer picture of the company's relative size. Western Union had a relative capitalization of 9.2, compared with a ratio of only 1.4 for American Telegraph and 5.98 for United States Telegraph. Being overcapitalized in this way meant that its return on investment was lower. By another measure, Western Union was capitalized at $485 per wire mile, with American Telegraph at $87 per wire mile and United States Telegraph at $250 per wire mile when the cost of constructing a line at the time was $220 to $250 per mile (upwards of $3,000 in 2005 values).[24] Such overcapitalization meant that Western Union would be unable to earn a comparable return on its investments, opening themselves for takeover, since the stock price would decrease until "reasonable" capitalization levels could be achieved.

Although overcapitalized, Western Union saw its primary hope in monopolizing telegraph service. Thus, in a series of stock swaps, first United States Telegraph and then American Telegraph were merged with Western Union. Western Union's headquarters moved from Rochester, New York, to New York City as part of the consolidation, setting the stage for the domination of the company by prominent New York City capitalists, most notably the Vanderbilt family. At the time of this consolidation, the company had 2,250 telegraph offices, control of 37,380 miles of line and 75,686 miles of wire.[25] This combination made Western Union the largest company (of any kind) in the country. It had also become the first telecommunications monopoly, establishing something of a model for subsequent technology-based firms in this and other fields.

In the decade after the Civil War, the telegraph monopoly became a vital aspect of industry and government operations. Because telegraphy offered the potential of immediate recorded communication, though at a steep per-word cost, business and government continued to be its chief users. In the decade following consolidation, Western Union greatly developed its system, added thousands of local offices, and expanded the number of messages it carried four-fold. This expansion, and constant technical improvement, served to suppress all but the best-financed potential competitors. In a sense the company was compelled to build the additional lines to follow the expansion of the railroads; had they not done so, an opportunity would have been presented to competitors. By building ahead of demand, Western Union minimized the demand-based entry incentives that occurred during the Civil War.

New entrants did nonetheless emerge, although only one really posed a major threat to Western Union. The most significant competitive threat came from a company and system supported by financier Jay Gould.[26] Gould's financial reserves—and often unscrupulous methods—made him a formidable threat to Western Union's dominance. Using the American Union Telegraph Company, Gould was able to construct lines on railroad rights of way despite Western Union's exclusive railroad contracts as the result of a new law passed by Congress that year. The competition lowered Western Union's stock price, giving Gould the opportunity to engineer a takeover of Western Union in 1881. After the takeover, he merged the American Union system with Western Union. By then, however, antimonopoly sentiment was developing, culminating in congressional passage of the Sherman Antitrust Act in 1890. As a result, Western Union began to shift from a policy of taking over competitors to interconnecting with them.

To maintain its technological leadership and to develop more patent barriers to entry by competitors, Western Union invested in research and development both internally (by contracting with known inventors) and by purchasing significant patents.[27] One significant way to increase capacity at a low marginal cost was to develop a method of *multiplexing* signals on the telegraph wire. Multiplexing allowed several distinct transmission paths to exist on the same wire.[28] Western Union purchased the rights to a duplex system (capable of transmitting two channels simultaneously over a single wire) for $250,000 (about $3.6 million in 2005), which entered service in 1872. Famous inventor Thomas Edison (1847–1931) and telegraph authority George B. Prescott (born 1830) developed a quadruplex system (capable of carrying four channels over a single wire), which Western Union put into service six years later.

Stimulating telegraph expansion throughout the 1870s was the rapid growth of railways. As rail networks spread through the West and into the South, telegraph lines followed in their rights of way, demonstrating that both transportation and communication relied on integrated networks to deliver their products. Another parallel between the two services was their monopoly pricing policy which brought howls of outrage from those (especially farmers) forced to use the systems. As neither rails nor telegraph faced any technical competitor, service could be priced at whatever the market would bear. There was no government regulation to temper corporate pricing strategies.

Network Economics

In evaluating the telegraph's development, what is striking is the early tendency toward *concentration*. A concentrated industry is one in which the production capabilities are owned by only a few firms—what economists often refer to as an oligopoly. Through the Morse patent, the early telegraph industry was controlled to limit competition on any established route. As patent-based control became less effective (through contractual disputes and technological rivalry with other patents), competition emerged among many of the routes that had been originally licensed by the Morse group. Indeed, after the Morse patent expired in 1861, competition flourished. Yet, despite this intense competition, Western Union achieved industry dominance by 1866, dominance which lasted until the telegraph became obsolete and thus irrelevant.

Why did this happen, and how did Western Union achieve dominance? The telegraph industry, like most (especially wire line) telecommunications industries, is very capital intensive. Such an industry has a low *capital turnover ratio*, defined as the ratio of gross revenues to capital investment. This means that the revenues received per dollar of capital invested is small. The capital *intensiveness* of these industries leads to regulation because few competitors can afford the price of entry to the market, and incumbent firms can use this capital requirement to create barriers to market entry.[29]

Hiram Sibley (as noted earlier, of the New York and Mississippi Valley Printing Telegraph Company) recognized the benefits of concentration and also managed to achieve it by avoiding significant initial capital costs because the railways that he contracted with incurred the considerable capital expense. Sibley's recognition that consolidation was the "cure" for the telegraph industry was, effectively, a recognition that "reasonable" returns on assets could not be achieved in a highly competitive marketplace. As the network grew larger, network externalities began to become important. Western Union used these demand-side externalities as leverage to isolate and eventually buy out its competitors. This leverage was exercised through interconnection and traffic-sharing agreements. As Western Union grew, failure to reach such agreements limited the effectiveness of the smaller telegraph firms, because the customers of competing firms could not reach as many cities as they could with Western Union. Eventually, it became more profitable to merge *with* Western Union than to compete *against* it. Western Union also raised the costs of (potential) new entrants by writing exclusive contracts with *all* railways. Not only did potential entrants have to acquire the necessary rights-of-way, they also had to invest in the construction of the lines with their own capital. Western Union achieved this by exchanging their stock for a direct capital investment in outside plant by the railroad companies. Thus, Western Union was able to impose a larger cost on new entrants than it incurred itself.

Impact of Instant Communication

Some of the earliest users of the telegraph network included gamblers, stock brokers, and journalists.[30] Each found the timeliness of the telegraph appealing, albeit for different reasons. As telegraph networks expanded and became more reliable, business messages of many kinds were transmitted, but press interest in and dependence upon this technology persisted.

Timely information is the lifeblood of journalists. Newspapers that could acquire information sooner than their competitors attracted a larger readership and thus more advertising revenue. Given the high costs of acquiring timely information by means of telegraphy, newspapers began to create press associations to share the cost of news gathering as well as to alter the nature of competition among themselves. The emergence in 1848 of the Associated Press (AP) as the dominant news agency in the United States was due in no small part to their skill in using (and circumventing artificial obstacles placed by) the telegraph network.[31]

When Western Union succeeded in effectively monopolizing the telegraph business in 1866, the viability of the AP was threatened, because, given its control of the wires, Western Union could have easily established a rival news organization that would have been a potent competitor. At the same time, the AP was a major customer of Western Union, one that could help justify the construction of a rival network. Faced with this potential standoff, in 1867 Western Union and the AP entered into an agreement in which the AP agreed not to use the wires of Western Union's rivals for its transmissions, and Western Union agreed not to enter the news gathering business and to offer special discounts to the AP. While this agreement stimulated a popular movement for telegraph reform, no effective dismantling of the agreement emerged.[32]

This was merely the first of many such "client-and-carrier" arrangements that appear throughout the history of telecommunications. The agreements and the debates they foster focus attention on the appropriate level of integration between the information content providers (i.e., wire services, the press, the movie studios, and broadcasters) and telecommunications carriers. Some argue that such artificial restrictions between content providers and carriers limit the efficiency and effectiveness of the information products that can be sold to users.

Submarine Telegraphy—Going International

Land telegraphy systems were expanding elsewhere as well. Europe had systems in most countries by the 1850s and by 1865 had established the International Telegraph Union to help coordinate technical standards, prices, and revenue sharing. Many European colonies (such as British India) also made use of their initial telegraph services. But except for some small bodies of water (rivers and lakes mostly), the telegraph was only a *land* service and stopped at the ocean's edge. Technology then available did not sufficiently insulate the telegraph wires under water or boost weak signals to allow for successful transmission of signals over any distance. Indeed, technologists had no idea of some of the problems they would face.

But that didn't stop people from trying. After small bodies of water had been successfully crossed (lakes and then the English Channel), the first attempt at a trans–Atlantic telegraph cable, developed by an Anglo-American consortium under the direction of American entrepreneur Cyrus Field, was made in 1857 but was not completed. A second and more successful attempt was laid down in the summer of 1858. For a tantalizing few months, operators exchanged signals between Britain and North America, but then signals became slower and eventually stopped altogether. An inquiry determined that undersea cable insulation was insufficient and that far too much transmission power had been used, literally overpowering the delicate instruments of the time. The developing American political crisis and Civil War prevented any replacement expedition for six years.

In 1865, using the carrying capacity of the huge hull of the *Great Eastern* (an iron ship intended to carry hundreds of immigrants to Australia and the largest ship in the world until the turn of the century[33]), a new cable was attempted—only to break just short of success. A third expedition in 1866 finally did the trick and was able to grapple for and repair the 1865 cable as well. From that day forward, cables have operated across the Atlantic and soon were laid across the Indian Ocean and parts of the Pacific as well. Soon multiple telegraph cables were in place connecting major population centers around the world.

The impact of submarine telegraphy, which some observers dubbed "the grand Victorian technology," was considerable.[34] The pace of international business and diplomacy quickened as replies now took perhaps hours rather than days or weeks. The flow of news was dramatically sped up. What happened in distant places no longer seemed so far away. Indeed, the very perception of time and distance changed as Europe was drawn closer to the rest of the world.

2.4 TELEPHONY—ADDING VOICES (1876–85)

Driven in part by telegraphy's high prices (an obvious reflection of user demand but also a target for price-cutting competition), many inventors sought to add flexibility to telegraph service. One approach was to supplement or even replace coded telegraphy with voice service. This would allow elimination of the trained telegraph operator. As practical knowledge of electricity increased, the desire to transmit tones over wires came ever closer to reality. Some "harmonic telegraphs" were invented that enabled multiple transmission channels to be constructed over a single set of physical lines (frequency division multiplexing, in today's terminology). Since the cost of constructing lines was very high, the economic value of the harmonic telegraph was high. These concepts were simple; developing a viable system was something else.

Many innovators were actively trying. Experimenters had been working on devices, notably Philip Reis, who nearly invented a telephone device as early as 1860. In the 1870s, Elisha Gray, an engineer for Western Union's manufacturing arm, Western Electric Company, recognized that the transmission of voice over a wire was a logical extension and filed a *caveat*[35] for a patent on such a device on February 17, 1876, just hours after a speech teacher named Bell filed an actual patent on a similar device.

Bell's Device—"Nothing Much New"

Among those seeking an effective multiple telegraph solution was Alexander Graham Bell (1847–1922), a teacher of the deaf at Boston University. In 1872 he began to work on his "harmonic telegraph," work that was inspired by experiments of the German scientist Hermann von Helmholtz. In July 1874, while visiting his parents in Brantford, Ontario, Bell witnessed a demonstration of Dr. Clarence Blake's "phonoautograph," which translated sounds to paper markings using a dead man's ear drum; this rather grim demonstration gave him the idea for the telephone. Bell was joined by assistant Thomas Watson in early 1875 and the two worked in a rented space in downtown Boston. Bell's financial backers in this effort were Gardiner G. Hubbard, a lawyer and teacher (who would later become Bell's father-in-law), and Thomas Sanders, a Boston businessman.[36] Bell's real research interest, however, and one closer to his own professional training, was in developing a means of transmitting the human voice over wires. Bell and Watson worked through 1875–76 trying to perfect a rudimentary telephone device.

Their first successful experiment in voice telephony did not take place until March 10, 1876, when Watson, under Bell's direction, had a wire attached to a diaphragm inserted into acid. By carefully choosing the type of acid and the wire material, the reaction caused by these two coming in contact created an electrical current. As the depth of the wire in the acid varied, the amount of current in the wire also varied. A person's speech caused the depth of the wire in the acid to vary proportionally to the speech energy, causing varying electrical current in the wire. This current was reconverted to sound energy by a simple horn loudspeaker. On that day, some of the acid spilled, causing an alarmed Alexander Bell to yell "Mr. Watson, come here, I want you!" into the telephone.[37] Watson heard the words over the instrument and rushed to Bell's assistance. The acid containers proved to be problematic. A year later and under contract to Western Union, Thomas Edison developed a carbon button transmitter which achieved the variable resistance effect without use of acid. Edison's transmitter also used a battery at the telephone instrument to improve the signal strength, which improved the audibility of the telephone. This was soon followed (in 1878) by an improved carbon transmitter invented by Francis Blake, rights for which were purchased by the Bell Telephone Company. Consequently, the need for acid in transmitters for variable resistance disappeared, although the telephones of the day contained lead-acid batteries. The conversion to central office-supplied power was not to come for another decade.

Although preoccupied by the tension and excitement of his invention's progress (to the point where most of his private students had abandoned him), Bell did not ignore practical business requirements. As his experiments with the telephone continued, Bell took steps to secure the rights to his invention. On March 7, 1876, Bell was granted U.S. Patent No. 174,465 for "Improvements in Telegraphy." Ten months later, in January 1877, he was granted a second patent covering the primitive telephone device that he had created. A Bell Telephone Association was formed by Bell, Watson, and their financial backers to exploit these patents, which decades later were often referred to as among the most valuable ever granted.

Search for Support

Although these patents would soon play a central role in telephone development, at first they held only potential value. Bell's more immediate concerns were financial, for without additional funding his continuing perfection of the telephone would be in jeopardy. Although the scientific community showed some interest in Bell's work, other investors (and the general public) remained unaware of its potential.

In one desperate attempt to raise funds for further work, Gardiner Hubbard offered all rights to Bell's patents to Western Union for $100,000 (roughly equivalent to $1.5 million in 2005 dollars). Uncertain (some sources say scornful) about the rightful place of a telephone in business affairs (for unlike the respected telegraph, the telephone provided no written record and could only transmit sound about 20 miles with existing technology) in an era when written communications was essential, Western Union rejected the offer.[38]

The rejection set back the search for funding and illustrated a key problem Bell's backers had to overcome. Because the telegraph already offered the practical service of instant distant communication, many potential investors saw the telephone as merely an improvement (or a local supplement to the telegraph's national reach), not a real breakthrough technology. Though it offered instant communication without need for experienced technical operators, the telephone seemed a luxury when the already-expensive telegraph would usually suffice.

Despite public indifference to the telephone, Hubbard and Sanders formed the Bell Telephone Company of Massachusetts in July 1877, to market their new device throughout the country. Fewer than 800 telephones were then in use.[39] Bell's patent rights were the firm's chief asset. Hubbard became principal officer while Sanders, with the largest financial investment, became treasurer. Watson was named superintendent of telephone instrument production, and Bell assumed the humble title of company electrician. Two days after creating this business union, Bell forged a personal union by marrying Hubbard's daughter, Mabel.

In truth the new Bell Telephone Company held two singular assets: the telephone patents and the technical and promotional genius of Bell himself. To stimulate further interest in his invention, Bell demonstrated his telephone to packed lecture halls in the spring and early summer of 1877. At the same time, Hubbard attempted, again unsuccessfully, to interest additional investors in the telephone's commercial prospects.[40]

An air of worry prevailed at the new company. Ads were placed offering to furnish telephones for the transmission of speech through instruments not more than 20 miles apart. The initial aim of the company, in other words, was to sell equipment, not service. It was assumed that customers would seek out their own service provider. Slowly, the public became curious, then interested, then intrigued by the potential of instant voice connections at a distance. Soon the nascent enterprise found itself a victim of success—too much demand and too few telephones.

Building a Telephone System

Perhaps following the example of the early telegraph industry, Gardner Hubbard developed the inspired idea of enlisting the services of local agents located across the country to expand the telephone system. Since the Bell Telephone Company was very short on capital, it would follow a policy of licensing local promoters to build and operate networks. Even though the telegraph had done much to facilitate a national capital market, it remained far from being efficient, so local promoters were often effective. These businessmen would receive a license to provide telephone service in their communities for a limited amount of time (usually five years) and agreed to certain conditions in its development, including system compatibility. Customers were prohibited from connecting any equipment to the network other than that provided by the local agent. Many of these entrepreneurs formed new companies to support and expand their efforts.

While this strategy was similar to the one pursued by the Morse interests several decades earlier, the Bell Telephone Company was able to avoid the many internal squabbles that marked the telegraph patent partnership. Furthermore, with its technical limitations the telephone was much more of a local operation at the time, so that the operational nature of a telephone company was quite different than a telegraph company, where long distance mattered. Thus, the business model that the Bell Telephone Company (and it successors) pursued consisted of the following elements:

- Offer limited local operating licenses to local entrepreneurs[41]
- Manufacture telephone related equipment; and
- Lease this equipment to the local operating companies.

Thus, the Bell Telephone Company was not at all involved with local operations (with the exception of the New England Telephone Company). The revenues derived from operations stayed with the local entrepreneurs, but rental revenue from the telephones and related equipment came back to the patent partnership. This form of operation has been referred to as a *patent franchise*.[42]

Telephones in 1877 were so primitive that the relatively few pioneering customers could only talk to each other if a direct line ran from the person placing the call to the person receiving it. Party (shared) lines[43] were standard, with as many as six to twenty customers on each line, making privacy impossible. An additional burden for potential customers was their responsibility for obtaining and maintaining these lines; the

Bell Telephone Company merely leased telephone instruments to sub-scribers.

The limitations of such a system—where each customer had to main-tain a physical connection with every other customer—were obvious. And thus the telephone system evolved from what we would today call a private line network to a switched public network very early on. It quickly became apparent to local promoters that users of the telephone wanted to be able talk to *any* other user, not just the other end of the tele-phone line that had to be constructed for them. This customer need could only be satisfied if the lines could be interconnected in an arbitrary way. Economically and technically, the most feasible way of doing this at the time was in a centralized telephone exchange. The first was a switch-board that served the 21 customers of the New Haven District Telephone Company in New Haven, Connecticut, in January of 1878. Three days later, another was opened in Meriden. It quickly became apparent that this was the appropriate mode of operation for telephony. The operators, boys with training as telegraph operators, called customers by their names. Female telephone operators soon replaced the boys because of their more courteous manner with customers.[44]

Early telephone switchboards consisted of a panel of input jacks, one for each customer. The operators would use a patch cord to connect two parties that wished to speak to each other. While this technique worked well for small exchanges, there remained a number of problems with this technology: it did not "scale" well and became cumbersome for large ex-changes, and it was difficult for the operator to detect when a connection had been terminated. Each board could accommodate approximately 100 connections; as a telephone company grew, it had to install additional boards and enable connections between them. The problem of scale was recognized as early as 1879, but was not solved until 1897, after a series of conferences finally recognized that a hierarchical structure of switches (a "divided exchange") could solve the problem. Thus, operators would transfer a call not on their switchboard to another operator, who would only concern him- or herself with interboard connections.[45] Soon ex-changes (essentially larger switchboards connecting many customers) appeared in major cities; the greatly expanded usability they offered en-abled the Bell firm to sell some 10,000 telephones by late 1878.

The use of telephone numbers was introduced late in 1878 during a measles outbreak in Lowell, Massachusetts, that struck down all four of the town's operators. Faced with operating a board with 200 names (the name associated with each input jack had to be memorized), telephone service was briefly paralyzed. But if each jack (hence each customer) was to be assigned a number instead of a name, then a reference list (eventu-ally a telephone directory) could be used to associate numbers with customers, thereby solving this problem.[46] Once initiated, the "obvious" idea quickly became widespread.

Patent Fight

By now, however, the Bell interests were not the only people interested in improvements in telephony. Western Union, now recognizing that it had committed a sizable blunder by earlier rejecting Hubbard's offer to purchase Bell's patents, decided to carve out a share of the telephone market. Thomas Edison created a telephone transmitter which had several advantages over equipment the Bell system used; among other things, Bell's transmitter required a speaker to bellow to make himself heard whereas Edison's transmitter permitted a more normal tone of voice. When Bell's own local agents began to clamor for Edison's transmitter, the battle was joined: the Bell forces sued Western Union for patent infringement. Some 600 telephone patent battles took place before the patents expired in 1894—every one of them won by Bell interests. This emphasis on patent control characterized the industry's early development. One set of cases reached the Supreme Court in 1888 and resulted in an important decision upholding Bell's rights.[47]

The fight between Bell and Western Union took place outside the courtroom as well. Each firm raced to sign up as many customers as possible while refusing to interconnect with the competitor's network.[48] Bell interests also pursued technical equipment improvements to overcome Western Union's temporary lead. Late in 1879, the two companies agreed to stop competing. Under an agreement designed to be in force for nearly two decades, Western Union sold its patents and network of 56,000 telephones in some 55 cities to Bell, withdrew from the telephone business, and Bell, in turn, pulled out of the telegraph business. It seemed a reasonable settlement as the telephone, then suitable only for short-distance communication, would continue to be ancillary to the long-distance telegraph network. Over the life of the settlement, Western Union received some $7 million from Bell.[49] More importantly, the two companies worked cooperatively, making it difficult for outside competitors to enter either market.

While this model was effective for quite some time, early signs of its breakdown came soon after the agreement, when limited long distance telephone service was begun in 1881 between Boston and Providence, Rhode Island, a distance of 45 miles. This line used two metallic conductors for each circuit, eliminating the earth-based return, which had been the source of much of the noise in the telephone connections of the day. In 1884 the first copper lines went into service between New York and Boston, a distance of nearly 300 miles. With its lower resistance compared to iron,[50] copper enabled longer distance communications circuits. Now the telephone could begin to compete against the telegraph's ease of long distance service.

Having resolved their troubles with Western Union, the Bell forces continued to press their advantage by acquiring a host of patents on different aspects of telephony. From control of only 66 patents in 1879, Bell's establishment of an engineering department (1881) led to control of some 900 patents by the time the original patents expired in 1894.[51]

2.5 DEVELOPING THE TELEPHONE INDUSTRY (1885–93)

Development of the switchboard hinted at the critical advantage, that is, access to a system of connections that would let a customer use the phone to call any other customer. Bell company organizers undertook to develop telephone service on this basis throughout the country.

Fractious Local Exchanges

One important step was to standardize the hundreds of small telephone outlets created by local agents under license from Bell. Traveling agents for the Bell company distributed technical manuals and provided instruction on the installation, use, maintenance, and repair of telephone equipment and exchange systems. Agents were also invited to annual switchboard conferences and meetings of the National Telephone Exchange Association to discuss and adopt uniform technical criteria and service standards. American Bell was also busy shipping equipment to the licensees for use on their networks.

Bell sources were no longer the only providers of telephone equipment and service when Bell's original patent expired in 1893–94. Hundreds of independent (non-Bell-owned) companies entered the business and created a host of problems for the fledgling Bell system, including varied and often incompatible technical standards and loss of local customer revenues. One of the most difficult problems was how (or even whether) to interconnect these disparate independent exchanges with Bell companies to achieve an integrated system. Similar problems existed within the Bell System where problems of management and cost sharing forced American Bell into a realization that efficient service between and among interconnecting exchanges demanded central management, rather than mere license agreements with a multitude of small companies.

Sometimes the problem was just getting the different companies to talk to each other. For example, in 1882 the Wisconsin and Chicago companies each contracted to construct segments of a Chicago–Milwaukee line to be connected at Racine, Wisconsin. Upon arriving at Racine, the companies discovered that they had used different gauge wires, and the resulting connections never afforded adequate service. Elsewhere, Southern New England Telephone sought to provide long distance service between New York and Boston by sublicensing local companies to construct portions of the line within their respective territories. The line was built (and then rebuilt using improved technology) but was never used because of squabbles between the partners over its ownership and operation.[52]

Western Electric

Founded in 1869 to manufacture telegraph and other electrical devices, Illinois-based Western Electric (the name adopted in 1872) became by 1880 the largest manufacturer of telegraph equipment in the United States. With two-thirds of its stock owned by Western Union and one of its senior officers, the company began to make telephone equipment in 1878–79 when the telegraph firm sought to compete with AT&T (American Telephone & Telegraph).[53] With the settlement with Bell, however, Western Electric had excess manufacturing capacity. And both the manufacturer and its parent company were tied up and in a somewhat confused management situation thanks to financier Jay Gould. Gould sought a quick financial return.

In mid-1882 American Bell purchased Western Union's one-third holding of Western Electric's stock (for $150,000 or about $2.7 million in 2005 values), expanding its stock holdings in the months to come until it held majority control. This purchase accomplished several things: it

- Eliminated price competition between the various suppliers and "piracy";
- Secured for American Bell additional key patents to further fortify its monopoly on telephony;
- Provided a reliable source of supply for its rapidly growing network; and
- Secured a perpetual license for Western Electric for the manufacture of telephones and telephone equipment.[54]

That ownership interest was soon increased to 60 percent, and by 1908 AT&T owned 80 percent (the last private shareholders were not bought out until 1966).[55] From the start of Bell ownership, Western Electric was granted sole rights to manufacture equipment under Bell patents, and would devote its research and manufacturing facilities only to Bell System company needs. By 1892, there were nearly a quarter-million telephones in use in the United States.

Developing AT&T

By the mid-1880s, a scant decade after Bell had obtained his initial telephone patents, the telephone business was characterized by individual market fiefdoms rather than any rational plan. Busy local agents recruited customers in their service areas and some cities (e.g., Boston, New York, and Chicago) had expanding networks. But most calls were placed through primitive switchboards operated by young men who often demonstrated an unreliable work ethic (women rarely worked outside the home at this time but would eventually take on this work). And almost no calls could travel more than 20 miles given the technology available. Equipment varied from community to community as standardization was unknown.

American Bell managers had grown frustrated at this disconnected and disorganized state of affairs. It was obvious that the Bell company's plan for providing service through separately owned and managed licensee companies would probably never work. Rather than trying to achieve cooperation among the small companies, American Bell took over the work of installing interexchange lines to connect local exchange companies, a scheme which still required the companies' cooperation, something not always easily achieved.

American Bell embarked on a plan to consolidate many local exchanges into larger operating units, taking an equity interest in the newly created enterprises. To attract local owners, the Bell company offered to exchange the original five-year contract for a perpetual license. One of the major problems of consolidating separate telephone companies was assuring compatibility of equipment; another was guaranteeing ownership of facilities; yet a third was defining a company's service area.

Hampered by certain aspects of the corporation law in Massachusetts, in 1885 American Bell organized a subsidiary in New York known as the American Telephone and Telegraph Company (AT&T). The mission of the new subsidiary was provision of a comprehensive system of long distance service to be achieved by tying individual franchises together with standard equipment. By the time the basic Bell patents expired in 1893–94, Bell engineers had developed a method for transmitting the human voice over distances greater than 20 miles (termed "long distance"). While the service was primitive and unreliable at best, it did allow people in Boston to talk to customers in New York, a feature which gave the Bell group a substantial advantage over its rivals.

Perhaps more importantly, the ability to transmit the sound of a human voice (not just a coded message) over long distance meant that telephone service would eventually eclipse telegraph service as the preferred mode of business and personal communications.

NOTES

1. See D. J. Woolliscroft, 2001. *Roman Military Signalling* (Charleston, SC: Tempus) for a well-illustrated discussion of Roman methods.
2. For an excellent treatment of the development and application of these mechanical semaphore systems in all parts of the world, see Geoffrey Wilson, 1976. *The Old Telegraphs* (London: Phillimore).
3. The standard historical source on the early American electrical industry is Harold C. Passer, 1953. *The Electrical Manufacturers, 1875–1900* (Cambridge: Harvard University Press), which discusses arc and incandescent lighting, and electric power including transport.
4. Ibid., p. 11.
5. See John D. Rider and Donald G. Fink. 1984. *Engineers & Electrons: A Century of Electrical Progress* (New York: IEEE Press) for the development of the AIEE and IRE, which combined in 1963 to become today's IEEE.
6. These resulted from the work of William Cooke and Charles Wheatstone, who later fell out with one another and initiated a decades-long battle over who played the key role. See Geoffrey Hubbard, 1965. *Cooke and Wheatstone and the Invention of the Electric Telegraph* (London: Routledge & Kegan Paul), and J. L. Keive, 1973. *The Electric Telegraph in the U.K.* (New York: Barnes & Noble).
7. The Morse Code is a *variable length code* because each character is represented by a variable number of transmission symbols (dots and dashes, in this case). It is actually a fairly good code by the standards of information theory in that the average number of symbols per character is small (smaller than a *fixed length code*, in which each character is represented by the same number of symbols). Morse represented the most frequently occurring characters of the English language by short symbol strings (for example, "E" is "dot" and "T" is "dash") and less frequent characters by longer strings ("X" is "dash-dot-dot-dash"). This results in a very efficient code.
8. Brock reports that this "eavesdropping" feature of the telegraph was used to advantage by the Mississippi Valley company. He also indicates that it was common practice to encode confidential messages. Gerald Brock, 1981. *The Telecommunication Industry* (Cambridge, MA: Harvard University Press), pp. 77–78.
9. For a good review of the subsequent history of Post Office attempts to re-acquire control of the telegraph and later telephone services, see the report of the U.S. Post Office Department, Postmaster General, *Government Ownership of Electrical Means of Communication.* 63rd Cong., 2nd Sess., Senate Document No. 399. (Washington: Government Printing Office, 1914; partially reprinted by Arno Press, 1974).
10. Brock, p. 58.
11. The partnership (Morse, Amos Kendall, F. O. J. Smith, and Alfred Vail) was subject to considerable internal disagreements, which later impaired their ability to develop a national system under their control.
12. This line actually terminated on the western shore of the Hudson River in New Jersey. It would take several years before the use of the natural rubber (*gutta percha*) for un-

derwater cable would be developed. In the meantime, messages were ferried across the Hudson River several times daily.

13. The receipts might have been higher, but the line was out of service for nearly one-quarter of that period. This underscores the new and experimental nature of the technology. Little was known about electricity; better construction techniques evolved over the next several years as more experience with system operation and construction was gained.

14. Robert Luther Thompson, 1947. *Wiring a Continent: A History of the Telegraph Industry in the United States, 1832–1866* (Princeton, NJ: Princeton University Press), p. 54.

15. Unlike the Morse system, which required operators who understood Morse Code, the House system used a keyboard that sent information in such a way that it could be printed by the receiving machine in Latin characters. This patent was used by many of the Morse competitors, even though it had some operational difficulties with respect to the Morse system. See Paul Israel, 1992. *From Machine Shop to Industrial Laboratory: Telegraphy and the Changing Context of American Invention, 1830–1920* (Baltimore: Johns Hopkins University Press), pp. 44–45.

16. The financing difficulties were due, in part, to the terrible reputation telegraph firms had among investors. Competition made these early systems only marginally profitable, making it impossible for them to pay the promised dividends. Furthermore, there was no shortage of scams involving telegraph systems.

17. U.S. railroads did not widely appreciate the synergy with telegraphy until 1855, although the first experiments began in 1851 with the New York and Erie Railroad. Railways in Europe were much quicker to adopt telegraphy.

18. These purchases, while bargains, often required significant investment to improve the lines and allow them to be integrated into the growing regional system, as these marginal operations often neglected maintenance and upgrading of their physical plant.

19. Thompson, p. 312.

20. Most notably, the competition came from Amos Kendall's Magnetic Telegraph Company, who built the first commercial line from New York to Philadelphia and F. O. J. Smith's widespread stock and patent rights. Kendall and Smith were among the earliest of Morse's partners in telegraphy. Thompson shows that one objective of the "Treaty of Six Nations" was to eliminate the influence of Kendall and Smith over the industry. See Thompson, p. 313.

21. These consolidations resulted in the removal of the last of Morse's original backers (Kendall and Smith) from active involvement in the telegraph enterprise.

22. Brock, p. 82.

23. Horwitz, p. 94.

24. Thompson, pp. 407–409.

25. Ibid.

26. Western Union was by this point controlled by the Vanderbilt family. Gould was engaged in several control struggles with the Vanderbilts, so this was simply an extension of that larger conflict.

27. Israel argues that technological advances did not provide a significant challenge to Western Union based on the numerous failures of technology-based competition to Western Union. Israel, pp. 146–151.

28. Today, multiplexing is frequently used by speeding up the bit rate on a transmission medium and dividing time slices between the distinct transmission paths (called *time division multiplexing*). This approach was not feasible with the telegraph, since it relied ei-

ther on human operators (the Morse system) or on mechanical printers (the House system). Thus, the capacity could only be increased by *frequency division multiplexing*, in which the distinct transmission channels are given unique tones (or frequencies) on which to transmit; the tones would be transmitted simultaneously. The receiver would then be sensitive to different frequencies so that the channels could be separated again.

29. Baumol and Sidak follow George Stigler in defining an entry barrier as an arrangement that imposes costs on an entrant from which the incumbent is immune. See William J. Baumol and Gregory J. Sidak, 1994. *Competition in Local Telephony* (Cambridge, MA: MIT Press), p. 129. The capital requirement can be a barrier to entry because an entrant faces a higher risk (due to strategic entry countermeasures by the incumbent), which results in generally higher capital costs.

30. Thompson, p. 47.

31. As is documented by Thompson in detail on pp. 225–239. See also Menahem Blondheim, 1994. *News Over the Wires: The Telegraph and the Flow of Public Information in America, 1844–1897* (Cambridge, MA: Harvard University Press).

32. See Horwitz, p. 94.

33. For the fascinating story of this white elephant, including the cable-laying missions, see James Dugan, 1953. *The Great Ship* (New York: Harper).

34. By far the best discussion of this is found in Vary Coates and Bernard Finn, 1979. *A Retrospective Technology Assessment: Submarine Telegraphy—the Transatlantic Cable of 1866* (San Francisco: San Francisco Press).

35. No longer used, a caveat was essentially announcement of an intent to file for a patent. A caveat indicated a line of work under way, but not yet fully developed enough to file an actual patent on specific devices.

36. Gardiner and Sanders had become acquainted with Bell as the teacher hired to instruct their deaf children.

37. Long thought by many to be a mere tale told long after the fact, these very words appear in Bell's hand on the pages of his laboratory notebook entry for this date. The notebook is now in the Library of Congress.

38. This decision may have been strongly motivated by prior interactions between Hubbard and Western Union President Orton. Hubbard had been a proponent of more governmental control over Western Union. Furthermore, the telephone had significant technical problems at the time the deal was offered. See Horwitz, p. 307, note 25.

39. Charles L. Brown, 1991. "The Bell System," in Fritz Froehlich et al., eds. *The Froehlich/Kent Encyclopedia of Telecommunications*, Vol. 2 (New York: Marcel Dekker), p. 66.

40. The Bell Telephone Company was reorganized as the National Bell Telephone Company on February 17, 1879. This reorganization was the result of a desperate need for additional capital and was urged by William Forbes, a man with a finance background. This reorganization reduced the influence of the original Bell backers, Hubbard and Sanders, who were in strong disagreement about financial matters, and elevated Forbes to the leadership position; it also raised additional capital to permit the continued operation and expansion of the Bell interests. Together, Forbes and system manager Theodore Vail strengthened the finances and operations of the firm. See George David Smith, 1985. *The Anatomy of a Business Strategy: Bell Western Electric and the Origins of the American Telephone Industry* (Baltimore: Johns Hopkins University Press), pp. 55–57.

41. This has business as well as technical reasons. The technological reason was that the telephone was a purely local device. The primary business reason is that the limited local licenses proscribed the activities of licensees, reserving new business opportunities

(such as interconnection between companies and "long distance") for the Bell Telephone Company.

42. Smith, pp. 5–7.

43. Party lines, now almost extinct in America, were those with two or more (and sometimes a dozen) customers on the same line. If one was called, all telephones on the line would ring at the same time. While obviously more efficient (rarely was more than one user in need of the telephone at the same time), party lines lacked the privacy assumed with single-party service.

44. John Brooks, 1976. *Telephone: The First Hundred Years* (New York: Harper & Row), pp. 65–66.

45. See Amy Friedlander, 1995. *Natural Monopoly and Universal Service: Telephones and Telegraphs in the U.S. Communications Infrastructure, 1837–1940* (Reston, VA: Corporation for National Research Initiatives), pp. 32–36.

46. The local company was concerned about the customer reaction to being called by a number instead of their name. Apparently, the objections were few. Numbering then quickly spread to other exchanges. Exchanges themselves remained named until the introduction of all-digit dialing in the late 1950s. This transition was done gradually, with the replacement of names by two-letter mnemonics (e.g., "Cedar" became "CE"), then the use of digits only in small towns. When this was expanded to a national scale, a significant public outcry ensued (unlike the conversion to telephone numbers) as people felt a sense of identity loss when the long-familiar exchanges disappeared into numbers. Digit dialing was not completely converted until the 1970s. See Brooks, pp. 74, 270–273.

47. See *The Telephone Cases: Cases Adjudged in the Supreme Court at October Term, 1887.* (New York: Banks & Bros) "United States Supreme Court Reports, Vol. 126," 1888, 531 pp. This is the only volume of Supreme Court decisions devoted to a single issue (and 10 interrelated cases).

48. In 1878, Western Union was an aggressive, well-capitalized competitor, having overcome its financial difficulties of the late 1860s and benefiting from its virtual monopoly. They undertook a policy of installing rival, non-interconnected networks wherever the Bell licensees installed one. Further, they threatened to remove the telegraph connections of institutions that connected with the system of a Bell licensee. At the time, losing the telegraph connection was a matter of vital commercial importance, whereas the telephone's utility was unproven. Brock, 1981, p. 94.

49. Brock, p. 95.

50. Iron was the common material used for telegraph wires because of its higher strength, even though its inferior electrical properties were well known as early as the 1840s. See Thompson, p. 46. The single wire with earth-based return was also common for telegraphy. Since the telegraph was essentially a digital medium, the problems with this approach were much less apparent. Furthermore, electromechanical telegraph repeaters were developed early on, so the higher resistance of iron was less of a problem. See Brooks, pp. 89–91.

51. Brock, p. 103.

52. *Defendant's Third Statement of Contentions and Proof*, Civil Action No. 74-1698 (U.S. v. AT&T), U.S. District Court for the District of Columbia, March 10, 1980, Vol I pp. 106–107.

53. AT&T, 1980. *Defendants' Third Statement of Contentions and Proof* in Civil Action No. 74-1698, *U.S. v. AT&T.* (Washington: AT&T, March 10), Vol I, p. 120 (hereinafter "Gold

Book"); and U.S. Department of Justice, Antitrust Division, *Plaintiff's Third Statement of Contentions and Proof* in Civil Action No. 74-1698, *U.S. v. AT&T.* January 10, 1980, Vol I, p. 27 (hereinafter "Green Book").

54. Western Electric would continue to supply many aspects of the market for electrical equipment, including magnetos and light bulbs until the mid-1910s. See the full story in Stephen B. Adams and Orville R. Butler, 1999. *Manufacturing the Future: A History of Western Electric* (New York: Cambridge University Press).

55. Ibid.

Chapter 3

Era of Competition
(1893–1921)

Expiration of Bell's basic patents in 1893–94 rapidly changed the structure of the telecommunications industry. Since telephone devices were no longer protected by patents, anyone could enter the business to manufacture or purchase telephones and equipment for their system. The resulting local exchange competition lasted until approximately 1921, with the passage of the Willis–Graham Act.

The emergence of competition introduced an important new issue into telecommunications—*interconnection*. Interconnection between service providers was not a major issue during the early competition between National Bell and Western Union; to the extent that it was an issue, it focused on the synergy between telephone and telegraph. With telephone competition emerging on a larger scale, the issue of interconnection became acute. Should telephone service providers be *required* to interconnect? If so, under what circumstances? The lessons of this quarter century provide insight into interconnection issues faced eight decades later.

Public sentiment also affected the nature of the telephone business. Beginning in the 1860s and continuing through the remainder of the 19th century, antimonopoly sentiment rose steadily. This took many forms, from the unsuccessful movement to "postalize" Western Union (turn it over to government ownership) in the 1870s to the more successful Grange agricultural movement. The notion of regulation began to enter the legal sphere at this time in the *Munn v. Illinois* decision of the Supreme Court (*see section 1.3*). The antimonopoly sentiment also permeated a few of the patent infringement cases that were brought against Bell prior to the expiration of the patents.

The attitude of American Bell and its affiliates toward its customers and its business during the monopoly period engendered an attitude that was often hostile:

> After Vail left, Bell adopted a policy which, if not "the public be damned," certainly was the "public be ignored" and before the turn of the century had severely damaged its position. . . . [Bell president] Hudson . . . thought the telephone use was a privilege, rather than a service. For seven years Bell remained at a standstill, adding few exchanges, permitting its equipment to deteriorate. . . . Bell installed telephones at its leisure and at the customer's expense, levied whatever fees it thought the market would bear, and made maintenance a sometime thing.[1]

This accumulated ill will would haunt American Bell and its affiliates at the outset of the competitive era, particularly when this experience was combined with the emergent populist antimonopoly sentiment.

3.1   IMPROVING TECHNOLOGIES

These three decades saw many important changes in telecommunications technology. Many of these technological changes had a significant impact on the structure of the industry and the dynamics of local exchange competition. The most important technological changes included establishment of a switching hierarchy, emergence of "common battery" systems, emergence of "automatic switching," emergence of viable long distance technology, and the inception of wireless telegraphy and telephony. Each of these would affect the competitive picture in different ways.

Switching—Connecting Everyone

Switching technology underwent three important changes in this period. The first was architectural in the sense that it allowed existing switching technology (manual exchanges) to be used to construct networks that were arbitrarily large. As noted earlier (*see section 2.4*), divided exchanges consisted essentially of the creation of a "switching hierarchy," in which new switches (or panels) were added that dealt solely with the interconnection of other switches. So important was this problem to telephone operators that as early as 1879, they began to note that flat rate service pricing (see appendix A) might not be economically feasible; that "per-switch" (per-call) pricing might have to be introduced.[2]

This lack of economy of scale in switching exists fundamentally because the complexity of the switching problem increases proportionally to the square of the number of lines being served.[3] This "dis-economy" of

scale was costly because of the size of the panels, the number of operators, and, most importantly, because of the costs of coordinating hand-
offs between operators. The switching problem consumed a good deal of
the technical and operational development of the telephone industry until
its resolution in 1897, when the switching hierarchy was introduced.[4]
Because of this, one historian notes that

> large urban exchanges were the most expensive and difficult to operate.
> Telephone service in the manual switching era was characterized by dis-
> economies of scale. In large systems, signaling was more complex, mainte-
> nance more expensive and labor less productive. The small-scale telephone
> switchboards needed by small towns and rural areas, on the other hand,
> were easy to manufacture and inexpensive to operate.[5]

Thus, the new entrants to the telephone business (*see section 3.2*) would
face a more favorable cost structure at the outset. With the introduction
of a toll switching hierarchy, this advantage became less apparent.

The introduction of "common battery" systems was another important change in the development of switching technology. In a common
battery system, the telephone office contains the battery that operates the
handset, line, and switchboard; that is, all elements of the local loop (the
line that connects the central office to the subscriber's premise). Prior to
the common battery system, each device had its own battery. In addition
to compounding maintenance problems, the distributed battery system
made signaling between the devices more difficult. Thus, common battery systems allowed an improvement in the efficiency of telephone operations, from both improved signaling and standardization. This reduced
the cost of operating exchanges and telephone systems.

Automatic Switching

The first widely accepted automatic telephone switch was invented in
1891 by Almon Strowger (1839–1902). He was an undertaker from
Kansas City who is said to have been motivated by "malicious or corrupt
telephone operators," who he claimed were steering business to his competitors. By replacing the operator with an automatic switch, this problem would be solved. Strowger received a patent for his switch, even
though it was not the first such device, because it could be scaled up to
handle 99 telephones, whereas earlier switches were limited to a relative
handful of users. Strowger founded the Automatic Electric Company
that year to manufacture and promote this technology.[6] Strowger's
"step-by-step" switch technology was widely enhanced and adopted for

telephone use. In many parts of the world, these durable switches are still in operation.[7]

AT&T resisted the adoption of the automatic switch at first, largely out of the belief that it was inappropriate to be involving the customers in the switching process. It was not obvious at the time how well customers would take to dialing their own numbers rather than simply announcing them to operators to dial for them. Furthermore, there was little standardization of telephone instruments capable of supporting automatic systems; early instruments required a separate button on the customer's telephone set that would initiate a ring at the destination. It was not until the 1920s that telephone call setup steps were completely automated in a transparent way.[8]

Not only was this antithetical to the "user transparency" service vision of AT&T, but the company had just "tamed the dragon of organizational complexity with manual technology, [so that] changing made little sense." Furthermore, "automatic switching around 1910 was not [cost] competitive with manual in making extensive toll and interexchange connections."[9] Even though AT&T began to develop automatic switching in 1902, when automatic technology was introduced in the Bell System, it was done to augment rather than to replace operators.[10] Bell was criticized for their late adoption of automatic switching,[11] even though their concerns were quite appropriate given the scale of their operation.

Long Distance Transmission

The first long distance line went into service in 1881 between Boston and Providence, Rhode Island. These lines used iron wire circuits (in which both the signal and return line were over iron conductors, instead of the return being through the earth), which improved the quality of transmission. These circuits could be and soon were improved and extended by using copper wires of increasingly greater diameter.

With the addition of *loading coils*, or Pupin coils (after Michael Pupin, 1854–1935, who received the patent on the device in 1900), the range could be extended further.[12] Loading coils were passive inductors (carefully constructed coils of wire) that altered the properties of the wire so that the electrical signals in the voice frequency range were emphasized at the expense of higher frequencies.[13] The long distance service, even with loading coils however, was of marginal quality, since it could not be amplified.[14] Higher quality long distance communications became practical in 1913, when the vacuum tube *repeater* was put into service. Repeaters require an *amplifier*, and no practical amplifier existed until Lee de Forest (1873–1961) invented his "Audion" three-element vacuum tube in 1906. AT&T bought partial rights to the patent and set out to develop a repeater.[15] Improvements in the Audion made by de Forest and a variety of scientists and engineers at AT&T, along with Edwin Howard Armstrong's (1890–1954) negative feedback amplifier, enabled the development of a practical repeater. With this device, a coast-to-coast link (New York to San Francisco) was placed in service on January 25, 1915.

The emergence of this true long distance network was to become a formidable competitive tool for the Bell System. Attempts by the independents to form a separate long distance network failed, largely motivated by AT&T's deepening business influence.[16] Control of the only effective long distance network meant that Bell customers could talk to many more people than the customers of independents, greatly enhancing the benefit of subscribing to the Bell System. As the competitive era wore on, the absence of a viable long distance alternative would work against the independents, gradually eroding their economic viability. This is one of the many reasons why the number of independent telephone companies began declining after 1905.

Wireless Telegraphy and Telephony

Although the focus of this account is not upon broadcast communications, development of long distance telephony and later mobile telephone communication cannot be fully understood without some background on the rise of wireless.

The theoretical foundation for the emergence of radio lies in Michael Faraday's (1791–1867) theory of electromagnetic induction in 1854, which was developed into a broad theory of electromagnetic fields by Scottish physicist James Clerk Maxwell (1831–79) in the 1860s and experimentally confirmed by Heinrich Hertz (1857–94) in 1887–88 experiments in Germany. As impressive a body of work as these three scientists produced, it was Italian Guglielmo Marconi (1875–1937) who developed a *system* of practical radio telegraphy in the late 1890s, and secured the necessary patents.[17] Marconi used "spark gap" techniques to create the signals for wireless telegraphy.

A spark gap transmitter works by increasing the voltage between two physically separated electrodes until a spark leaps across the gap (functionally, this is the equivalent of an automobile spark plug). One electrode was attached to a transmitting antenna and the other to ground. This caused the electromagnetic disturbance of the spark to radiate from the antenna. The spark gap system would cause a rapid succession of sparks (at a rate of eight per second in the Marconi system) to occur for shorter (Morse DOT) or longer (Morse DASH) time intervals.[18] These sparks would cause a resonant "ringing" in the antenna apparatus that would cause energy to be broadcast. The frequency of the ringing was a function of the length of the antenna and the properties of the other components. Receivers could then be tuned to detect these emanations because the rate of sparking within an interval corresponded to the "carrier frequency" of the signal. While this technique worked for binary telegraph signals, it could not be easily extended to more complex needs of voice transmission. Further, Marconi soon discovered that anyone could receive the signals; this caused privacy and security concerns for both commercial and military clients.[19]

Marconi's primary vision of wireless telegraphy was essentially as a point-to-point medium, one that found a particular application to communications with ships at sea and between stations across oceans—the latter in competition with telegraph cable companies. The Marconi company soon acquired a dominant position in radio communications, handling 90 percent of radio communications in the United States between 1910 and World War I.[20] Early wireless, however, was "a lonely place" because of its low usage. This developing situation formed the context of two Berlin wireless conferences in 1903 and 1906 and a London conference in 1912. These international meetings were concerned with agreement on techniques and protocols for senders using one system to find and communicate with receivers using another.

Growing in part out of these international gatherings, the first American legislation concerning the use of wireless telegraphy at sea passed in 1910 and was updated in 1912. With the separate Radio Act (also passed in 1912), the Department of Commerce gained the authority to license American wireless transmitters, though the law did not give the Secretary of Commerce the discretion to *deny* a license to any citizen.[21] The legislation established government control over the use of frequencies and placed some restrictions on operators' behavior.[22]

Michael Pupin had suggested the advantages of using "continuous wave" systems as early as 1899, but it took the later efforts of Fessenden, de Forest, Alexanderson, and others to make such signals a practical reality.[23] With the development of Lee de Forest's improved vacuum tube, new technology could be applied to radio. Inventor Reginald Fessenden (1866–1932) was able to demonstrate broadcast speech by as early as 1906. Instead of using a spark gap to create a "carrier wave" (a system that was not precisely controllable, consumed a lot of spectrum, and was not easily adaptable to voice), the vacuum tube enabled the transmission of a continuous wave carrier, consisting of a sine wave. Such a signal could be more easily modulated with a voice signal, more precisely controlled, and more spectrum efficient. This was clearly the technology of the future, and American Marconi (Marconi's affiliate in the United States) began several attempts to capture the necessary patent rights.[24]

Patent Pool

Commercialization of any radiocommunication system, however, was inhibited by diverse ownership of key patents. AT&T held some patent rights to use of de Forest's triode vacuum tube, Westinghouse had the rights to Armstrong's negative feedback amplifier, and General Electric owned the rights to the Alexanderson alternator, a sizable and expensive transmitter that could produce continuous waves efficiently at high power. When the United States entered World War I in April 1917, the Navy, which had been conducting wireless experiments since 1904, became sensitive to the potential conflict with the foreign-owned American Marconi Company. Furthermore, being witness to Great Britain's use of its control of international telegraph undersea cables to isolate Germany,[25] the Navy decided to take action to develop an American wireless infrastructure. Using the wartime emergency as a lever, the Navy persuaded the major patent holders to create a pool of relevant patents for the duration of hostilities.

The need for such a pool naturally continued after the war, and private negotiations forged a new patent pool as a series of agreements among such licensees as General Electric (GE), Westinghouse, AT&T, and United Fruit (which held valuable crystal wireless tuning patents). The pool was arranged under the auspices of the newly formed (1919) Radio Corporation of America (RCA), which also oversaw a market segmentation agreement—essentially a manufacturing cartel. RCA took over the facilities of the former enemy countries (primarily German wireless stations) and purchased the stations of American Marconi. With all of this, RCA was to be an operating company for ship and intercontinental traffic and a sales agent for radio equipment manufactured by GE and Westinghouse. The segmentation agreement created a "radio group" (consisting of GE and Westinghouse) and a "telephone group" (made up of AT&T and Western Electric). This agreement was reached before the concept of broadcasting was widely understood and made no provision for such a service.

Regular broadcasting began with Westinghouse's KDKA station in Pittsburgh in November 1920. Within two years, radio broadcasting would take on unforeseen popularity with hundreds of transmitters and tens of thousands of receivers. This popularity would lead to new problems (*see section 4.1*).

3.2 COMPETITION APPEARS

On the expiration of Bell's two original patents in 1893–94, competitors jumped into a rapidly expanding telephone market. Literally thousands of independent telephone companies were organized to serve less populated areas not yet reached by Bell licensees or, in rarer cases, to engage in head-to-head competition with a Bell-affiliated company for local subscribers. Most of the new firms (there were some 3,000 commercial companies by 1902) were small and undercapitalized. "The proportion of telephones controlled by the independents rose from 19 percent in 1897 to 44 percent in 1902, and then slowly increased to 49 percent in 1907."[26] But AT&T was stronger than the percentages suggested.

Rise of the "Independents"

When the basic Bell patents expired, the public was ripe for the emergence of competing systems due to several converging factors. For example, "agricultural states were . . . chosen for the greater number of independent telephone companies."[27] While this was in no small measure due to Bell's strategy of building systems in larger cities and higher density areas, the antimonopoly sentiment of the Grange movement, which was focused in agricultural areas, no doubt played a role as well. Price mattered too: "Lower rates, combined with the appeal of their underdog status, made the independents popular."[28]

This popular sentiment had its effects in the patent arena as well, where a challenge to American Bell's Emile Berliner (1851–1929) carbon microphone patent[29] was interpreted so narrowly as to effectively nullify it; the patent had been delayed in the patent office for several years, ostensibly for administrative reasons (applied for in 1877, it was not granted until 1891).[30] With this 1897 reversal of the Berliner patent, American Bell's strategy of protecting its monopoly through patent litigation began to fail. Several other key patents were overturned in short order following this case. By 1901 AT&T abandoned its patent-based strategy against the independents and began a policy of outright competition. This will be discussed in more detail below.

A U.S. Department of Commerce and Labor study in 1902 identified three general types of new entrants to the telephone business:

- Commercial telephone systems that were established with the intent of making a profit;
- Mutual telephone systems that were established for the purpose of providing service where Bell licensees did not. These were often by means of cooperatives, where the subscribers capitalized the systems; and
- Rural (or farmer) systems that were very informally organized and were designed to interconnect farms in a region. These were inexpensively built, often without a switchboard, and occasionally using barbed wire as lines.[31]

Of these, only commercial systems were likely to come into competition with Bell licensees. The mutual and rural systems often had no accounting system and sometimes put the burden of line maintenance and repair on the affected subscriber(s).[32]

The independent systems were predominantly of the commercial variety (94 percent of the wire miles served by independents in 1902). These commercial systems had, on average, approximately 700 subscribers. By comparison, the tiny mutual systems had only 2.3 percent of the independents' wire miles and served, on average, 90 subscribers in 1902. Data on rural systems are scarce, although it is generally accepted that they reflected only a small percentage of the total telephone subscribers and they were smaller still (11 subscribers on average in 1902). Nonetheless, establishment of farmer lines illustrated the strength of the demand for telephone service and "the feasibility of unsubsidized telephone service in sparsely populated areas."[33] The politician and independent telephone businessman D. Elbert Reynolds pointed out:

> The Bell [Company is] giving excellent service, and the independent or farmer company gives service from about 7 or 8 o'clock in the morning until 8 or 9 at night, provided the lines are in order, and the service entirely unsatisfactory. But inasmuch as one service costs $24 and the other less than $4 a year, [the farmers] seem to be satisfied with what they are getting.[34]

The Bell System owned hundreds of patents still in force (and was adding more all the time), possessed a large manufacturing facility, offered long distance capability, and controlled many large local exchanges. By the time competition came to the industry, Bell was virtually the equivalent in size and economic power with the old Western Union. Outsiders seeking a share of the market faced a formidable and entrenched rival.[35]

Vail and AT&T Centralization

For the next two decades, the twin problems of acquiring or consolidating local telephone companies while integrating new technology would consume the time and attention of AT&T's management. From its position as owner of increasingly large segments of the network, AT&T was able to function effectively as the national network's central planner and manager. However, not all local companies were happy to come under AT&T's jurisdiction and control—some companies strongly resisted AT&T's acquisition overtures, while others were happy to sell out.

In the 1890s, American Bell continued to develop its internal network, including converting its lines to two wire, expanding the "long lines" long distance network, and moving to the "common battery" switchboard system.[36] With competition, AT&T began to compete on the basis of network size and service quality in conjunction with patent litigation. American Bell had effectively passed out of existence at the end of 1899, with its assets being assumed by the New York-based American Telephone and Telegraph Corporation (*see section 2.5*). This allowed AT&T to increase its control over its licensees—which had been restricted under Massachusetts law—and to increase its capitalization without governmental interference.[37]

Competition from the growing independents took its toll on AT&T earnings, especially given the capital required for expansion. AT&T was forced to borrow to finance construction, and saw its debt grow to $200 million in 1907 (equivalent to about $3.8 billion in 2005), from just $60 million five years earlier. In 1907, after a lengthy battle with Boston financial interests, those associated with financier J. P. Morgan assumed control of AT&T and returned Theodore L. Vail (1845–1921) to its leadership.[38] Vail became the single most important AT&T leader in company history. Persuaded by Gardiner Hubbard to leave the government's Railway Mail Service, Vail had first joined Bell as general manager in 1878. He was the first professional manager of the operation and served as president from 1885 to 1887 before resigning because of policy differences with the Boston financiers who then controlled the Bell firm. Vail was brought back into AT&T as a board member after Morgan took financial control of the company. Vail served again as president from 1907 to 1919, and was Chairman of the Board when he died in April 1920. Vail pushed AT&T ahead on three fronts: merging or taking over other telephone companies, welcoming and thus helping to shape developing regulation, and continuing and even expanding basic research.

Building on a degree of public dissatisfaction with the confusion of the competitive 1890s, Vail soon established "one system, one company, universal service" as an organizing credo for AT&T. He is perhaps best known for his statement in the company's 1909 *Annual Report* concerning "universal service," a phrase which appears to have originated with him:

> The position of the Bell System is well known. It is believed that the telephone system should be universal, interdependent and intercomunicating, affording opportunity for any subscriber of any exchange to communicate with any other subscriber of any other exchange. . . . It is not believed that this can be accomplished by separately controlled or distinct systems nor that there can be competition in the accepted sense of competition.[39]

Under Vail and into the 1920s, AT&T reorganized both its New York-based general departments and the operating companies. The former helped to streamline and make more efficient company-wide technical and financial functions. The latter were reorganized and consolidated by the early 1920s into the Bell Operating Companies familiar for decades to follow. "Vail's revitalized Bell System was now characterized by uniformity of practice and policy, standardization of operations, and cooperation in planning and coordination in the commercial development of the business."[40] Vail's consolidation of research and development functions (the number of Western Electric engineers alone had increased five-fold to nearly a thousand men from 1910 to 1915) evolved into the Bell Telephone Laboratories in 1925.

From a structural point of view, American Bell and, later, AT&T gradually took controlling interest in their licensees beginning in the 1880s with the profits of the monopoly period. This enabled the development of a system that was consistently engineered and operated, so that a highly uniform level of service could be supplied. Financially, American Bell and AT&T received rental payments on equipment, license fees for the privilege of being a Bell licensee, and toll revenues from their operating companies.

Vail also was concerned with strengthening AT&T's monopoly. In addition to more aggressively purchasing independents, Vail wanted to merge with a telegraph company. Such a merger would have two principal benefits for AT&T:

- The physical infrastructure used by the telegraph network would be under AT&T's control so that if a new technology that enabled telephony over the telegraph plant were developed, AT&T's exposure to this risk would be small.
- The company could benefit by integrating voice and record communications.

To this end, Vail approached Postal Telegraph with a merger proposal. Postal Telegraph was attractive because it had an extensive international cable system. Clarence Mackay, the head of Postal Telegraph, was also interested in such a merger, but *he* wanted to control it. When this disagreement over control caused this merger to fail, Vail approached Western Union. At the time, Western was in financial difficulties, due to management problems and competition from Postal Telegraph, so they welcomed AT&T's offer. As a result, in 1909 AT&T acquired effective control over Western Union.

Bell's Response to Competition

The Bell System's initial response to the growth of telephone independents was to pursue a strategy of patent-based protection of their monopoly. Beyond this, Bell management effectively ignored the competitors.

Prior to the company's 1901 *Annual Report*, the directors did not see fit to even mention the existence of competitors.[41] It is useful to separate Bell's response to competition into early (1894–1906) and late (1907–20) competitive periods.[42] The dividing point, 1907, marks the return of Vail to the presidency of AT&T and the influence of the Morgan interests (there is also a gradual increase in governmental regulation). Through 1906 Bell responded to competition in four important ways.[43]

Pricing.   The Bell companies decreased prices to meet the competition's prices. These price decreases came for both the telephone instrument rental and the service charge. While the independents often charged less, this was often a result of their smaller size. These "[e]xchange diseconomies also created a neat little trap that snared many an independent competitor."[44] Essentially, switching becomes more costly per subscriber as the number of subscribers grows. Thus, independents who offered low rates initially found those rates unsustainable as their subscriber base grew. By the time the independents arrived on the scene, many municipalities were requiring franchise agreements that, in many cases, limited the rates that the telephone company could charge. The industrial era had schooled customers to believe that economies of scale should be expected, so that requests to increase prices were met with skepticism and allegations of price gouging, while, in fact, it was diseconomies of scale at work rather than price gouging.

Network Size.   American Bell worked hard to expand the size of their subscriber base by interconnecting the exchanges of their affiliates (via their long distance network) and refusing interconnection to independents. They also worked aggressively to purchase the "rural lines," with the rationale that if they gained the farmers, then they would also get the services that the farmers relied upon for support (e.g., the merchants and banks), which would soon encompass an entire community. Thus, intense competition among service providers for the rural lines ensued.[45]

AT&T began to buy out or "sublicense" independent exchanges. With this, they purchased an immediate network that could be attached to other licensees. A sublicensing agreement was an interconnection arrangement with an independent in a region that was not served by Bell. In these agreements, Bell agreed not to construct exchanges in their territory, and the independent agreed not to interconnect with another independent. Efforts by the independents to organize a separate long distance network failed on several occasions, allegedly because of AT&T.[46] J. P. Morgan and his affiliated interests had made a sizeable financial investment in AT&T by the mid 1900s. Much of AT&T's competitive advantage was in its ability to connect many exchanges, creating a large "network" of people to call for AT&T's subscribers. The most spectacular example of this involved the merger of the Telegraph, Telephone, and Cable (TT&C) Company and the Erie Telegraph and Telephone (ET&T) Company in Erie, Pennsylvania. TT&C was formed to create a rival long distance network. In acquiring ET&T, it was essentially replicating the horizontal integration of AT&T. Allegedly at the behest of J. P. Morgan, however, the financial backing of TT&C was withdrawn, sending the company into a financial crisis. Through a front, AT&T purchased control of TT&C and dissolved the company, reducing it to a local affiliate.[47]

Undermining the Independents. AT&T developed a variety of strategies to undermine the independents. These included advertising its own services and using its influence to restrict the availability of capital to the independents to block competitive entry. Bell had mixed success using this strategy, with its greatest successes coming in the financial and political area. By the 1890s, Bell had become a prominent organization with many contacts in the business community, particularly after the Morgan interests gained control. AT&T's strategy was to use its influence to restrict the independents' access to desperately needed investment capital. This had the effect of slowing the independents' rate of expansion.[48] Bell's strongest political influence was with municipal authorities. Many cities required telephone companies to obtain operating licenses. In some cases, Bell's influence with prevailing politicians motivated establishment of onerous terms for the independents.[49] The independents' naivete at telephone operations in some cases prompted them to promise lower rates than they could deliver and still stay in business.

Refusal to Sell. Bell refused to sell Western Electric equipment to the independents, stimulating the emergence of competing supply firms, such as Stromberg-Carlson, Automatic Electric, and Kellogg Switchboard. These new manufacturing entities were responsible for many technological developments in this period. Some of these innovations led to compatibility problems as independents later either integrated with or interconnected to the Bell System.

Taken as a whole, these four policies were partially successful—the responses "slowed [AT&T's] loss of market share, but did not stop it."[50] After 1907, entry by new carriers became more difficult because of the increasing importance of long distance service and the lack of communities without telephone service. With Morgan's takeover, AT&T's stance toward its competitors changed. In 1909 AT&T began to acquire a "substantial interest" in Western Union, ultimately resulting in a controlling share.[51] This financial interest was augmented by close cooperation between the two systems, resulting in significant economies of scope

with regulating the nation's expanding railway system by issuing "certificates of necessity" which allowed railroads to build and operate.

In addition to the federal effort, by the early 1900s several states had established public utility commissions to regulate railways (and some other businesses) within their borders. The first state commissions appeared in Wisconsin and New York in 1907, growing out of the progressive political movement controlling both states. By then five states had specifically established some degree of control over telephone companies, though none took control over rates charged.

The first legislation specifically subjecting telecommunications companies to federal regulatory supervision, the Mann–Elkins Act of 1910, added the words "telegraph, telephone, and cable companies" engaged in interstate commerce to the list of railway common carriers already regulated by the ICC. Mann–Elkins was concerned with rate regulation, however, and did not address acquisitions and mergers in the telecommunications industry. Thus, both AT&T and the major independent companies (all of which had supported the legislation) continued to acquire smaller telephone carriers. Further, the ICC was not an activist regulatory agency in telecommunications, responding only to complaints brought before it.[59] Mann–Elkins at least provided a precedent and mechanism for interstate regulation, even if it was not exercised effectively.

1913 Antitrust Suit

Three years after passage of Mann–Elkins, independent companies, in a state of growing alarm at the rate of industry consolidation, complained more insistently to the Justice Department. The Attorney General had successfully broken up Standard Oil and American Tobacco using the Sherman Antitrust Act, and AT&T feared that it might be next. Vail had reason to be concerned, for AT&T's competitors had been making complaints that generally fell into three categories:

- The Postal Telegraph Company complained that it had been discriminated against by AT&T/Western Union because those telegraph messages that originated on telephone lines were directed to Western and not to Postal;
- The independents claimed that they could not get access to AT&T's long distance network; and
- Some independents claimed that the merger policy was illegal under the Sherman Antitrust Act.[60]

There was also concern that the incoming Wilson administration, with its advocates for "postalization" (nationalization), might act aggressively against AT&T.

These complaints intensified government alarms first raised when AT&T purchased control of Western Union in 1909, thus concentrating most telegraph and telephone service in the hands of a single firm. The Justice Department responded with an investigation and ultimately filed an antitrust suit against AT&T on July 24, 1913, in the U.S. District Court for Oregon. The suit alleged an unlawful conspiracy to monopolize and restrain trade in the transmission of telephone service in Oregon, Washington, Montana, and Idaho. It was the first of what would turn out to be three federal antitrust actions filed against AT&T.

Kingsbury Commitment

Faced with growing pressure from the government, AT&T decided to negotiate. In the fall of 1913, the Attorney General of the United States held a series of meetings with AT&T Vice President Nathan C. Kingsbury in an attempt to find a solution to the independent companies' complaints about the Bell System's consolidation efforts. The result of these conversations was an agreement in the form of a letter dated December 19, 1913, known as the "Kingsbury Commitment," wherein AT&T agreed to:

- Refrain from acquiring additional independent telephone companies and submit any pending acquisitions to the Justice Department and the ICC for approval;
- "Promptly dispose of its entire holdings of stock" in Western Union (AT&T sold its controlling interest the next year); and
- Make "arrangements . . . promptly under which all other telephone companies may secure for their subscribers toll service over the lines of the companies in the Bell System."[61]

The Kingsbury Commitment can be seen as an attempt to maintain dual service while creating an integrated long distance system, but there is no evidence that independents "availed [themselves] of its costly and non-reciprocal toll interconnection arrangements."[62] Many independents were unfavorably disposed toward the Kingsbury arrangement, as it left them unable to sell their systems on favorable terms. Economically, selling out to the dominant player would have been an efficient market exit strategy for the independents if they had realized that the access game was lost. Rather than risk losing their entire investments, the independents would just as soon have joined with AT&T to seek the passage of a law which would have permitted mergers and consolidations of competing telephone companies.[63]

The principle (if not all the details) of the Kingsbury Commitment—that AT&T would cooperate with government oversight in return for recognition of its dominant industry position—would define the relationship between the company and government for the next several decades. Despite its intentions, however, the Kingsbury Commitment was only partially successful. By restricting AT&T from purchasing *any* independent companies, the government had deprived those independents who wished to sell out of their likeliest purchaser. At the same time, interconnection of independent companies gave many of them a new lease on life (as well as an interest in the well-being of the Bell System which now contributed to their revenues), making them more likely to continue their independent status. Less than a year after its creation, the Kingsbury Commitment was partially set aside when the Attorney General determined that the document was acting as something of a barrier to a consolidated telephone system.

Wartime Government Control

With American entry into World War I, Congress mandated that the Post Office take over "supervision, possession, control and operation" of AT&T, effective August 1, 1918. The action was taken partly in response to deterioration in service as a result of trained personnel entering military service and extraordinary increases in telephone traffic.[64] The major elements of the resulting contract between AT&T and the Postmaster General were:

- The property was to be maintained by the government and turned back in comparable condition as when it was received;
- The license and rental contracts between AT&T and the licensees were to be continued as before; and
- The government guaranteed payment of interest and existing amortization charges on all Bell System obligations and securities (including dividend payments).

Telephone rates were soon increased by the Postmaster to cover expenses and dividend payments. Government control ended after one year on July 31, 1919.[65]

By agreeing to the terms of the contract, the government provided *de facto* support for some controversial elements of AT&T's rate structure, namely the 4.5 percent license fee that the operating companies were required to pay to AT&T. The manner in which the Postmaster chose to increase rates was via a $3.50 "service connection charge" that the Bell companies had been trying to impose for some time; thus, government placed an imprimatur on a controversial Bell System policy.[66]

Public outcry over these rate increases helped doom serious attempts to perpetuate government ownership.[67] For at the same time, the Navy had taken control of wireless stations. After the war Congress addressed the question of industry structure when it rejected permanent government ownership of the telecommunications industry, despite pleas from the Postmaster General and Navy officials that only public ownership could ensure low prices and adequate service.[68] This was merely repeating an argument that officials had been making in official reports for decades—indeed, ever since selling the telegraph back to private hands in 1847. The congressional hearings on telecommunications ownership in 1919 were the high water mark of possible permanent government ownership.

Willis–Graham Act

Instead, Congress passed the Willis–Graham Act of 1921, which freed the telephone industry to again pursue acquisitions and consolidation. Both Bell and independent owners had supported passage of the bill. Passage of Willis–Graham stimulated a spate of acquisitions by Bell. While these acquisitions included territorial expansion, they were focused on eliminating competing companies. The independents, recalling the pre-Kingsbury era, entered negotiations with AT&T. These negotiations resulted in the Hall Memorandum (signed in June 1922), in which AT&T agreed "to make no purchases of, or consolidations with, independents unless demanded for the convenience of the public or unless special reasons existed making the transaction desirable for the protection of the general public service or Bell System property."[69] The Hall Memorandum essentially circumscribed the boundaries of the Bell System and the structure of the telephone industry by voluntary agreement.

Thus, the early competitive era in telephony drew to a close. Industry structure stabilized in the early 1920s and remained in largely the same form until the 1960s, when interexchange competition reemerged. Local competition would only reemerge in the 1990s, but this time under strict interconnection requirements.

* * *

The passage of the Willis–Graham Act in 1921 and its aftermath brought the early competitive era of telecommunications to a close. While sector competition would emerge in radio broadcasting, the industry would soon settle into established market segments of voice (the Bell

System), record (Western Union), and broadcasting (many companies). There would be little competition across these boundaries until the 1970s, even though competition might occur within each segment.

NOTES

1. Joseph C. Goulden, 1968. *Monopoly* (New York: Putnam), p. 59.
2. Amy Friedlander, 1995. *Natural Monopoly and Universal Service: Telephones and Telegraphs in the U.S. Communications Infrastructure, 1837–1940* (Reston, VA: Corporation for National Research Initiatives), p. 33.
3. For a system of N subscribers, there are N(N – 1)/2 ways in which the subscribers can be interconnected. When N is large (e.g., N > 20), the complexity can be approximated by $N^2/2$. For example, a network with 25 subscribers must be able to support 313 connections, while a network of 100 subscribers must be able to support 5,000 connections.
4. The switching hierarchy was possible because of detailed traffic studies that had been carried out. It was successful because it created an organizational structure that was capable of dealing with technical complexity. See Milton Mueller, 1989, "The Switchboard Problem: Scale, Signaling, and Organization in Manual Telephone Switching, 1877–1897," *Technology and Culture*, 30(3), July, 534–560.
5. Milton L. Mueller, 1993. "Universal Service in Telephone History," *Telecommunications Policy*. 17: 352–369 (July), p. 357.
6. John Brooks, 1976. *Telephone* (New York: Harper & Row), pp. 100–101.
7. Coauthor Sterling saw them being actively used in a Polish village outside of the city of Cracow early in 1993. He had seen similar switches in a Budapest museum shortly before—suggesting the great variance even in the comparatively backward Eastern European systems in the aftermath of decades of Soviet domination.
8. M. D. Fagan, ed. 1975. *A History of Engineering and Science in the Bell System* (New York: Bell Telephone Laboratories).
9. Mueller, 1989, pp. 558–559.
10. Amos E. Joel, ed. 1982, *History of Engineering and Science in the Bell System*, Vol. 3, *Switching, 1925–1975*. Warren, NJ: Bell Laboratories, p. 7, and Fagen, p. 547.
11. See, for example, Goulden, 1968.
12. For a good study of this period and the role of loading coils, see Neil H. Wasserman, 1985. *From Invention to Innovation: Long-Distance Telephone Transmission at the Turn of the Century* (Baltimore: Johns Hopkins University Press), especially chapters 4 and 5.
13. These higher frequencies were irrelevant in this era. Later, when multiplexers were introduced, many loading coils had to be removed. In fact, the introduction of digital ISDN was delayed in part because of the presence of undocumented loading coils in long subscriber loops.
14. Brooks, pp. 121–122.
15. Brooks, pp. 137–141.
16. AT&T and its financiers were able to cut off the financing for the long distance ventures.
17. See J. J. Fahie, 1899. *A History of Wireless Telegraphy 1838–1899* (Edinburgh: William Blackwood & Sons; reprinted by Arno Press, 1974) for a contemporary summary of the theory and experiments by these and other innovators.

18. Hugh G. J. Aitken, 1985. *The Continuous Wave: Technology and American Radio, 1900–1932* (Princeton, NJ: Princeton University Press), p. 33.

19. For further discussion of this drawback, see Gleason L. Archer, 1938. *History of Radio to 1926* (New York: American Historical Society; reprinted by Arno Press, 1971).

20. Horwitz, 1989, pp. 105–106. See also Peter J. Hugil, 1999. *Global Communications Since 1844: Geopolitics and Technology* (Baltimore: Johns Hopkins University Press).

21. Hugh G. J. Aitken, 1994. "Allocating the Spectrum: The Origins of Radio Regulation," *Technology and Culture.* 35: 686–716 (October), pp. 690–691.

22. "Willful and malicious interference" with radio communication was prohibited and penalized, as was "uttering or transmitting false or fraudulent signals." Archer, p. 106.

23. Aitken, 1994, offers the best modern treatment of this process.

24. Horwitz, pp. 109–111.

25. Britain achieved this dominance by controlling the supply of *gutta percha*, a natural rubber that was essential for the insulation of undersea cables. They actually cut Germany's cables, which restricted Germany's ability to raise the funds necessary for their war effort, and forced them to use wireless, which played into the hands of British codebreakers.

26. Brock, p. 121.

27. J. Warren Stehman, 1925. *The Financial History of the American Telephone and Telegraph Company* (Boston: Houghton Mifflin, 1925; reprinted by Augustus M. Kelley, 1967), p. 55.

28. Brooks, p. 109.

29. This patent was instrumental in providing operational parity with Western Union's telephone operation in the late 1870s. Recall that Thomas Edison had developed a carbon microphone for Western Union that was superior to the magnetic or liquid system of National Bell and that provided, therefore, a significant competitive advantage.

30. Goulden, pp. 60–61.

31. U.S. Bureau of the Census, 1906. *Census of Electrical Industries: Telephones and Telegraphs: 1902* (Washington, DC: Government Printing Office).

32. Stehman, pp. 52–55.

33. Brock, p. 111.

34. Goulden, p. 62.

35. "Bell's revenue was insignificant in 1878, 54 percent of the Western Union revenue in 1885, 68 percent . . . in 1890, and 98 percent . . . in 1895." This provided sufficient capital to hold off, or take over, the most serious competitors. Brock, p. 108.

36. Brooks, p. 105, and Friedlander, pp. 34–37. The common battery switchboard was a significant change that solved a number of signaling and operations problems, allowing the integration of the licensees into a single system. See also Mueller, 1989.

37. This transfer was done in two steps. First, American Bell turned over to AT&T its assets and contracts so that the former company was left holding AT&T stock and some patents. After this was accomplished, American Bell's shareholders were given two shares of AT&T stock for each share of American Bell stock they held. With that, American Bell was left only with the patents, which it licensed to AT&T. See Stehman, pp. 74–75.

38. Brooks, 1976, pp. 120–124. See also Albert Bigelow Paine, 1921. *Theodore N. Vail: a Biography.* New York: Harper.

39. AT&T, *1909 Annual Report*, pp. 22–23.

40. Robert W. Garnet, 1985. *The Telephone Enterprise: The Evolution of The Bell System's Horizontal Structure, 1876–1909* (Baltimore: Johns Hopkins University Press), p. 154.

41. Brooks, p. 112.

42. Gabel, 1969, "The Early Competitive Era in Telephone Communications, 1893–1920," *Law and Contemporary Problems.* 34: 340–359 (2), Spring, p. 349.
43. Prior to 1900, American Bell was the parent company. The actions in the early period prior to 1900 should therefore be attributed to American Bell.
44. Mueller, 1989, p. 558.
45. Mueller, 1993, p. 361.
46. Brock, p. 119 and Goulden, p. 66.
47. Brock, pp. 119–120.
48. Goulden, pp. 66–67.
49. Ibid.
50. Brock, p. 121.
51. Stehman, pp. 147–152.
52. Ibid., pp. 152–153.
53. See Horwitz.
54. Gabel, p. 356.
55. The Sherman Act had been the key tool in the government's breakup of the Standard Oil and American Tobacco trusts in 1911.
56. Mueller, 1993, p. 358.
57. Ibid.
58. Ibid., p. 359.
59. Gabel, pp. 357–358.
60. Brock, p. 155.
61. AT&T, *1913 Annual Report*, p. 24.
62. Mueller, 1993, p. 365.
63. Gable, p. 353.
64. Stehman, p. 174.
65. Ibid., pp. 178–179.
66. Horwitz, p. 101.
67. Ibid., p. 101.
68. Op. cit., note 1.
69. Gabel, p. 353.

Chapter 4

Regulated Monopoly (1921–56)

Beginning in the 1930s, the integrated structure and increasing size and power of AT&T (as well as the increasing importance of telephones in all aspects of business and government) were subject to a series of investigations. Telecommunications was not alone in this Congressional concern; the report on the telephone industry was but one in a number of parallel Congressional studies of large holding companies in different lines of business brought about because of the often poor performance of such firms in the early years of the Depression.

4.1 FURTHER TECHNOLOGICAL ADVANCE

This era saw some remarkable technological advances in both transmission and switching. The advances in customer premise equipment were less radical, consisting primarily of the standardization of telephones including use of dial mechanisms. The operational technology in this period remained strictly analog.

Transmission—Multiplex

The fundamental innovation in this period was the development of multiplexed transmission systems. These systems allowed a single physical cable to carry many simultaneous voice channels by carefully applying modulation techniques. Improvements in modulators and demodulators for broadcasting in the 1920s and 1930s could be applied as well to interoffice transmission systems. Improvements in cable technology helped as well; coaxial cables developed during and after World War II had a far higher bandwidth than twisted pair cables, and so were able to carry many more voice channels.

The systems used frequency division multiplexing (FDM), which divided a high capacity channel into several subchannels. At the lowest level, each of these subchannels was a voice communications channel. While this enabled the efficient use of high capacity transmission media, the technique suffered from two major problems that limited the clarity of long distance communications. Analog repeaters amplified both the intended signal *and* accumulated noise, resulting in deteriorated quality over long transmission paths (*see section 1.1*). And electrical terminations and adjustments of filters matching the electrical characteristics of the incoming signal changed over time due to aging of components and changes in the transmission medium such as heating/cooling and humidity. The effect of this is *crosstalk*, when the signal energy from one voice channel impinges on another channel. When crosstalk becomes sufficiently severe, it is impossible to clearly hear and understand a conversation on the interfering channels. Neither of these problems would be resolved until digital transmission systems were introduced in the early 1960s.

Available transmission media ranged from twisted pair wires and high frequency radio systems early in this period to coaxial and microwave systems after 1945. Coaxial cable systems provide a significantly increased capacity over twisted pair. Like twisted pair, however, these systems required a continuous right-of-way between sender and receiver. The control of such rights-of-way has long been an important issue, as illustrated by the exclusive contracts that Western Union wrote with railroads.

Transmission—Microwave

Microwave technology was first theoretically conceived in the years before World War I, although the first important experimental link was only established in 1931 from France to Britain.[1] Marconi himself did some experimental work in the years prior to his death in 1937. Bell Labs was active in microwave research by the mid-1930s.[2] What all these researchers were working with was very short radio waves, transmitted very high in the spectrum (800 MHz and higher), that appeared able to carry considerable traffic and could be aimed in any direction.

Work continued during World War II in several countries and by several companies, including Philco, RCA, and Bell Labs (based in part on research work on microwave equipment undertaken by Harold T. Friis and his colleagues).[3] Wartime requirements (including the perfection of radar) delayed civilian application of microwave technology but greatly expanded the understanding of spectrum's upper reaches in which microwaves operated. The first nonmilitary experimental microwave networks, primarily for video transmission, were placed into service by Philco and RCA right after the war.

Engineers soon discovered that, coupled with directional antennas, transmission systems could be built that did not require a continuous right-of-way between sender and receiver. Instead of using a cable with periodic repeaters, *microwave* transmission systems use high frequency electromagnetic propagation through the atmosphere in place of a cable. Periodically, this signal would have to be amplified just as in cable systems. In microwave networks, these repeaters were far more expensive than for cable systems, since the repeater sites required a tower, two antennas, and the necessary receivers and power transmitters. Thus, the economy derived with such systems is due to cheaper repeater sites and from sufficiently powerful transmitters and sensitive receivers so that the repeater spacing could be long enough, typically 20 to 30 miles.

AT&T's initial commercial application of microwaves in 1947 was also to distribute video, in this case between New York and Boston. The 220-mile system was built with five "hops" averaging just over 27 miles and utilizing FM transmission.[4] The Bell System was soon extending its microwave (and cable) systems for use in television networking; they reached Chicago and St. Louis by 1948. A coast-to-coast television service became possible in 1951 with completion of a microwave link to the West Coast that could carry 600 calls or a single television program.[5] Microwave called for relay towers built within sight of one another, typically 20 to 35 miles depending on terrain. The national system was cheaper to build than any wire or cable network; the coast-to-coast microwave link of 107 towers and amplifiers cost $40 million (some $290 million in 2005 dollars), about half the expense of a coaxial cable network.[6] By 1953 the microwave network was transporting network television signals to most parts of the country.[7] AT&T soon dominated microwave service, providing nearly 80 percent of television links and more than a fifth of the country's telephone trunk connections. Some 40 percent of AT&T's total intercity circuits were carried on microwaves by the early 1960s.[8]

AT&T maintained their policy of refusing interconnection to private networks. This policy was effectively legitimized by the Federal Communications Commission (FCC) in a sequence of decisions during the late 1940s and early 1950s. However, in these decisions, the FCC never *explicitly* declared video carriage a monopoly service, unintentionally leaving the door open for future challenges. Indeed, the FCC gave temporary authority to broadcasters to construct private networks to alleviate the effects of AT&T's bandwidth shortage.[9]

But broadcasters were not the only firms that wanted to construct private line networks. By the mid-1950s, a collection of powerful industry groups, including the Automobile Manufacturer's Association, the National Association of Manufacturers, the American Petroleum Institute, and the Newspaper Publishers Association, began to lobby on behalf of their respective members for the permission to construct private networks.[10] These industry organizations, in addition to Motorola, lobbied in support of applications before the FCC by Minute Maid Corporation and Central Freight Lines of Texas, both of whom wanted to build private networks to support their business operations. This would lead to definitive FCC action late in the decade (*see section 5.3*).

Switching

As noted earlier (*see section 3.1*), automatic switching was introduced in 1892 by Almon Strowger, although the Bell System resisted its use until the 1920s. As the organization of the switching function stabilized and automatic switching technology matured, it was introduced more systematically into the Bell System's network. One practical hurdle that had to be overcome for automatic switching was the standardization of customer premise equipment (telephones). The dial mechanisms of early telephone sets were quite variable. It wasn't until the 1920s that the rotary dial mechanism became standard.

Early automatic switches were step–by–step devices.[11] These switches were characterized by a switching fabric that was separated into distinct "cells," each of which had Strowger's electromechanical switch. This switch used progressive control that was hard wired. In progressive control, the call is transferred from cell to cell within the switch until the call is connected to the next line or trunk. These switches proved to be effective and durable, though relatively slow, and were most cost-effective for small offices.

In 1930 Bell System engineers installed the first panel switch. This differed from the step-by-step by its switching fabric, which consisted of a large matrix of crosspoints instead of several "cells" of relays. Contacts were made by a set of wipers that attached to vertical bars. A call would cause a wiper to move to the appropriate crosspoint and touch it for the duration of the call. This allowed for faster call setup and also supported larger installations more economically than step-by-step technologies. The mechanisms used by the switch turned out to be unreliable and required frequent maintenance.

This unreliability motivated Bell Laboratories to develop *crossbar* switch technology, originally invented in Sweden. The architecture of the crossbar switches was similar in many ways to the panel switches, except with improved design. Their improved reliability came from the small motions required by the mechanical elements, in contrast to the large motions required of the mechanical elements of the panel switch. The Number 1 Crossbar first went into service in 1938, but it was the Number 4 Crossbar, designed for toll offices, that were most successful and widely implemented during the post-World War II economic boom in suburban end offices.

Radio Broadcasting

While improving its telephone system, AT&T was faced with a wholly new industry that seemed closely related. As noted earlier (*see section 3.1*), regular radio broadcasting began when Westinghouse expanded the amateur transmitter of one of its engineers to become station KDKA. The purpose of this 1920 upgrade was to promote the sales of its newly available receivers, just as amateur broadcasts had stimulated the sales of crystal sets. By May 1922, 218 licenses had been issued; by the following year, this more than doubled and the first serious interference or signal congestion emerged. To address this, in 1922 the Secretary of Commerce allocated a second frequency (400 meters, or 750 kHz) for radio broadcasts in addition to the existing 360 meters (833.3 kHz) that had been allocated in 1920. These new frequencies were available only to stations with a transmission power greater than 500 watts. By mid-1923, amid steady growth of the new business, the entire 500 to 1500 kHz band had been allocated to broadcasting.[12]

From 1922 to 1926, AT&T operated two radio broadcasting stations, one in New York and the other in Washington, DC. The company briefly viewed radio as but another aspect of its telephone business—"phone booths of the air."[13] It initiated experiments with networking (connecting two or more stations with its telephone lines) and radio advertising (dubbed "toll radio"). Its conception of radio programming paralleled telephony: the company would provide the facilities but content would be offered by others. Finding that it did not work in practice for the new medium, AT&T was soon learning how to develop broadcast programming and was selling advertising time to help support its stations.

The notion of a chain or network of more than one station began experimentally in 1922–23, as first one and then several other stations were linked with AT&T's New York station WEAF by telephone wire so that multiple stations could carry programs more cheaply than any one station could produce them. By December 1925 AT&T could interconnect 26 stations on a fairly regular basis. When other stations wished to create their own networks, however, AT&T refused to provide them the long distance lines needed. Competitors were forced to use telegraph lines that lacked sufficient frequency response for voice and music broadcasts. RCA, one of several station owners, considered constructing its own wireline network to interconnect its stations.

Faced with this potential competition and increasingly uncomfortable with the dichotomies inherent in a widely respected public utility operating an entertainment service, in mid-1926 AT&T sold its radio interests to RCA.[14] By this agreement AT&T's monopoly in point-to-point transmission was preserved and RCA's substantial role in broadcasting was ensured. The Bell System subsequently withdrew from an active role in the broadcasting business with the exception of some sponsorship of prestige programs such as radio's later *The Bell Telephone Hour*.[15] AT&T also retained one other critical, though behind-the-scenes, role.

AT&T retained contractual rights with RCA's new NBC network—and later with CBS and ABC—to interconnect radio stations by telephone cables for the purposes of networking. In 1948, use of microwaves and coaxial cable extended this role into television.[16] The first coast-to-coast television interconnection entered service in 1951.[17] Until the mid-1980s' inception of satellite links for terrestrial broadcasters, AT&T's broadcast interconnection service (operated as a part of Long Lines) was a valuable source of continuing income for 60 years. Yet AT&T's essential role was all but invisible to the consumers of network radio or television.

Transistors

Emergence of the vacuum tube as a standard amplification device by about 1920 profoundly changed both wired and wireless telecommunications services. Despite their utility, however, vacuum tubes had some undesirable technical aspects: they were relatively bulky, fragile, and power hungry. One of the most significant sources of power consumption of vacuum tubes occurred in the generation of an electron beam, which was achieved using a heated filament, much as with an incandescent light bulb. Their fragility was due to the glass tube and the sometimes delicate structure of the internal control mechanism. These limitations drove the search for a replacement.

Seeking that something, AT&T's Bell Labs began an intensive project in "solid state" physics even before the end of World War II.[18] In part, the project sought possible replacements for vacuum tube technology.

But ideally, results might lead to a replacement for the troublesome tubes *and* many of the mechanical relays and other devices central to telephone switches. Less than three years later, the project's most important result, the transistor, dubbed the "major invention of the century" was first demonstrated at Bell Labs on December 23, 1947.[19] The transistor "effect" discovered was "a process of electronic amplification based on a semiconducting crystal (and no longer by a vacuum tube)."[20] Researchers John Bardeen (1908–91), Walter Brattain (1902–87), and William Shockley (1910–89) set in motion a still-continuing flow of solid state technology improvements and shared the Nobel Prize in physics in 1956.

Instead of regulating the flow of current by manipulating an electron beam (as in vacuum tubes), transistors worked by regulating the flow of electrical current through junctions of dissimilar silicon. Transistors did not have to generate an electron beam (thus saving energy), and they did not require a vacuum enclosed in glass. These characteristics made them more rugged. However, the physics of the junctions on which transistors relied was highly dependent on temperature. Despite introducing some new limitations, engineers began adopting the transistor for devices that required active electrical amplifiers. Devices that had to be small, rugged, and/or had low power consumption were the leading candidates for conversion to solid state devices.

After initial patents had been filed, Bell Labs went public with its new device in mid-1948,[21] though practical commercial transistors were still several years off. Only in 1951 did manufacture of the improved device begin and a year later it was first used in the telephone network. That same year Bell Labs held a series of high-level seminars for the industry, for which companies paid $25,000 to attend and learn how to manufacture improved transistors for different applications. Consumer products—the first were a small portable radio and a tiny hearing aid—first appeared in 1953–54. "As a memorial to Alexander Graham Bell and to his interest in the deaf, Bell did not require royalties on transistors produced for hearing aids."[22]

But just as it resolved two drawbacks of vacuum tubes—heat and size—the transistor raised one of its own. To achieve its full potential, any given device would take dozens, and soon hundreds, of discrete transistors all of which had to be handwired on a circuit board. Their tiny size made this a difficult proposition. Truly complex devices, such as the early mainframe computers then being built, would require thousands of transistors all of which had to be tediously (and expensively) interconnected by hand. Something better than hand assembly was required for the solid state revolution to continue (*see section 5.1*).

4.2 ECONOMICS

Building on these technical innovations, 20th-century telecommunications regulation involved three basic economic assumptions:

- Telecommunications was a natural monopoly, and, as such, was subject to regulation;
- The proper method for balancing the interests of investors and subscribers was the cost-plus, rate-of-return regulatory approach; and
- Because the same equipment was used to transmit intrastate and interstate calls, a method to allocate the cost of that equipment between intrastate and interstate jurisdictions was necessary.

During this period of regulated monopoly, these assumptions dominated the efforts of both state and federal commissions, and, on occasion, involved the courts as well. In all of these activities, the Bell System played an active role in seeking to protect its own financial and strategic interests.

Concepts of "Natural Monopoly"

With the signing of the Willis–Graham Act in 1921 (*see section 3.3*), Congress signaled its acceptance of the idea that monopoly, not competition, was the proper industry structure for telecommunications. After a period of what could be regarded as destructive competition, in which competing telephone companies refused to interconnect with one another, and the creation of competing telephone companies required the construction of additional, and seemingly duplicative, facilities, regulators were willing to accept the notion that telephone service was a natural monopoly.

The concept of a natural monopoly is based on the idea that, in some industries, "the cost, demand, and behavioral conditions are such that allowing a single firm to serve the market is most efficacious."[23] In other words, one firm can provide a service more efficiently and effectively than can any combination of one or more firms.[24] The reason for this greater efficiency has to do with economies of scale and scope, as well as with cost and demand characteristics.

Natural monopoly arises from economies of scale and scope across the total range of market demand.[25] Economies of scale exist when the average cost of producing a good or service decreases as output increases—the more units produced, the lower the cost of each one. If this is true for all of the units demanded in the marketplace, then one firm can satisfy all of market demand at the lowest cost. An underlying reason for scale economies is the level of fixed costs involved. The higher fixed costs are, the more scale economies apply as the fixed costs are spread over more units of output. As an example, fixed costs of $1 million over an output range of one million units result in unit costs of $1.00. Fixed costs of $1 million over an output range of two million units result in unit costs of $.50. If the units of output coincide with market demand, then a natural monopoly may exist. In other words, if total market demand is two million units, one company can satisfy that demand at the lowest possible cost— 50 cents.[26] Competition in such a situation would be inefficient.

Economies of scope exist when two or more goods or services can be produced by one firm at a lower total cost than if the goods or services were produced separately. As an example if one firm can produce outputs X and Y at a total cost of $10, while another firm can produce X for $6 and a third firm can produce Y for $7, scope economies exist, since $10 is less than $13.

Since it is possible to have economies of scale without economies of scope, and economies of scope without economies of scale, economists determine whether a natural monopoly exists by referring to the concept of "subadditivity" of costs. Subadditivity takes into account a firm's economies of scale and scope. Quite simply, subadditivity exists when the cost of producing one or more outputs is cheaper when done by one firm than when done by two or more firms. As Train puts it, "A cost curve is said to exhibit subadditivity at a given level of one or more outputs if the cost of producing these outputs is lower with one firm than with more than one firm, regardless of how the output might be divided among the multiple firms."[27]

The theory of natural monopoly was not a new concept even in 1921. Several economists discussed various aspects of the theory before and during this time period.[28] John Stuart Mill spoke of natural monopolies in 1848, noting that water and gas could be provided more cheaply in London by one company, rather than by several. Henry Carter Adams, in 1887, stated that industries in which there were increasing returns to scale were by nature monopolies and so needed to be regulated by the government. In 1902, Thomas Farrer offered five characteristics of natural monopoly, including the provision of an essential service, the existence of economies of scale, and the idea that "customers of the industry must require a certainty and well-defined harmonious arrangement of supply which can only be attained by a single supplier."[29]

An economist writing during this period, Richard T. Ely (1937) noted that the conditions of natural monopoly make competition self-destructive; he listed these conditions:

- The commodity or service is one for which purchasers will select producers based on a small difference in price;
- The business must be one in which the creation of a large number of competitive plants is impossible; and
- The proportion of fixed to variable costs must be high.[30]

It is easy to see how regulators and legislators would regard the telephone industry as reflecting Farrer's characteristics, as meeting all three of Ely's conditions, and as being an industry that had tried competition and found it to be less than efficient.

Telecommunications would continue to be regarded as a natural monopoly in all of its aspects until the 1970s when the FCC allowed competition in customer premise equipment and in long distance services:

> Whether or not an industry is a natural monopoly is not an immutable fact. Technology and tastes (demand) are the fundamental influences, and, as these change, optimal industry organizations can change; industries which once were in this category may be removed from it, and new industries may become natural monopolies.[31]

During this period, however, regulators were convinced that telecommunications was a natural monopoly and that the most effective way of dealing with the industry was to regulate AT&T and the remaining independent telephone companies as monopoly service providers.

Regulation was necessary to assure that AT&T and the independents would not abuse their monopoly power:

> The sole producer who survives the competitive struggle while efficient—in that it can produce the output demanded at a lower cost than two or more firms—could restrict output, raise prices, and reap monopoly profits. Hence regulation may be needed to thwart this temptation.[32]

The dominant form of regulation designed to prevent companies from abusing their monopoly position was rate-of-return regulation. If the workings of a competitive marketplace were not available to balance the interests of the supplier and the consumer, then regulators would control a company's earnings in order to balance the interests of subscribers and investors.

Rate-of-Return Regulation

Rate-of-return regulation (RoR) appears to be a relatively uncomplicated approach to controlling business activity. A regulatory body establishes, through a rate hearing, the amount of revenue a company needs to generate in order to cover its service costs, pay its debts, provide earnings for its stockholders, and maintain its credit rating so it can continue to attract investors. This amount of revenue, the revenue requirement, is the amount of money the company is allowed to target when it sets its rates. While the basic concepts appear to be simple, the implementation of this form of regulation has often been a contentious process, with regulators and the regulated company arguing over each step in the process.[33]

The RoR method provides regulators with a method for balancing the interests of the subscriber with those of the service provider. On the one hand, it is the regulator's duty to assure that the prices charged by the monopoly provider are "just and reasonable." On the other hand, the regulator has to assure that the prices charged are not "confiscatory"; in other words, the prices charged must be sufficient to cover the provider's costs so that the provider continues to be a viable company. A revenue requirement that exceeds the provider's valid needs results in prices that are too high to be "just and reasonable." A revenue requirement that does not meet the provider's valid needs results in prices that are "confiscatory." Obviously, the key is to establish exactly what constitutes "valid needs."

The RoR formula is:

$$RRQ = E + (RB - D) \times R$$

Where: RRQ = *Revenue Requirement*
E = *Allowable Expenses*
RB = *Rate Base*
D = *Accumulated Depreciation*
R = *Allowed Return*

The values assigned to each of these elements have been the subject of much debate. In a RoR hearing, the regulated firm presents its figures for E, RB, and D in the above formula. The figures presented are usually based on the actual transactions that took place during a previous 12-month period, the "historic test year." Since the purpose of the rate hearing is to identify the amount of revenue a provider will attempt to achieve through future service prices, it is in the interests of the regulated firm to try to include in its formula any known and expected changes, such as planned salary increases. It is up to the regulator to determine how valid such changes are and whether they should be included. The calculation of revenue requirement on historical figures has led to a phenomenon called "regulatory lag." As an example, assume a firm's revenue requirement is calculated on the expenses and rate base incurred in year one, and that the revenue requirement is targeted to be recovered through prices billed in year two. Assume further that the firm greatly expanded its staff and added significant equipment during year two, none of which was included in the revenue requirement calculation and therefore not in the service prices charged during year two. These increases in staffing and equipment will not be included in the firm's revenue requirement calculation, and therefore its prices, until a later rate hearing. It is this timing difference that is called regulatory lag.[34]

It is the responsibility of the regulator to determine the levels of E, RB, and D that are allowable in calculating RRQ. Any expenses or investment disallowed in calculating RRQ are considered "below the line." They are not recovered through the prices charged to ratepayers and so are, in effect, charged to the stockholders. Expenses and investment deemed to be appropriate for inclusion in the RRQ calculation are "above the line" and so are built into service prices to be recovered from the ratepayer.

The major concern regarding expenses has been which expenses are "allowable." The regulated firm can include everything from employees' wages to taxes to utility costs to country club memberships; however, the regulator can determine that some of these expenses should not be covered by the rates charged to subscribers. For example, regulators often found that advertising costs were not valid expenses to be incurred by the ratepayers of a monopoly service. Such expenses were then considered "below the line."

The rate base consists of the buildings, equipment, land, and other facilities that have been used in providing regulated service. It is an extremely important element in the rate-of-return equation because it is the basis upon which the regulated firm can generate revenues to provide earnings for its shareholders, to build retained earnings to be used for future plans, and to generate a profit. Determining the proper rate base has been a major concern for both the regulator and the regulated firm. Perhaps the most important aspect of the rate base is the valuation basis used. For all practical purposes, the rate base can be valued in one of two ways: (1) on the basis of historical costs (the actual dollars expended at purchase), or (2) reproduction costs (the dollars it would cost to replace the items at current prices). Prior to World War I, regulators tended to use reproduction costs because they seemed to be more reliable; the lack of standardized accounting practices made it difficult to determine exactly what historical costs had been. However, when construction costs rose sharply after World War I, regulators began to adopt historical costs as a valuation base. The regulated firm, as is to be expected, advocated whichever method would result in a larger rate base.

Court involvement was, therefore, inevitable. Until the early part of the 1940s, the courts were actively involved in trying to determine how best to value the rate base, with judges split between reproduction costs and historical costs as the method more likely to yield a return that would be fair to the ratepayer and also fair, that is, not confiscatory, to the regulated firm. In 1944, in a case involving natural gas, the Supreme Court adopted the "end result" doctrine. Rather than the courts attempting to determine what constituted the "fair value" of the rate base, the courts determined that the regulatory commissions could use whatever valuation method they desired so long as the end result was just and reasonable to both the ratepayers and the firm. The FCC and the majority of state commissions were then free to enforce historical cost as the required valuation method.[35]

In addition to valuation base, another major concern regarding rate base is that only those facilities that are "used and useful" are to be included. Only those buildings, equipment, and facilities that are used in the provision of service are to be included in calculating the revenue requirement that will result in the prices charged for those services. A major issue involved in the "used and useful" standard is the treatment of plant under construction. Regulators must determine at what point, and how, to reflect the costs of constructing new facilities that are not yet in use at the time of the rate hearing.

While the issue of rate base valuation may have dominated the regulatory landscape during the early part of this period, the issue of the allowed return has been a dominant point of controversy ever since. The allowed return is the amount that the regulated firm is permitted to earn on its investment. Just one or two percentage points make a significant difference: "Given the rate base, earnings are 25 percent higher under a 10 percent return than under an 8 percent return."[36] The larger the rate base, the greater the impact of even a fraction of a percentage in allowed return. Determining the appropriate allowed return is an important regulatory challenge.

Regulators in setting the allowed return have tended to use the "cost of capital" approach. The cost of capital "may be defined as the annual percentage that a utility must receive to maintain its credit, to pay a return to the owners of the enterprise, and to ensure the attraction of capital in amounts adequate to meet future needs."[37] In implementing the cost of capital approach, regulators first identify the firm's capital structure, and then determine the costs of debt and equity.

A firm's capital structure is its proportion of debt and equity. As an example, a firm's capitalization may consist of 40 percent long-term debt and 60 percent common stock. The cost of debt is usually fixed; however, the cost of equity is difficult to determine. Regulators must decide the level of return on equity that is required to make the firm's common stock attractive to investors. While several methods have been used to determine the cost of equity, the most often used methods have been a market-determined standard and a comparable earnings standard.[38] The market-determined standard attempts to capture the level of earnings and dividends an investor expects before investing in a firm; the comparable earnings standard looks at what an investor can earn by investing in other firms with a comparable level of risk.

Once the capital structure and the costs of debt and capital are determined, the allowed return is calculated by summing the weighted costs of debt and capital. Returning to the earlier example of a firm with 40 percent debt and 60 percent equity capitalization, if the cost of debt is assumed to be 10 percent and the appropriate cost of equity is determined to be 15 percent, the resulting allowed return is calculated as 13 percent, as follows:

	Capital Structure		Cost		Weighted Cost
Long–Term Debt	40%	×	10%	=	4%
Common Stock	60%	×	15%	=	9%
Cost of Capital (Allowed Return)					13%

It is important to remember that the RoR method did not guarantee the regulated firm that it would achieve the allowed return. Instead, RoR regulation provided the firm with the opportunity to earn the allowed return. Fluctuations in service demand, changes in investment or expense levels, or changes in the cost of debt could result in earnings below, or above, the allowed level.

The RoR method was refined during the period of regulated monopoly from 1921 to 1956. The fine points of valuating the rate base and calculating cost of capital were developed, and the RoR methodology became the dominant form of telecommunications regulation until the 1980s when commissions began to examine alternative, incentive-based regulatory models.[39]

Separations

Regulators during this time period were not just concerned with refining RoR methods; they were also concerned with issues of jurisdiction. Telecommunications operates under a system of dual jurisdiction. While the FCC has authority over interstate services, along with the associated investments and rates, the state commissions oversee intrastate services, along with the attendant rates and investments. A complicating factor in this arrangement is that the same buildings, the same equipment, and the same technicians are used in the provision of both interstate and state services. Some method of allocation is necessary in order to assure that federal and state regulations are being applied to the appropriate services and investments. That method is called jurisdictional separations. The need for separations was recognized, and the initial steps to develop a separations system were taken during this time period.

The implications of the separations process for the pricing of services are clear when the RoR method of regulation is considered. The more investment and expense that is included in the calculation of revenue requirement, the larger the revenue requirement figure and the greater the service prices needed to attain that revenue target. The more investment and expense allocated to the interstate jurisdiction, for example, the higher the prices for interstate services. The more investment and expense included in the state rate of return calculation, the higher the prices for services falling into the state jurisdiction, notably local service rates. This aspect of separations was not lost on either the regulators or the Bell System. The arguments about separations rules hinged on the desire of regulators and the Bell System to shift investment and expenses, and so price levels, between the state and federal jurisdictions.

The separations process is time consuming and cumbersome, requiring significant record keeping; however, the process is based on fairly simple principles. Each investment dollar is to be allocated to the appropriate jurisdiction based on some estimate of use of that investment. The book costs of investments that are used in only one jurisdiction are directly assigned to that jurisdiction. For example, the book costs of an interstate dedicated line can be directly assigned to the interstate jurisdiction. The book costs of investments that are used in the provision of both state and interstate services are allocated between the jurisdictions based on some measure of relative use. For example, a local telephone switch routes local calls, state toll calls, and interstate toll calls. The book costs of that switch are allocated to the appropriate jurisdiction based on the percentage of traffic that is state (local and state toll) and the percentage that is interstate.[40] If 12 percent of the minutes going through the switch are interstate, 12 percent of the book costs of the switch should be allocated to interstate.

The first step in the separations process is to determine the book costs for the categories of investment that are to be allocated. Prior to the full deregulation of customer premise equipment (CPE) in the 1980s, there were five primary categories: subscriber station equipment (or CPE), local distribution plant (the local loop), local switching equipment, toll switching equipment, and interexchange facilities.[41] Once the book costs for these investment categories are determined, the costs are allocated to the appropriate jurisdiction, either through direct assignment or through a relative use factor appropriate to that category of investment. Expense dollars are allocated based on investment allocations. For example, maintenance expenses associated with interexchange facilities are allocated to state or interstate based on the allocation of interexchange facilities investment. When the process is complete, the investment and expenses that a firm devotes to providing regulated services should be apportioned between the state and interstate jurisdictions, so that the firm's revenue requirement for each jurisdiction can be calculated.

The Supreme Court, in its 1930 *Smith v. Illinois Bell Telephone Company* decision, required cost allocation that separated all costs, including the cost of station equipment or CPE. This was called the station-to-station approach. The case had started when Illinois Bell took the Illinois state commission to court, claiming that the local exchange rates specified by the Illinois commission for the city of Chicago were confiscatory. The District Court found for Illinois Bell, and the Illinois commission appealed to the Supreme Court. The Supreme Court noted that the District Court had erred in arriving at a decision without doing some form of cost allocation. According to the Supreme Court, "separation of intrastate and interstate property, revenues, and expense of the company is important not simply as a theoretical allocation to two branches of the business; it is essential to the appropriate recognition of the competent governmental authority in each field of regulation." As a result, the Court found that the "validity of the commission's order in this case can be suitably tested only by an appropriate determination of the value of the property employed in the intrastate business." Such a determination would involve "a reasonable apportionment of the telephone exchange property used in both [state and interstate] classes of service."[42] Despite the *Smith v. Illinois* decision, the station-to-station approach was not immediately accepted.

Prior to the 1930 decision, little had been done to develop any form of cost allocation. The Interstate Commerce Commission (ICC) did not address the issue at all, and did little to regulate interstate toll rates. State commissions during this time period were more concerned with matters of rate base valuation than with cost allocation.[43] Meanwhile, the Bell System found board-to-board a more lucrative approach. In board-to-board allocation, only switch and interexchange costs are allocated. The costs of CPE and the local loop are assigned to the state jurisdiction to be recovered through state and local rates. As Gabel has pointed out, the almost nonexistent regulation of interstate toll rates, when coupled with the greater regulatory oversight exercised by the state commissions, made the board-to-board approach the more financially sensible method for the Bell companies.[44] Interstate rates were based on discussions with the ICC, not on rate hearings. However, investments and expenses allocated to the state jurisdiction could be recovered through rate hearings and so result in higher state rates. Even after the *Smith* decision, the Bell companies continued to follow the board-to-board approach.

State commissions did not become interested in separations matters until the creation of the FCC in 1934. The FCC proved effective in negotiating a series of significant interstate toll rate reductions. Between 1936–40, interstate toll rates were reduced by about $27.5 million annually.[45] State toll rates, on the other hand, did not decline; indeed, the state commissions were faced with a barrage of rate hearings in which the Bell companies asked for increases in state rates. The resulting difference in rates became known as the toll rate disparity problem. By 1951, state toll rates were 35% higher than interstate rates.[46] State commissions, fearing that they would look ineffective in comparison to the FCC and its seeming ability to lower interstate rates, began to look for ways to remedy the situation.

The period from the FCC's creation through the 1950s was an interesting time during which the state commissions argued for the development of separations procedures that would shift more costs to the interstate jurisdiction; the Bell System came to see the financial benefits of the station-to-station approach; and the FCC failed to take the initiative in formulating separations rules. The state commissions, acting through the National Association of Railroad [later Regulatory] and Utility Commissioners (NARUC),[47] attempted to solve the toll rate disparity problem by trying to dissuade the FCC from mandating toll rate reductions, and by drafting a series of cost allocation proposals designed to shift more costs out of the state jurisdiction.

In 1942 the FCC began a formal inquiry to deal with separations issues but never made any definitive decisions. Instead, the FCC would agree not to block state separations procedures, or would adopt separations rules that were often proposed by the Bell System. In 1943 AT&T agreed to a $50 million rate decrease and finally agreed to a separations procedure.[48] This procedure was focused only on *revenues*. The result of these procedures and rules would be to shift significant amounts of revenue requirement to interstate toll, thus staving off any planned interstate toll rate decreases.[49] In 1947 the FCC agreed not to object to the use of a *Separations Manual* issued by NARUC. The use of the 1947 manual resulted in an increase of $19 million in interstate revenue requirement and the withdrawal of a proposal for a further interstate toll rate reduction. In 1951 the FCC accepted the "Charleston Plan," a plan originally proposed by the Bell System. This changed the method used to allocate local exchange facilities so that a greater percentage would be allocated to interstate. As a result, some $90 million in investment and $22 million in expenses were shifted to interstate toll; in response, the Bell System filed to raise interstate toll rates by $14 million. In 1956 the FCC allowed the use of the so-called Modified Phoenix Plan, which had already been adopted by NARUC; the result of the plan was to shift almost $40 million in revenue requirement to interstate toll. The FCC continued to approve separations changes suggested by the Bell System, accepting changes in 1962 that shifted $46 million to the interstate revenue requirement and accepting the "Denver Plan" in 1965 which shifted almost $100 million in revenue requirement.

During this period, the Bell System changed its philosophy. The Bell System greeted the creation of the FCC by continuing to espouse the use of the board-to-board method, arguing that *Smith v. Illinois* had been superseded by the Communication Act of 1934, which precluded the FCC from having jurisdiction over state rates, services, or facilities. As the FCC continued to contemplate large interstate toll reductions, the Bell System came to see the usefulness of increasing interstate revenue requirements in order to prevent rate reductions. This strategy appeared all the more successful since the Bell companies were able to prevent rate decreases for state rates, citing increased costs at the state level despite the shift of revenue requirement to interstate toll. Indeed, in the period from 1942 to 1965, there was a shift of $280 million in revenue requirement to interstate toll; only 22 percent of the corresponding reduction in state revenue requirement resulted in state rate reductions.[50]

Through the 1950s, the issue of how to accomplish separations was an extremely contentious one. The most basic, and significant, controversy underlying the whole cost allocation question was the appropriate treatment of the investments that comprised local service: subscriber station equipment, the local loop, and local switching equipment. Proponents of a board-to-board approach argued that none of the local investment should be allocated to the interstate jurisdiction, and so to interstate rates. According to advocates of the board-to-board approach, toll service was incremental to local service, a mere add-on. Local facilities such as the loop and the local switch were necessary to provide local service and would be in place even if no toll calls were ever placed. The only facilities used for toll service were toll switches (or toll operator boards) and the interexchange facilities that connected them; therefore, toll prices should bear only the cost of those facilities. The other side of the argument, the station-to-station approach, held that toll calls would not be possible without the local plant. A toll call used local facilities; therefore, toll prices should bear some portion of the cost of the subscriber station equipment, the local loop, and the local switch.

The separations process proved to be a valuable tool for both regulators and the regulated firm. As will be discussed (*see section 7.3*), the shifting of local costs to interstate toll became a crucial component of the universal service policy to keep local residential rates low and therefore widely affordable. The separations process also became a useful tool for the Bell System in managing its earnings and prices. The development of the separations process also provides an interesting example of the challenges involved in regulating a firm in a complex industry. During this period, decisions about separations rules were often driven by the regulated firm rather than by the regulators.

4.3 REGULATION—SETTING A PATTERN

Driven by problems with existing legislation and the new broadcasting industry, which required some system of governmental oversight to avoid chaos, this period saw the implementation of vital legislation, most of it still in force.

Radio Act

The growth of radio broadcasting by the mid-1920s was becoming increasingly chaotic. In the face of only weak Department of Commerce attempts to regulate under the Radio Act of 1912,[51] many stations began independently changing their frequencies and powers, causing rapidly worsening interference in most major cities. The 1927 Radio Act asserted governmental ownership of and the right to regulate the spectrum. The statute created the Federal Radio Commission (FRC) with power to license stations, assign frequency bands and station wavelengths, fix operating times and power levels, and ensure that the use of the airwaves was "in the public interest, convenience, and necessity." Broadcasters were *not* common carriers: their rates would not be regulated, and they would be responsible for the use made of their facilities.

Originally intended to be a temporary body, the FRC was made permanent in 1929. The five-commissioner operation was seen as the best and most effective means of regulation for radio. But some aspects of radio regulation remained with the Department of Commerce, and telegraph and telephone regulation was still largely in the ineffective hands of the Interstate Commerce Commission. This divided authority struck many as an inadequate approach to effective telecommunications policymaking. Indeed, there were dozens more players involved, that is, the individual states.

Splawn Report

As part of the search for change, a massive telecommunications industry investigation was begun in 1932 under the supervision of Dr. Walter M. W. Splawn, special counsel for the House commerce committee which had jurisdiction over communications issues. Splawn submitted his 4,200-page, three-part report in segments from April to June 1934. It was the most extensive study of the structure, finances, and operations of the telegraph, telephone, and wireless industries ever conducted.

Among other things Splawn demonstrated that renewed acquisition activity had pushed Bell System ownership of the nation's telephones to nearly 80 percent—the same level that would prevail a half century later. Yet Splawn felt additional research was needed before specific conclusions could be reached about possible changes in the structure of the industry. In the meantime, he recommended that:

- federal regulation be unified under a single communications commission;
- the commission be funded by assessing the regulated industries themselves;
- use of the holding company as a means of control in telecommunications be examined closely; and, more specifically
- all "big companies and their subsidiaries"—especially AT&T—be subject to a "thorough and complete study" that would probably require at least a year and a million dollars.[52]

Just a few months earlier, an interagency study under the direction of Secretary of Commerce Daniel Roper had reached somewhat similar conclusions.[53] A mere 14 pages long, it identified four major firms, AT&T chief among them, which dominated wire and wireless means of communication.[54] The report expressed concern about the lack of any national plan for telecommunications development. Despite pressures of the Depression, the report favored continued private ownership and operation of telecommunications "at least for the present."[55] Most importantly, however, it recommended "the transfer of existing diversified regulation of communications to a new or single regulatory body, to which would be committed any further Federal control of two-way communication and broadcasting."[56]

Communications Act

A key result of these two reports and a specific request by President Franklin Roosevelt was the passage of the Communications Act in June 1934. The Communications Act was not a true landmark policymaking effort, for in effect, it was an administrative consolidation of regulatory functions into a single independent regulatory agency.[57] In an alternative formulation, one researcher summarizes the Communications Act of 1934 as "essentially empower[ing] a modern agency to carry out the traditional provision of the common law of public callings."[58] A common misconception about the 1934 legislation is that it codifies the idea of "universal service" as that term came to be understood beginning in the 1970s (*see section 9.4*). Another argues persuasively that neither the language nor the intent of Congress in writing the act can be seen in this way.[59]

The new act established the FCC to provide definition to the sometimes hazy congressional policy statements in the Act. Formation of the Commission, which centralized most regulation of wired and wireless communications, marked the first time that the federal government had created the basis for comprehensive oversight and regulation of telecommunications. "The [FCC's] power to compel interconnection, to suspend rates pending an investigation, to allocate frequencies, and to require prior approval for any expansion of facilities provided the framework for moving long-distance communications from a market-oriented industry" to one largely defined by politics and regulation.[60] While the FCC was granted important powers, its creation essentially froze the telephone industry and the dominant role of AT&T in place for years. (Ironically, however, telephone issues were of relatively minor importance to the FCC—well into the early 1970s, the FCC would spend far more time and energy over broadcasting problems than with telephone concerns.)

Since the Act was federal legislation, the power of the FCC was limited to interstate communications. The states retained authority over intrastate communications. While this system of dual regulation has allowed for experiments in regulatory forms at the local level, it has also made it difficult for the United States to have a consistent telecommunications policy at all jurisdictional levels. This objective of a uniform and consistent policy has resulted in numerous jurisdictional conflicts over the years.

Early in its existence, the FCC demonstrated a bias favoring monopolistic service provision. In lobbying Congress to enable the FCC to consolidate the two competing telegraph companies (Western Union and Postal Telegraph), the FCC cited the advantages it perceived of the consolidation of the telephone industry. Thus, while the FCC would soon turn a very critical eye on AT&T, it was in the context of a monopolistic industry organization; the question of competition (which was not foreclosed by the Act) was not then seen as a viable alternative.

FCC Telephone Investigation

The FCC's first action (one mandated by Congress to build on Splawn's findings) was to launch an intensive study of the telephone industry's structure and operations. From 1934 to 1939, using a special staff of some 300 economists, accountants, and attorneys, the FCC conducted a series of detailed studies as FCC Docket No. 1 under the direction of FCC Commissioner Paul A. Walker, a former Oklahoma state commissioner. The study eventually cost some $2 million in the midst of the Depression (roughly $24 million in 2005 dollars).[61]

Every aspect of AT&T's operations was meticulously investigated, including the relationship between AT&T and its Western Electric manufacturing subsidiary, and particularly Western's exclusive right to supply telephones and related equipment to the rest of the Bell System. Dozens of detailed FCC research studies were issued, and AT&T published rebuttals to nearly every one. Among AT&T's complaints, the company argued it was not being given any chance to comment on the FCC research and findings. AT&T was most concerned about this investigation, and issued rebuttal booklets as each of the inquiry's special studies was released in limited numbers by the staff. But the mass of detail was more than the daily press could handle and the study received little media attention until the first summary report of its findings and recommendations was issued.

Walker Report

A lengthy initial report by Walker was released in April 1938. Sharply critical of AT&T management, the Report made the following conclusions:

- Western Electric overcharged the Bell Operating Companies (BOCs) for equipment. This inflated the rate base and cost AT&T customers an estimated $51 million per year;
- AT&T's accounting and depreciation practices were variable and erratic, and also resulted in a higher rate base; and
- The AT&T licensing fee charged to operating companies was arbitrary, since it was not based on the cost or value of services provided to the BOCs by AT&T. The report claimed that the cost of these services was either undocumented or unknown.

Based on these and other findings, the Walker Report made the following recommendations:

- The FCC should be granted the authority to review and approve a wide range of Bell System practices in advance of their implementation. This authority would extend to issues such as depreciation calculations, plant expenditures, and intercompany contracts.

- The Bell System should be explicitly prevented from participating in competitive businesses to protect its financial stability from these risks.
- The FCC should choose one of two possible options for equipment manufacturing: either mandate competition in procurement by the operating companies or regulate Western Electric as a public utility.

In all, Walker's report recommended dozens of changes in the industry and in regulation including a far stronger supervisory role for the FCC over telephone industry operations.

After considerable complaint and some effective lobbying by AT&T officials, Walker's conclusions (but not the detailed analysis itself) were toned down for the final June 1939 report submitted to Congress.[62] The final FCC report called attention to the continued high profits of AT&T in the midst of the Depression as one indicator of the company's heavy-handed operations. Yet despite the huge FCC effort, the investigation had little impact in the short run, and no formal federal effort was launched to change AT&T's structure. Only a decade later would the investigation seem to bear fruit (*see section 4.4*).

Since Congress had given the FCC regulatory authority over AT&T, the Justice Department had fairly little to do with the company on a day-to-day basis. In November 1940, Assistant Attorney General Thurman Arnold, writing to a citizen seeking an antitrust action against AT&T, noted that AT&T "has a monopoly of telephone service in the United States, owning 80 % of the telephone stations and 95% of the toll lines." Arnold went on to describe the company as a natural monopoly with the lawful status of a public utility subject to regulation by state and federal regulatory bodies. Thus, Arnold wrote, the letter-writer would have to appeal to one of these regulatory bodies for relief. In essence, Arnold concluded that the comprehensive regulatory scheme employed under the Communications Act and state regulatory laws was sufficient to address problems of any size or power inherent in AT&T.

4.4 ANTITRUST: THE 1949 SUIT

With wartime concerns resolved, the government's indifference to AT&T had changed markedly by 1949. Drawing on the results of the FCC's 1939 investigation, conferences with FCC staff, and the results of its own internal study, the Department of Justice's antitrust division concluded that substantial revision of AT&T's corporate structure was the only way to resolve continuing ownership and competition problems in the industry.[63]

The Truman administration antitrust division was a strong and active one. Several cases were filed: against GE in a lamp patent case, A&P foods, the tobacco companies, and American Can, among others. The motion picture studios were being divorced from their theater chains, a decision upheld by the Supreme Court in 1948. In 1947, the Supreme Court had upheld an earlier government decision to break up the Pullman sleeping car combine, and many saw close parallels between the railway case and the telephone business.[64]

Many were surprised, however, when the government moved on AT&T in 1949, for there had been no complaints about its service or pricing and the company had done yeoman service in a variety of ways during the war. Further, the telephone giant was clearly pressing ahead with technology (the transistor had been announced in 1948, and coaxial cable and microwave links were being actively developed). Financial analysis showed AT&T's rate of return was well below that of the 50 largest U.S. manufacturing firms.[65]

On the other hand, regulators at both state and national levels were still heavily influenced by Depression-era thinking and theory. A vertically integrated company like AT&T (or Pullman or the movie studios) had the potential to dominate a business and control prices, and to many regulators of the time, that danger was sufficient rationale to take action. Further, of course, the Justice Department already had a wealth of telephone industry data on hand from the 1936–39 FCC telephone investigation. That it was a decade old seemed not to matter.

1949 Complaint

In January 1949, the Department filed a 73-page antitrust suit to break up the Bell System.[66] Filed in the U.S. District Court for New Jersey (where the Western Electric manufacturing subsidiary was headquartered) as *United States v. Western Electric Co.*, the government charged that AT&T and Western Electric had engaged "in a combination and conspiracy to monopolize" and had

> in fact monopolized . . . the production, manufacture, distribution, sale, and installation of telephones, telephone apparatus, telephone equipment, telephone materials, and telephone supplies . . . in violation of . . . the Sherman Act . . . by [among other things] eliminating . . . competitors engaged in the rendition of telephone service.[67]

The complaint went on to allege that AT&T had engaged in the "elimination] of all substantial competition in the manufacture and sale of" telecommunications equipment to Bell System operating units by means such as

- "vesting in Western Electric the exclusive right to manufacture and sell" such equipment;
- "requiring" Bell operating units "to purchase their equipment exclusively from Western"; and
- "vesting in AT&T the power to control the operation of all branches of the Bell System, including research and development, and the manufacture, sale, installation, and operation of Bell System telephone plant and facilities."[68]

To remedy all this, Justice sought to sever Western Electric from AT&T, or in Attorney General Tom Clark's words: "The chief purpose of this action is to restore competition in the manufacture and sale of telephone equipment now produced and sold almost exclusively by Western Electric at noncompetitive prices."[69] Western might be divided into at least three independent companies to encourage competition. Further, Western would be required to sell its half-interest in the Bell Labs research arm to AT&T. The Labs, in turn, would have to license on a nondiscriminatory basis its patents to non-Bell System users.

In answering the complaint in April 1949, AT&T defended the integrated nature of the Bell System as one vital to a high quality telephone service. Further, and foreshadowing an argument to be made a quarter-century later, AT&T argued at length that previous agreements and settlements with regulators obviated both the need for this suit and its legal basis.

Derailing the Suit

For reasons that have always been somewhat hazy, there was no contact between the two sides in the suit for the next two years. Only in August 1951 did the government serve its initial discovery[70] request on AT&T. In the meantime, setbacks in the Korean War (which had begun in June 1950) led President Truman to declare in late 1950 a state of national emergency in which many civilian concerns were subordinated to national security needs. Building on its many roles in assisting the defense buildup, AT&T sought in early 1952 to freeze the antitrust process for two years. The company claimed that key executives were being distracted from national needs by the antitrust process. The Defense Department, relying totally on the company (Defense undertook no investigation of its own on the possible impact of an antitrust finding on national security needs), concurred and strongly supported the company's request for delay. Later Congressional hearings produced considerable evidence of the close cooperation between Bell and Defense Department personnel in this request for a delay.

The effect of all this was to prevent any effective action on the case for all of 1952. In the meantime, the facts underlying the government position—most dating back to the FCC investigation of the 1930s—grew colder. Then partisan politics intervened. In the 1952 national elections, the government changed from two decades of Democratic control to a Republican President (Eisenhower) and Congress. A Republican Justice Department seemed far less inclined to pursue the nation's biggest company and, predictably, the huge case never went to trial.

Although the new administration seemed lukewarm about continuing the case, AT&T was determined to bring the process to a close as the continuing legal uncertainties affected the company's ability to raise needed capital. The company's incentive, combined with government indecision led eventually to discussions about a settlement, or consent decree, where the government would agree to terminate legal proceedings and settle its differences. Several meetings were held in the months after the Eisenhower administration came into office where Bell System and government figures informally discussed how to bring the case to some kind of conclusion.

In mid-1953, these initial meetings led President Eisenhower's Attorney General, Herbert Brownell, and AT&T's general counsel, T. Brooke Price, to confer privately at a judicial conference in White Sulphur Springs, West Virginia. Although persuaded by Defense Department concerns that an integrated AT&T was vital to the nation's national security in the midst of the Cold War, Brownell was reluctant to simply dismiss the case given the record that had been developed that made clear AT&T's resistance to competition. He suggested to AT&T's top lawyer that some relatively minor concessions by the company could form the core of a relatively modest settlement of the matter without doing serious damage to AT&T's business.

AT&T officials turned again to the Defense Department for further support to end the case. The head of Bell Labs, Dr. M. J. Kelly, prepared a statement which was used by Defense officials to press the Department of Justice to settle the case. For nearly a year, nothing further happened. Early in 1954, settlement negotiations re-opened between AT&T and Justice officials who jockeyed back and forth on what the company would concede and what the government would accept. Antitrust Division officials continued to press for some kind of divestiture. Once again, AT&T returned to the Defense Department and all but wrote the department's plea to Justice to drop the case once and for all.[71]

Discussions dragged on into 1955 with some in the Antitrust Division pressing for at least minimal "divorcement" of Western Electric from AT&T. The FCC was asked by the Department of Justice for its views and after considerable internal debate, offered a statement generally supporting settlement of the case without such divorcement. Then things began to move rapidly. The parties finally reached agreement on the terms of the settlement in late December 1955, and a Consent Decree was accepted by the U.S. District Court for the District of New Jersey on January 24, 1956.[72]

1956 Final Judgment

The 1956 Consent Decree (also referred to as a Final Judgment) that ended the case restricted AT&T to the provision of regulated common carrier telephone service and manufacturing of equipment that supported that purpose. Put another way, AT&T was not allowed to enter the fledgling computer industry which was an unregulated field; for that matter, AT&T was not permitted to enter any other field unrelated to telephony. Yet exceptions were granted—AT&T could continue to provide ancillary services for federal government needs (a Defense Department requirement), experimental purposes, and furnishing circuits to other carriers, among others.[73] So the restriction was somewhat flexible.

Further, AT&T was required to license all of its 8,700 existing patents (including those on the valuable transistor) to any applicant (except RCA, General Electric, and Westinghouse) on a royalty-free basis, though it could charge reasonable royalty rates for use of all subsequent patents the Labs would develop. Any company so applying, however, had to license Western Electric (at reasonable fees) to make use of its own patents. The three excepted firms, major communications manufacturers, would have to pay royalties for all AT&T patents unless they allowed the Bell System free use of their own patents.

By limiting the business sectors which AT&T could enter, instead of changing AT&T's structure (the divestiture requested in the government's original complaint never took place), the decree essentially meant that AT&T retained the dominant share of the telephone business while continuing to submit its rates to state and federal regulators for approval. Yet the 1956 decree did not force AT&T to open its business to competitors; the document contained

> no provision (1) requiring the defendants to sell at nondiscriminatory prices to independent telephone operating companies or prohibiting or limiting sales of such equipment to such companies; [and] (2) requiring sales at nondiscriminatory prices to common carriers competing with Bell of equipment, other than telephone equipment, used by the Bell companies.[74]

Because the government's attempt to sever Western failed, AT&T concluded (as company officials later told Congress) that "in effect, [the] decree constitutes an admission for [the] government that the way we are doing our Bell System job is a legal and proper way as far as antitrust laws are concerned—[it is] in effect a blessing of [the] present setup."[75] Described a different way, AT&T gave up sole access to its patent empire—for 80 years a central feature of its strategy—in return for governmental recognition of its dominant role in the industry.

Aftermath

As AT&T had dodged the government's bullet, naturally the negotiations leading to the Consent Decree were almost immediately a source of some suspicion and controversy. After the Democrats regained control of Congress in 1956, extensive hearings on the AT&T case were held for 17 days before a House of Representatives antitrust subcommittee, resulting in a 2,800-page printed record. Congressional investigators had to depend on the FCC, Department of Defense, and AT&T to provide key documents when the Department of Justice refused to do so. The investigations revealed the close relationship between the Department of Defense and AT&T as the latter fought to avoid dismemberment. House investigators concluded that, overall

> the present decree not only fails to effectuate the purposes of the Sherman Act, but was arrived at without adequate consideration of the issues by those responsible for the protection of the Government's interests. From the time when AT&T first voiced its objections to standing trial, no top-level official of the Government appears to have given these issues the serious and searching consideration which their gravity demanded in the public interest. . . . This blot on the enforcement history of the antitrust laws can only be erased by those who are responsible for it.[76]

In other words, little had changed other than the opening of Bell Laboratories patent vaults. Despite government claims to the contrary, conventional wisdom was that AT&T had come out of the antitrust case very well: the company was intact and prevented from entering a business (mainframe computers) that was then of little interest to telephone company managers or anyone else. Yet as with many agreements intended to define an industry or its main players for all time, this one began to unravel in a very short time.

NOTES

1. John Bray, 2002. *Innovation and the Communications Revolution from the Victorian Pioneers to Broadband Internet* (London: IEEE), p. 131. Regular commercial links followed, but only some two decades later.

2. Joseph H. Vogelman, 1962. "Microwave Communications," *Proceedings of the Institute of Radio Engineers* 50:907 (May).

3. That microwave technology served as one of the instruments that paved the way for the later undoing of the Bell System bespeaks a certain irony. As one historian of the system has observed, "Seldom, outside warfare, can there be such examples of spectacular technology [the transistor, communications satellite systems, and microwave] being used for the destruction of the institution which produced them." H. M. Boettinger, 1983. *The Telephone Book: Bell, Watson, Vail and American Life, 1876–1983* (New York: Stern), p. 8.

4. "Boston to Washington Coaxial Opened," *Broadcasting* (November 17, 1947), p. 100.

5. A Canadian coast-to-coast microwave link followed in 1958. See Robert J. Chapuis, 1982. *100 Years of Telephone Switching (1878–1978)* (Amsterdam: North-Holland), p. 292.

6. "A Promising New Era Begins for Television," *Life* (date not known, 1951), p. 66.

7. *The Bell System's Role in the Development of Nationwide Network Television* (New York: AT&T, 1967), pp. 6–7.

8. George C. Southworth, "Survey and History of the Progress of the Microwave Art," *Proceedings of the Institute of Radio Engineers* 50:1206 (May 1962).

9. Robert Britt Horwitz, 1988. *The Irony of Regulatory Reform: The Deregulation of American Telecommunications* (New York: Oxford University Press), pp. 147–148.

10. Ibid., p. 225.

11. Chapius, part II.

12. See Christopher H. Sterling and John Michael Kittross, 2002. *Stay Tuned: A History of American Broadcasting*, Third edition (Mahwah, NJ: Lawrence Erlbaum Associates), p. 95, for a chart illustrating the expansion of the AM spectrum from a single frequency in 1920 to the expanding band after 1923.

13. For a definitive discussion of AT&T policy with its pioneering WEAF in New York from 1922 to 1926, see William Peck Banning, 1946. *Commercial Broadcasting Pioneer: The WEAF Experiment* (Cambridge, MA: Harvard University Press).

14. For more information on this now-forgotten episode, see Banning (for the most complete account), Archer (1938), and Sterling and Kittross, pp. 74–77.

15. The program began on the NBC radio network in April 1940 and ran for 18 years, presenting concert music by the Bell Telephone Orchestra conducted by Donald Voorhees, as well as guest performers. See John Dunning, 1976. *Tune in Yesterday: The Ultimate Encyclopedia of Old-Time Radio 1925–1976* (Englewood Cliffs, NJ: Prentice Hall), pp. 58–59.

16. See *Broadcasting Network Service* (New York: Long Lines Department, 1934), and *The Bell System's Role in the Development of Nationwide Network Television*.

17. "A Promising New Era Begins for Television," *Life* (September, 1951), pp. 63–66 includes photos of the coaxial and microwave facilities. Ninety-four stations carried an address by President Truman to a potential audience of about a million viewers.

18. The Editors of *Electronics*. *An Age of Innovation: The World of Electronics 1930–2000* (New York: McGraw-Hill, 1981), p. 71.

19. Ernest Braun and Stuart Macdonald, 1982. *Revolution in Miniature: The History and Impact of Semiconductor Electronics* (New York: Cambridge University Press), p. 33.

20. Robert J. Chapuis and Amos E. Joel, 1990. *Electronics, Computers and Telephone Switching: A Book of Technological History* (Amsterdam: North-Holland), p. 131.

21. In a famous and uncharacteristic moment of underestimation, the *New York Times'* report on the device on July 1, 1948, took up only four inches buried on page 46 after

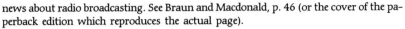

news about radio broadcasting. See Braun and Macdonald, p. 46 (or the cover of the paperback edition which reproduces the actual page).

22. Ibid., p. 48.

23. James C. Bonbright, Albert L. Danielsen, and David R. Kamerschen, 1988. *Principles of Public Utility Rates*, Second Edition (Arlington, VA: Public Utilities Reports), p. 33.

24. The concept of a natural monopoly is a common one in utility regulation. Two sources that cover the concept of natural monopoly in some depth are William W. Sharkey, 1982. *The Theory of Natural Monopoly* (New York: Cambridge University Press), and Kenneth E. Train, 1991. *Optimal Regulation: The Economic Theory of Natural Monopoly* (Cambridge, MA: MIT Press).

25. Train, pp. 10–11.

26. If, on the other hand, the market demand is actually more than two million units, but the firm's cost per unit begins to rise after two million units (because of having to increase its production costs), then a natural monopoly does not exist.

27. Train, p. 11.

28. Sharkey, pp. 13–20.

29. Ibid., p. 15.

30. Ibid., p. 16.

31. Michael Waterson, 1988. *Regulation of the Firm and Natural Monopoly* (New York: Basil Blackwell), p. 145.

32. Bonbright et al., p. 33.

33. While new incentive forms of regulation have been developed during the past 10 to 15 years, rate-of-return regulation is still the method (at the time of this writing) that is applied to many small telephone companies at both federal and state levels.

34. Regulators and the regulated have long been aware of this aspect of rate-of-return regulation. The alternative to the historical approach to rate-of-return is to calculate the revenue requirement based on forecasted expenses and rate base and to then perform a true-up calculation after the resulting rates have been charged for a period of time to determine whether the firm over- or underearned. This approach was adopted when access charges were established at the onset of long distance competition. (*See section 7.3*.)

35. Phillips, pp. 321–331. This discussion of valuation of rate base, as well as the following discussion regarding allowed return relies heavily on chapters 8 and 9 of Phillips.

36. Ibid., p. 375.

37. Ibid., p. 388.

38. Ibid., pp. 394–400.

39. Policymakers came to regard RoR regulation as providing the regulated firm with the wrong set of incentives. Regulators feared that the cost-plus nature of this method led to "gold plating" and inefficiencies. This aspect of the RoR method will be discussed more fully in section 4.3. Train also explores in some depth the potentially adverse effects of RoR regulation, pp. 19–113.

40. Separations studies actually allocate costs among local, state toll and interstate toll calls. Prior to World War II, it was common practice for local telephone companies to do such three-way studies. After World War II, however, in most states (California being the exception) local companies provided state commissions with the separations results that showed state numbers in the aggregate, not broken down between local and state toll. See Richard Gabel, 1967. *Development of Separations Principles in the Telephone Industry* (East Lansing: Michigan State University Press), pp. 6–7.

41. Ibid., p. 15.

42. 282 U.S. 133; 51 S. Ct. 65; (1930); 75 L.Ed 255, pp. 148–150.

43. Gabel, pp. 16–17. Gabel also points out that in 1910, the New York Public Service Commission tried unsuccessfully to separate the costs of exchange and toll services, while the Kansas Commission tried, with a little more success, to apportion expenses between exchange and toll in 1918, pp. 20–23.

44. Ibid., p. 18.

45. Ibid., p. 36.

46. Ibid., p. 61.

47. NARUC later changed its name to the National Association of Regulatory Utility Commissioners.

48. Horwitz, p. 134.

49. See Gabel, pp. 47–118 for more specific information regarding the discussion of NARUC and Bell System proposals. See also Carol L. Weinhaus and Anthony G. Oettinger, 1988. *Behind the Telephone Debates* (Norwood, NJ: Ablex), for a discussion of the development of the separations rules.

50. Gabel, pp. 127–129.

51. The 1912 Radio Act gave the Secretary of Commerce no discretion to turn down a license application, nor the power to develop enforceable regulations with which to license stations. Court decisions and a finding by the Attorney General in 1926 made the situation intolerable, forcing Congress to take action.

52. U.S. House of Representatives. *Preliminary Report on Communication Companies.* 73rd Cong., 2nd Sess., House Report 1273, April 18, 1934, p. xxix.

53. *Study of Communications by an Interdepartmental Committee.* 73rd Cong., 2nd Sess., Senate Committee Print, 1934 (reprinted in John M. Kittross, ed. *Administration of American Telecommunications Policy*, Vol. 2, 1980. New York: Arno Press).

54. The others were Western Union, International Telephone & Telegraph, and the Radio Corporation of America.

55. *Study of Communications*, p. 5.

56. Ibid., p. 6.

57. Milton L. Mueller, 1997. *Universal Service: Competition, Interconnection, and Monopoly in the Making of the American Telephone System* (Cambridge, MA: MIT Press), pp. 156–157.

58. Horwitz, p. 130.

59. Mueller, chap. 13.

60. Brock, p. 179.

61. It was initially funded by Congress for $750,000 at a time when the FCC's entire budget for its first year of operation was only twice that ($1,525,000). See Fred W. Henck and Bernard Strassburg, 1988. *A Slippery Slope: The Long Road to the Breakup of AT&T* (Westport, CT: Greenwood Press), p. 5.

62. U.S. Federal Communications Commission, 1939. *Investigation of the Telephone Industry in the United States.* 76th Cong., 1st Sess., House Document 340 (reprinted by Arno Press, 1974).

63. One indicator of the close connection between the 1939 report and 1949 suit: the FCC's counsel for the former, Holmes Baldridge, directed the Justice Department's antitrust case until 1951. See House of Representatives, 1959. *Report of the Antitrust Subcommittee . . . on Consent Decree Program of the Department of Justice.* 86th Cong., 1st Sess. (January 30), p. 33, note 18.

64. *United States v. Pullman Co.* 330 US 806 (1947).

65. This paragraph draws on Alan Stone, 1989. *Wrong Number: The Breakup of AT&T* (New York: Basic Books), pp. 67–72.

66. One of the attorneys at Justice was Holmes A. Baldridge, who had directed the FCC Telephone Investigation (as an FCC staff attorney) during its final stages, and it was his familiarity with AT&T that enabled him to advise the Attorney General to file the suit. The filing decision was aided by state PUCs who "were unable to determine the reasonableness of Western Electric's charges. Brooks notes the irony of this in light of the financial crisis that enveloped the Bell System due to their lack of effectiveness of AT&T in gaining rate increases, Brooks, p. 234.

67. *Complaint* in Civil Action 17-49, United States of America v. Western Electric Co. Inc., and AT&T, pp. 20–21, paragraph 59 (as reprinted in Christopher H. Sterling, et al., eds. 1986. *Decision to Divest: Major Documents in U.S. v. AT&T, 1974–1984* (Washington: Communications Press), Vol. 1, pp. 68–69.

68. Ibid., pp. 22–23, para 60 (b).

69. Ibid., p. 33.

70. *Discovery* is the legal term for the process of exchanging of information by plaintiff and defendant that takes place in the pretrial phase of a lawsuit.

71. For details on this settlement story, see *Report on Consent Decree Program*, pp. 45–95.

72. For the full text of the *Final Judgment in Civil Action No. 17–49*, see Sterling et al., eds. 1986. *Decision to Divest*, Vol. 1, pp. 123–142.

73. *Report on Consent Decree Program*, p. 37.

74. Ibid., pp. 38–39.

75. Ibid., p. 39.

76. Ibid., p. 293.

Chapter 5

Competition Reappears (1956–74)

The two decades following the 1956 Consent Decree witnessed a revolution in the telecommunications marketplace. AT&T's hallowed status as virtually the only provider of telephone equipment and service—a role long subject to only superficial regulation—slowly began to erode. By the late 1950s, advances in technology (some developed, ironically, at Bell Labs) had created alternative ways to deliver telephone services. At the same time, regulators were providing unprecedented opportunities for new companies to enter the telephone equipment and services industries.

Yet none of this was part of any governmental or corporate grand plan, for until the early 1970s, there was no plan, grand or otherwise. What now seems a logical and planned sequence of decisions opening up telephone markets to competition was in fact a haphazard set of separate actions responding to individual requests, complaints, inventions, and ideas. Taken together, they changed the industry and laid the foundation for the transformation of the industry.

5.1 TECHNOLOGY—ADDING COMPETITIVE OPTIONS

The pace of electronics technology picked up noticeably in this period, with major progress in both systems and means of transmission. What had been a fairly staid industry subject to steady but fairly slow changes began to experience some revolutionary breakthroughs.

Solid-State Electronics

Demand was rising from many sources for simplified circuits that could be more easily and flexibly designed, built, and used. Many researchers were working on the problem. One British engineer made a prescient comment in a conference paper delivered in 1952:

> With the advent of the transistor and the work in semiconductors generally, it seems now possible to envisage electronic equipment in a solid block with no connecting wires. The block may consist of layers of insulating, conducting, rectifying and amplifying materials, the electrical functions being connected directly by cutting out areas of the various layers.[1]

The key innovation, the silicon *chip*, occurred to two different men within months of one another in 1958–59.[2] Robert Noyce (1927–90), then working in Mountain View, California, for Fairchild Semiconductor, and Jack Kilby (1923–2005), working in Dallas for Texas Instruments, independently developed different approaches to the *integrated circuit (IC)* that still defines electrical progress today. What both men had in mind was completely integrating the dozens or hundreds of individual transistors and their countless interconnections in miniature form embedded in a chip of silicon. Once designed, these chips could be fabricated in mass production.

The chip was first revealed by Kilby at the Institute for Radio Engineers convention in 1959.[3] As had happened at the first announcement of the transistor more than a decade earlier, skepticism greeted his announcement. Manufacturing such tiny high-capacity devices would be incredibly difficult, and a large proportion of each batch would likely be defective. And the plant in which they would be made would be expensive to build and operate. The designing process would have to be microscopic.

But demand from both military and commercial space markets was huge. Early government contracts helped fund further research and development and improved means of chip manufacture. The first commercial ICs were announced in 1960, mass production began in 1962, and more than 25 companies were making them in 1965.[4] Many were spun off of existing companies or were new start-up firms helping to give Silicon Valley its name and chief industry.

By the late 1970s, the manufacture of solid state devices was a $6 billion industry; this was before introduction of the personal computer. As important was a unique feature of this new and expanding technology: the continual drop in per-chip cost just as per-chip capacity was increasing. One of the most significant applied advances came when engineers discovered that they could fabricate circuits containing multiple transistors (and other components) on a single piece of silicon. As invented in

1959, these integrated circuits further reduced the space and power consumption required by circuits, making transistors even more attractive.

Computer Revolution

Computers as we know them today developed from automatic tabulating and calculating machines used in business and government early in the 20th century. Those machines were electromechanical and were instructed by the use of punched cards to undertake repetitive tasks, freeing people to do other things.

During World War II, military needs led to the initial breakthroughs to true computing. Fifty miles northwest of London in Bletchley Park, British codebreakers trying to keep up with changing German machine-coded communications got a hand from British Post Office engineers who developed the "Colossus" computing device, the first of which went into service in late 1943. It was programmed to help reduce the number of code possibilities that human codebreakers had to deal with. The British had a dozen of the machines by the end of the war, but their existence remained a secret for three decades. In Philadelphia, engineers at the University of Pennsylvania's Moore School developed a room-filling device called ENIAC for the Army Ordnance Corps. Completed in 1945, it could (with difficulty) be programmed to develop firing tables for artillery—and was so used for a decade. ENIAC was powered by 18,000 vacuum tubes, creating problems of heat and constant replacement. "Software" as we know it today did not yet exist—these huge machines were "programmed" by changes in their physical settings, a complex and time-consuming chore.

The development of commercial business computing began in the late 1940s as several separate teams, and then companies, sought to build on these pioneering efforts to derive an economically viable machine.[5] The public first learned of this on Election Night in November 1952 when a UNIVAC computer (developed by the ENIAC team, now working for Remington Rand) was used by CBS News to predict the outcome of the presidential race. The machine did so well that its human operators found the results (projection of an Eisenhower landslide) hard to believe and held back the predictions for several hours! The Census Bureau began to use UNIVAC to speed up tabulation of 1950 census data. Given how complex and hugely expensive these early hand-built machines were, some predictions suggested there might only be a need for a dozen or so in the world.

Only slowly did other companies enter what would become a race to dominate the world of "electronic brains." Aiming at the defense market, IBM released its 701 stored-program computer in the early 1950s, but for several years remained undecided about pursuing a costly developmental push to perfect smaller but more powerful machines. The IBM 650 was their first business success with more than 1,000 built after its introduction in 1954. At the same time, IBM was the chief contractor on the huge Air Force SAGE (semi-automatic ground environment) project to place two-dozen automated aviation control centers around the country. SAGE demands for computing power and capability led to rapid leaps in development of magnetic core memory devices.

General Electric, RCA, AT&T's Western Electric and several other firms were all actively pursuing computer projects by the mid-1950s, though together they did not match IBM's performance. By 1960 IBM dominated the growing American computer market. "Computers" at this point meant mainframe devices—room-filling machines with memory units, control boards, printers, and other ancillary parts. The first widely used software programs, FORTRAN and COBOL were an important key to making the machines more flexible in application.[6] Introduced in 1964, the IBM System 360 became the first widely adopted machine for industry use. More than 1,000 were sold in the first month. (As part of a Consent Decree in an earlier antitrust case, IBM now sold as well as leased its machines.)[7]

Combining the growing capabilities of the commercial computer, and later solid-state electronics, marked a huge breakthrough for telecommunications.[8] What had been two quite separate industries with very different business cultures began to develop common interests as both commercial and government users sought ways to interconnect computers for what became known as distributed (or remote) data processing. But the telephone business was regulated while, aside from occasional antitrust forays, the computer industry was not. How to overcome that essential policy difference took almost as much time and effort as the technical research that created the issue (*see section 7.4*).

Communication Satellites

Combining microwaves with communication satellites in Earth orbit (an idea that had appeared only in science fiction) was first seriously proposed by Arthur C. Clarke, the British science and science-fiction writer, in a 1945 article.[9] Clarke theorized that if a satellite could be placed in orbit some 22,300 miles above the earth, it would appear from earth to stay in the same position at all times (a *geostationary orbit* or GSO). It could thus act as a super-tall antenna, allowing signals to be sent back to earth (*downlinked*) to a coverage area (*footprint*) covering about a third of the planet.

U.S. rocket and satellite development was slow at first, early experiments using leftover wartime German V-2 rockets and suffering from low budget priorities. Only in the mid-1950s did the Air Force begin to make heavy investments to perfect an intercontinental ballistic missile rocket. The Soviet launch of *Sputnik* in October 1957 shocked the West. Impact in Washington was substantial, leading to increased funding for the National Science Foundation, formation of the National Aeronautics and Space Administration (NASA), and creation of the Defense Advanced Research Projects Agency (DARPA). The race to perfect huge rocket vehicles, soon dubbed "the space race," became a central part of the Cold War. Money was no longer a limitation. After the spectacular and very public 1957 failure of the U.S. Navy's *Vanguard* rocket, a U.S. Army team headed by former German rocket scientist Wernher von Braun finally got the first American satellite into orbit in January 1958.[10]

Experimental communications satellites launched in the late 1950s and early 1960s used low-earth orbits (*LEOs*) a few hundred miles above the Earth's surface. Available rockets could only launch satellite payloads into LEOs despite knowledge of the GSO's benefits. These pioneering satellites could only transmit for part of each orbit around earth—usually about a half hour—when the satellite could be "seen" from a ground station. Of these early "birds," the most famous was *Telstar*, designed and built by AT&T and launched in 1962. It was a hugely expensive project:

> As for the space side of the bet, even the greater expense of the low-orbit system with its proliferation of satellites and complex tracking ground stations was to AT&T's advantage. American television company profits [were then] pegged to investments and therefore the more expensive Telstar turned out to be, the better.[11]

But the more than 50 trans-Atlantic television programs and thousands of telephone calls carried by *Telstar I* dramatically demonstrated to the general public what could be accomplished with satellite technology. It was followed by a near-twin *Telstar II* in 1963.

A substantial policy debate argued the benefits of *passive* satellites (cheaper, available now) that would act as super-tall antennas, merely reflecting signals back to Earth, and *active* satellites (more expensive and requiring more time to develop) that could store signals with onboard *transponders*. *Telstar* was an active satellite and its success and public reception helped to decide the issue, along with the clear recognition that only active satellites offered the potential for substantial further evolution.

Because of their costly design, construction, and launch, communications satellites were first used primarily for international government and military communications. As experience led to greater efficiencies, however, interest in potential domestic applications rose. As had already been the case with microwave and coaxial cable networks, the needs of broadcast television pushed technology. In 1965 the ABC television network proposed a domestic satellite (*domsat*) orbiting in the GSO to the FCC as a means of saving millions of dollars compared with the cost of terrestrial interconnection service provided by AT&T. Such a "bird" would allow great flexibility in network operations; the network also promised free use to the nation's struggling educational television stations. Though the application was turned down, it prompted a long FCC policy study (concerned more with questions of ownership and control than technology) that ended only with the 1972 domsat decision (*see section 5.4*).

Coaxial Cable

Experimental attempts to develop a telecommunications cable with two conductors (and thus far greater carrying capacity) took place early in the century. The principle had already been applied to submarine telegraph cables. A "wire in which one conductor surrounds the other . . . drastically reduces interference, and allows a much wider frequency range."[12] Extension of the idea to telephony took place largely at Bell Laboratories, leading to the first patents in the field in 1929.[13]

Initial Bell System application of coaxial cable focused on telephone use: each cable route (made up of several individual cables) could carry 1,800 calls.[14] The history of coaxial cables is one of steadily increasing capacity. By 1953 improved cables raised capacity to 9,300 and by the late 1960s that had increased to 32,400 voice circuits per route.[15] By the 1970s, improved systems could carry up to 132,000 conversations.[16] Coaxial cable technology and effective amplifiers made possible the first transatlantic telephone cable in 1956. TAT-1, a shared project of AT&T and the British Post Office, required a huge outlay:

> It was a stupendous pioneering undertaking. Some 4500 miles of coaxial cable had to be made to the most exacting specification ever devised, and new machinery had to be designed for laying the cable in waters up to 2.5 miles deep. Surveys of the transatlantic route had to be carried out to select the most suitable. One hundred forty-six repeaters had to be built to withstand the rigors of laying and the extreme water pressures in the deep ocean, and cable of functioning without attention for at least 20 years.[17]

Coaxial cable helped the early growth of "community antenna" television (CATV) systems in rural and some suburban areas. CATV began in the 1950s as a means by which isolated communities could share a large ("community") antenna, and its signal could be redistributed over a co-axial cable to individual homes. The high bandwidth infrastructure required by CATV would eventually make it an interesting competitor. During this first period, it was largely ignored by the FCC, which was substantially occupied with over-the-air television.

5.2 TERMINAL EQUIPMENT COMPETITION

What would become a terminal equipment revolution began very quietly, virtually unnoticed save by those directly involved. Starting in 1921, the "Hush-a-Phone" (a plastic cup-like device that fit over the speaking end of a telephone receiver and allowed a caller to speak privately in a noisy location) had been sold to some 125,000 telephone users. In the late 1940s, however, the Hush-a-Phone Corporation began to receive troublesome indicators that local exchange companies, most owned by AT&T, were informing customers that attaching such "foreign" (non-telephone company provided) equipment to their telephones was a violation of tariffs and was thus illegal.[18]

Hush-a-Phone

Concerned with this threat to its business, the Hush-a-Phone Corporation's chief, Harry A. Tuttle, appealed to the FCC in late 1948. Moving at a stately pace, the commission held hearings, and, seven *years* later, upheld AT&T, finding that use of a Hush-a-Phone could distort the voice of the person speaking into the telephone. The FCC agreed that prohibiting subscribers from attaching their own equipment to telephone company facilities was necessary to preserve the network's technical viability. Furthermore, noted the Commission, such restrictive tariffs had been allowed to remain in force over several decades by both the FCC and state regulatory commissions.[19] As long-time FCC Common Carrier Bureau Chief Bernard Strassburg later wrote,

> The Hush-a-Phone case is a classic illustration of the regulatory values that dominated the entire telephone regulatory community for generations. They were embraced by the FCC from its beginnings . . . into the 1960s. Thus it was the conviction of the FCC and its staff that they shared with the telephone company a common responsibility for efficient and economic public telephone service and that this responsibility could only be discharged by the carrier's control of all facilities that made up the network.[20]

The Hush-a-Phone Corporation, taking what was then an unusual action (today appeals are routinely expected), appealed the FCC finding to the Court of Appeals for the District of Columbia Circuit. On November 8, 1956, the court overturned the FCC in a short decision with a soon-to-be widely quoted sentence. The court held that, lacking a showing of demonstrable public harm, such restrictive tariffs were an "unwarranted interference with the telephone subscriber's right reasonably to use his telephone in ways which are privately beneficial without being publicly detrimental."[21]

As a result of the court order, the FCC ordered the Bell System to modify its tariffs accordingly. But *Hush-a-Phone* was only a limited precedent that appeared to allow connection of nonelectronic devices to telephone company facilities. Little had really changed in the telephone equipment industry. Virtually all telephones and related terminal devices continued to be obtained on a rental basis from telephone companies. Virtually none were sold outright.

Carterfone

By the late 1950s, Texas inventor and oil prospector Tom Carter was marketing his "Carterfone," an acoustic coupler (much like early computer modems) allowing private mobile radio telephone systems (such as those used on oil rigs in the Gulf of Mexico) to be connected to AT&T's wired telephone network. As with the Hush-a-Phone, no direct electrical connection to telephone company facilities was involved; a caller using AT&T facilities could be "patched through" via the Carterfone to a person on a private system. Nonetheless, the local AT&T operating company objected and went so far as to disconnect subscribers who continued to use Carterfones. Yet by 1966 Carter had sold some 3,500 of his devices, often reimbursing customers cut off by AT&T. When attempts at compromise with the Bell System broke down, Carter, not at all intimidated, slapped AT&T with a private antitrust suit in federal court in Texas. The suit was soon dismissed under a legal principle that required the matter to be heard first by the regulatory agency (the FCC) that had jurisdiction over the matter.[22]

After considerable reflection, the FCC in 1968 held that while Carterfone violated AT&T's tariffs, those foreign attachment tariffs were themselves unlawful since they failed to differentiate between harmful and harmless attachments.[23] Carter was free to sell his device. The FCC ordered AT&T to eliminate these tariff provisions nationwide, though how that might best be accomplished was left to the company to resolve. The FCC decision, quite a shift from its original holding in *Hush-a-Phone* that was later overturned by the federal court, was an indication of the growing importance of marketplace economic thinking within the agency's Common Carrier Bureau, an approach arguing for competition instead of regulated monopoly as the scheme best suited to encourage technical innovation and lower prices in the telephone business.

Bell System PCAs

AT&T accommodated the FCC mandate and its own purposes by replacing foreign attachment tariffs with "protective connecting arrangements" (PCA): a device provided on a tariffed basis (i.e., at a monthly rental charge to customers) to make the connection between any foreign device and the public switched network. Even if the foreign equipment to be attached to the AT&T network operated exactly like a telephone company-provided device (and most did), a PCA was still required. The PCA's added monthly lease charge to the customer effectively discouraged use of most non-telephone company devices. And because the devices were (according to a later FCC finding) overengineered and -designed, they were even more expensive to make and thus to lease.

The result of the *Hush-a-Phone* and *Carterfone* cases was to slowly open up the terminal equipment market to competitive entry by non-Bell System providers. The two decisions were limited in scope, their effectiveness blunted by initial government acquiescence to AT&T's restrictive PCA system. But the Bell System may have been too clever for its own good: the company's insistence on the use of PCAs over the next few years would translate into dozens of "episodes" of alleged anticompetitive behavior that were recited by the government in support of its 1974 antitrust case.

FCC Certification

The terminal equipment controversy continued for several more years until it became plain to the Commission that AT&T's real purpose was protecting its equipment monopoly, not protecting network viability. Fed up with AT&T's continued reliance on PCAs in denial of the spirit of the *Hush-a-Phone* and *Carterfone* decisions, and after a three-year investigation (including consultations with state commissions), the FCC in 1975 instituted its own policy of equipment certification.

Adopted only after considerable opposition from the Bell System, the Commission's certification program allowed any manufacturer to seek FCC approval that its device would cause no technical harm to the network and could be interconnected without need for a PCA. From now on, the FCC would rely on a system of standard plugs and jacks (rather than direct wiring, then standard in the Bell System) to allow use of devices from various manufacturers. Further, the FCC held that *all* equipment, including that manufactured by the telephone companies, would from now on have to meet the Commission's technical standards. A year later, the decision was extended to private branch exchanges, main station telephones, pay phones, and party line equipment. The Court of Appeals upheld the Commission when the telephone industry appealed, and when the Supreme Court declined to review, the decision went into effect late in 1977.

A small competitive industry selling telephones and related devices began to develop as a result. For the first time, consumers could purchase a telephone in a style and color they liked and no longer pay monthly lease charges. But local Bell operating companies continued to assess a (smaller) monthly charge for use of a customer-supplied telephone. Eventually state public utility commissions disallowed such charges.

5.3 OPENING UP TRANSMISSION

At the same time that the courts and FCC were chipping away at the edges of AT&T's dominance of the terminal equipment market, a potentially greater competitive threat to the Bell System was developing in the services arena. As with the terminal equipment market, the process began very slowly and was based on technological developments—in this case the beginnings of data transmission and the availability of microwave transmission.

As noted earlier in this chapter, remote computer processing was very slowly developing in the late 1950s and seemed likely to become ever more important. The analog Bell system had been designed for voice—not digital data—transmission. Some larger companies with widespread operations began to complain that they could not obtain the needed data connections to support their growing computer networks. This apparent inability (or unwillingness) of the Bell System to meet some user needs was only one of the motivating factors for change. Another was the practice of rate averaging employed by the Bell System to subsidize the cost of local connections. For large users, actual costs were often below average costs, so an important motivation behind the construction of private line networks was a desire to lower their expenses.[24] From the point of view of these firms, they were only asking for what they were already permitted in transportation: the ability to run a private system if it was economical for them to do so.

Above 890 Decision

In November 1956, the FCC opened a study of the use of microwave frequencies above 890 MHz, the upper limit of the UHF television allocation. It is notable that this was framed not as a *common carrier* issue, but rather as a *frequency allocation* matter. The consequence of this was that the standard for decision making was simply whether or not adequate frequency space in that portion of the spectrum existed. Common carrier decision standards, such as public necessity and considerations of economic harm to common carriers would therefore not apply here, although such points were raised by AT&T and other public network operators.[25] AT&T and the other public network operators argued strongly against allowing firms to construct private networks, as was proposed by the commission. They raised three primary concerns:

- Private networks would effectively be "cream skimming" by major users, who, because they would no longer be contributing to the cost of the public switched network, would raise the cost for the residential and small business users who remained (and could not afford their own network);[26]
- This could make communications deployment during times of national emergency more difficult (because there would now be multiple networks, raising concerns about interconnection); and
- Scarce frequency resources should be deployed to benefit everyone, not just the few firms that could afford to build private networks.

AT&T argued strongly that the FCC should retain its traditional approach—restricting microwave licenses to established carriers and a few specific service providers—rather than broaden the microwave franchise. This would avoid cream skimming activities by large users. Bell attorneys also asserted that AT&T certainly should not be required to interconnect such potential competitors.

In August 1959 the Commission concluded that the public interest would best be served by, and the frequencies available could readily support, limited licensing of private microwave systems.[27] The Electronic Industries Association had presented a detailed study showing that a 20-fold increase in the use of this spectrum was possible without causing interference; in this frequency allocation case, the finding created a presumption in favor of permitting the construction of limited private networks. "Limited" meant that such systems were to be used only for the licensee's own purposes (as a kind of closed-circuit communication for major business or government users), not resold to others (a restriction lifted in 1966). The FCC perceived that demand for such service was going unmet by existing carriers, chiefly AT&T, and that licensing private microwave systems would have no adverse economic effects on existing carriers. The practical implication of this ruling posed a threat to the system of nationwide price averaging, because a firm could operate a private network more cheaply than it could lease the equivalent number of lines. But the FCC did *not*

- Require AT&T to interconnect these networks with their network,
- Authorize companies to compete with AT&T in providing general long distance service, or
- Pool traffic in cost-sharing cooperatives.[28]

Given the limited purpose of these networks, the Commission decided the question of interconnection between these new private networks and AT&T's switched network could be defined in subsequent tariff filings. Despite this important limitation, the *Above 890* decision began to open a door. With its decision, the FCC allowed large companies to build their own private-line networks. And though the decision was limited in scope and participation, a very important point was made: AT&T and other common carriers were no longer the only service providers.

TELPAK

Having lost the main battle to prohibit incursions into some of its private-line business service offerings, AT&T countered the threatened loss of big customers in 1961 with a radical set of tariffs known as "TELPAK."[29] Offering significant volume discounts on AT&T's microwave service, these tariffs were aimed directly at potential private network operators and were designed to persuade them to use (or continue to use) Bell System services rather than going to the expense and trouble of building their own. The TELPAK tariffs were set substantially below the pre-*Above 890* tariffs—so low, in fact, that most companies had little incentive to construct private networks. Faced with a loss of business (construction and operation of private networks), both Motorola and Western Union complained to the FCC that TELPAK tariffs were set below the cost of providing the service—an indicator of anticompetitive predatory pricing.

These complaints triggered a sequence of litigation that lasted 20 years—the last TELPAK tariffs were only terminated in 1981. The contentious issues in TELPAK were ones that went to the very heart of definitions of common carriers and telecommunications pricing, issues that are still being discussed today. Constantly at issue was the degree to which the discount tariffs amounted to unfair dealings by AT&T against its competitors. Over time, the TELPAK tariffs were both simplified and increased in price, usually in response to specific FCC or court requirements.

The prolonged controversy highlighted a developing problem in FCC supervision of industry tariff filings. Critics accused AT&T of marshaling its considerable economic and legal talent to slow things down in response to proceedings or decisions with which it disagreed. Critics also claimed that AT&T used its considerable resources to overwhelm a small government agency, the FCC, with enough documents to keep an army of analysts busy for years. The FCC's practice of generally accepting AT&T's tariffs as filed (with the exception of the TELPAK controversy), unless a competitor or user raised substantial and specific questions, only added to the critics' complaints.

In rebuttal AT&T claimed that it was merely complying with FCC directives or competitors' filings at the requested or necessary level of detail. Complex tariffs called for complex filings of sophisticated documents and if the filings overwhelmed the FCC's ability to regulate the company, at least no one could accuse AT&T of inattention to detail.

Both *Above 890* and AT&T's subsequent TELPAK rates showed large business users what the price possibilities were. Both of these alternatives were based on *incremental costs*, and so did not include the cost transfers that were the result of the separations process. Despite this activity, NARUC and the FCC continued to negotiate separations manuals that transferred more and more capital from the state to the interstate jurisdiction. The result was that AT&T's regulated interstate prices continued to escalate, limiting attempts to move toward incremental cost-based prices.

Seven-Way Cost Study

By the mid-1960s it was increasingly evident that the FCC did not know enough about AT&T's cost structure to adequately judge the validity (and thus the legality) of proposed tariffs for any of AT&T's services. The last in-depth government analysis, the FCC's own 1930s investigation (*see section 4.3*), was now decades out of date. So, as part of a larger study of both telegraph and telephone tariffs and service, the FCC ordered AT&T to undertake a comparative analysis of its rate of investment, expenses, and revenues associated with seven classes of communication service, which made up the Bell System's interstate operations. TELPAK tariffs were included in the study which assessed AT&T's financial situation for the year ended August 31, 1964.

The findings of this "Seven-Way Cost Study" were either a shock or mere confirmation of what many had long thought: the study showed wide variations in rate of return among the compared services. For example, regular subscriber telephone service, a monopoly offering, was earning 10.2 percent on investment, while TELPAK, offered to meet competitive services, returned only 0.3 percent (internal AT&T studies suggested the company actually took a loss on the service[30]), a figure so low that AT&T could hardly have been said to have earned any money on the service. The conclusion seemed clear: AT&T was subsidizing its developing competitive offerings with revenues from monopoly services.

One result of the Seven-Way Cost Study was the FCC decision to abandon its policy of rubber-stamp approval of most AT&T tariff filings. From now on, in a process which grew steadily more confrontational, AT&T tariffs would be examined closely and many would be delayed or modified to the degree allowed under the Communications Act.

5.4   CREATING SPECIALIZED CARRIERS

Huge changes often begin with tiny steps. As AT&T was boxing with the FCC over the Seven-Way Cost Study, a future AT&T nemesis made its first tentative appearance. On the last day of 1963, an application was filed with the FCC seeking authority to construct and operate a private microwave telephone system between Chicago and St. Louis.[31] The system would involve flexible use of microwave frequencies for a private radio service supporting either voice or data needs of customers who would supply their own terminal equipment. Years would elapse before the application was finally approved and service inaugurated. The applicant was Microwave Communications Inc., soon known simply as MCI.

Formation of MCI

In its original form, MCI was the brainchild of John D. "Jack" Goeken, an insightful innovator (who would later initiate telephone service on commercial airliners). A salesman for radio equipment in central Illinois, Goeken hit upon the idea of providing truckers with radio-telephone service on the busy 290-mile route between Chicago and St. Louis.[32] Use of telephones in their trucks would greatly ease use of the route, helping to speed up pick-ups and deliveries.

He proposed allocating half the bandwidth used by the Bell System to the voice channels so that the capacity of the system would be larger. Although this would make speaker identification more difficult, he felt that people who worked together frequently would be willing to tolerate the lower quality for a lower cost. He also proposed less redundancy, which would lower costs as well as reliability.[33] What Goeken initially proposed was essentially a "shared" private microwave network. In *Above 890*, the FCC had proscribed such arrangements except in certain circumstances. To allow Goeken's plan was to change this policy. In addition to this,

Goeken would request interconnection with AT&T's network so that MCI's customers could access the microwave link. That was yet another departure from *Above 890*, one that would force the FCC to deal with issues of interconnection and appropriate interconnection pricing.

Use of microwave links would greatly expand the limited range of such systems. Early in October 1963, Goeken and two associates organized MCI "to sell communications equipment and to construct communications systems."[34] Two months later, just before higher FCC licensing fees went into effect, MCI rushed to file its application. Only Goeken and one of his partners had any knowledge of microwave technology and capabilities.[35]

While Goeken was full of ideas and enthusiasm, he lacked capital. MCI was woefully underfunded from the start—merely a few thousand dollars from the original partners—and thus pursued its FCC microwave license application with credit and hope. Save for Goeken, the few original investors dropped out, unable to meet the mounting bills. Still, stock in the little company with an idea but no business slowly became more widely held. Goeken made appearances before Congressional committees and developed a "Jack the Giant Killer" image for taking on AT&T's legal army fighting to crush his application. For nearly six years, the MCI saga before the FCC was a combination of regular modification of MCI's original filing, counter-filings from AT&T and others urging the FCC to deny the proposed new service as unnecessary and duplicative, and occasional Commission actions slowly moving the process along.

Formal hearings before an FCC examiner began in early 1967, by which time MCI had developed a detailed marketing plan that demonstrated a market need for its service, now estimated to cost about a half-million dollars to establish. Goeken was already suggesting how the system would expand if successful. In a reversal of its original thinking, the Common Carrier Bureau recommended approval of MCI's application as a useful test of market competition. In October, the hearing examiner recommended approval of the application. But MCI's constant search for sufficient investment funds continued as legal and other bills were delayed. And Goeken's relations with the MCI board were increasingly strained under the financial pressure.

The need for investors led in 1968 to William G. McGowan, a millionaire New York management expert. He and Goeken did not get along from the start,[36] but after extensive negotiations with MCI's board, McGowan came in as chief executive officer (Goeken would be president) of what was now called Microwave Communications of America Inc. McGowan's new business plans called for several affiliated networks to tie in with the original Chicago–St. Louis route. The regional firms would make it easier to raise investor interest.[37]

Approving MCI

In April 1968, the full FCC held a hearing on the MCI application, with representatives from AT&T, Western Union, and General Telephone of Illinois among MCI's detractors. The commissioners appeared divided over MCI's merits. The months that followed saw two events that helped sway the final decision. Late in the year, the Presidentially appointed Rostow Commission issued a report calling for, among other things, more open entry into private-line services.[38] And AT&T, in a classic case of poor timing, proposed to the FCC a new tariff to provide almost exactly the kind of service MCI had initially suggested, even though AT&T had earlier argued that such a service wasn't needed.

The FCC finally approved MCI's application by a vote of 4 to 3 on August 13, 1969.[39] While affecting only one company and its limited service, this decision was viewed as an important step in opening up telecommunications competition. The weekly newsletter *Telecommunications Reports* concluded "the domestic telecommunications business will never again be the same."[40]

Interconnection Negotiations

With an FCC license in hand to offer private microwave service between Chicago and St. Louis, MCI entered into negotiations with AT&T and two of its local operating companies for the one portion of the service MCI could not provide: the interconnection from MCI's microwave towers to the local loop or "last mile" part of AT&T's network that entered customers' homes or businesses. Negotiations dragged on for more than a year through no less than 12 complex and changing draft contracts.

MCI was forced to keep postponing inauguration of its service (which finally began in January 1972). Even when MCI service got under way, AT&T insisted on connecting MCI to Bell System facilities in a way that was arguably more involved and expensive than necessary, a situation MCI felt compelled to accept in order to begin service.[41] While MCI had fought for its license for seven years (final appeals were resolved only in 1971), actual system construction took only about seven months. Further, the legal battle before the FCC and with AT&T had cost some $10 million while system construction on the original Chicago–St. Louis line totaled less than $2 million.[42] Such were the costs of pioneering.

As the struggle to get the original MCI system up and running continued, McGowan established his 16 regional MCI companies, all of which soon applied to the FCC for new network routes, including connections between Chicago and New York, Boston and New York, St. Louis and Dallas, and Dallas to Los Angeles.[43] He had in mind a national telephone system, developed on regional lines much as Vail had pushed original Bell System growth.

More Applicants

Approval of the original MCI application opened the door to a number of similar specialized microwave service applications—more than 30 (for some 1,700 microwave base stations) in less than a year.[44] Collectively, they represented a threat to the established carriers. The private-line market, largely shared by AT&T and Western Union as the 1970s began, was worth $1.1 billion in revenue. Most of that accrued to AT&T, though Western Union's $135 million from this service contributed 27 percent of the company's total revenue.[45] Customers such as the federal government, airlines, and manufacturers made up two-thirds of the specialized service market.

In addition to MCI's affiliates, which constituted about half the applicants for private-line microwave services, another major player was the Southern Pacific Communications Corporation (SPCC), part of the California-based railway. SPCC wanted to expand its existing private network (begun in 1962) to sell specialized common carrier services to others. After negotiations with McGowan on a possible merger, SPCC went its own way.[46]

One unique applicant was the Data Transmission Corporation (Datran), which proposed something wholly different: the first digital switch national network designed specifically for data transmission.[47] Founded in 1968 (the same year McGowan began revitalizing MCI), Datran was the brainchild of Sam Wyly who had made a fortune with his University Computing Company in Dallas. Armed with a $50 million line of credit, Wyly planned a national network from the start, with some 250 towers to connect 35 cities. It could use microwave frequencies more efficiently due to its digital operating mode. The initial late-1969 application, however, was for a digital network connecting Chicago to St. Louis (the same route for which MCI originally applied), then to Kansas City, down to Dallas, and finally to Houston. To reach both coasts, Datran would lease analog capacity from other carriers, including, of course, AT&T.

SCC Decision

Several other firms were awaiting FCC decisions on their applications as well, and more seemed likely to come in. Slowly realizing what it had unleashed, in mid-1970 the FCC initiated a general investigation on whether to create a new category of *specialized* common carriers (SCCs) to supplement existing telephone carriers. Among others, and as might be expected, AT&T opposed the idea, arguing that existing carriers could and would meet any future needs and that such specialized firms would merely "cream skim" major customers from heavy-traffic routes while ignoring smaller customers served by existing telephone carriers.

The new applicants countered that they would offer specialty services not generally available from existing carriers and would do so at rates at least competitive with and often far lower than those charged by AT&T and other traditional companies. After reviewing the substantial record in the proceeding, the FCC agreed with the applicants, and in June 1971 concluded that

> there is a public need and demand for the proposed facilities and services and for new and diverse sources of supply, competition in the specialized communications field is reasonably feasible, there are grounds for a reasonable expectation that new entry will have some beneficial effects, and there is no reason to anticipate that new entry would have any adverse impact on service to the public by existing carriers such as to outweigh the considerations supporting new entry.[48]

In short order, the SCCs began to develop their service. By late 1973, both MCI and SPCC were offering coast-to-coast analog voice private-line services, using both their own facilities and those leased from AT&T. Datran was providing all-digital service for computer data transmission. And yet another transmission technology was on the horizon.

Domsats

Based on initial success with global satellite communications (*see section 5.1*), the FCC initiated a study of the possibilities for domestic communications satellites (domsats) in 1966. The commission was faced with a growing number of domsat applications from satellite manufacturers and other noncarriers. Naturally AT&T (which felt satellites were merely another means of improving the services of existing carriers) and the new Communications Satellite Corporation opposed these applications.[49] Further, Congress held a number of hearings on the issue, and the White House became involved. Satellite communications had been, and was going to continue to be, of high-level policy concern. Both branches of government came out strongly for a competitive commercial system of domestic satellites. The FCC took the indicated lead.

In its final decision in June 1972, the FCC established what was quickly dubbed an "open skies" policy allowing open and competitive entry into this new domestic satellite service by providers other than established carriers. AT&T was restricted to only using satellites for its existing monopoly services for a period of three years to allow this new class of carrier to get started.[50] As it happened, the limitations on AT&T continued until mid-1979 by which time competition was rife. Further, in proceedings that dragged over several years, the company was required to file satellite interconnection tariffs to allow new domsat carriers access to AT&T's network.

In early 1973 an internal report to AT&T's top officers was projecting that satellite services and MCI (and, presumably, other SCCs) threatened 70 to 80 percent of AT&T's voice-grade private-line business.[51] Significant though far lower losses were calculated for other voice services, including both MTS (message toll service, or regular long distance basic telephone service) and WATS (wide area telephone service, the discount high volume bulk calling offering). Another huge threat was posed to AT&T's lucrative broadcast network connection service which the company had dominated since 1926. Satellite delivery of video signals opened a potentially vast new market, as became clear in the late 1970s.

The first American domsat, owned by Western Union, was launched in April 1974 as *Westar I* followed by a sister satellite just six months later. Its 12 transponders could provide 7,200 telephone circuits or 12 color video channels. Other satellites quickly followed. In 1975–76, the first announcements came of cable program services' plans to use domsats to create "instant" national network distribution. Domsats would soon change the face of American telecommunications, beginning with electronic media. In 1975 the first plans were announced for domsat distribution of pay cable and "superstation" signals nationwide, effectively creating instant national networks.[52] The announcement began a rush to build more domsats, and the cable and ground antenna industries thrived. By the mid-1980s, traditional broadcast networks began to use domsats to distribute their signals, and the number of new cable networks was expanding monthly. Most voice and data networking also used domsats, or international communication satellites, for global links.

5.5   AT&T RESPONDS TO COMPETITION

For the most part, competitive equipment or service providers in the 1970s were small and often underfinanced. But AT&T strongly resisted all competitive thrusts with a variety of legal delays before regulatory agencies, complicated competing tariffs, and more effective marketing of its own service.

Often AT&T's various responses to competition were perfectly legal. Antitrust laws do not prevent a company from responding to competition with vigorous countermeasures. The laws also allow a company to urge regulators to reject a competitor's offerings. On the other hand, early announcement of a service (to hold on to customers who might otherwise defect) or filing tariffs that price services below the cost of providing those services are examples of unlawful behavior. Where to draw the defining line between legal and unlawful reactions to competition has always been a difficult issue.

Changing Image

Under H. I. Romnes, AT&T's CEO from 1967 to 1972, the company often took a conciliatory approach to both potential competitors and general government concerns. This had been the AT&T approach for decades— steady, conservative, and closely identified with the power establishment in government and business. After all, most AT&T employees (and many had spent their entire careers in the Bell System) felt that they were not merely selling a product; they were carrying out a public trust.

As a corporation, AT&T did not rock the boat and did not stir up animosity in Washington. In general, it behaved like what it was: an old-line, blue-chip corporation whose stock was the most widely held in America. But Romnes decided that change was needed. The company was facing competition from without and confusion from within in a rapidly changing telecommunications world. He chose as a successor a man very different from himself.

On becoming AT&T's chairman early in 1972, John D. deButts took to aggressively facing down competitors and government regulators alike with a "how dare you" tone, as if AT&T, the largest company in the world, could possibly be considered distinct from the national interest. Charging AT&T with bad behavior was like charging the U.S. military with disloyalty. As a result, AT&T's management adopted a tone of moral indignation concerning growing government "experimentation" with competition. For, in the forces of competition, AT&T saw nothing less than the end of universal service and the end of the company's historic mission to interconnect the world. For nearly 100 years, AT&T had been the dominant provider of telephone service. This status had been achieved, in part, because state and federal regulators had supported AT&T's centralized ownership and management of telephone network. Regulatory support was based on the belief that an integrated structure led to better service and affordable prices for local and long distance calling.

NARUC Speech

Perhaps nowhere did this mode of thinking—a blend of arrogance, entitlement, high moral purpose, and reluctance to change—become clearer than in a major policy address by deButts in September 1973. Invited to the annual meeting of the National Association of Regulatory Utility Commissioners (NARUC) in Seattle, deButts, the first AT&T CEO to appear there in decades, reminded his audience of state and federal regulators of the challenge to viable and affordable local telephone service that AT&T (and thus the regulators) faced. He asked his listeners to consider the source of this great challenge:

> [This challenge] comes from entrepreneurs who see opportunities for profit in serving selected segments of the telecommunications market and who, not unnaturally, want a piece of the action. It comes from newly authorized purveyors of communications services who, unburdened by any obligation to the whole body of customers, address their attention to those it costs least to serve and profits most. . . .
>
> At issue is the degree to which competition should obtain in a field that has been brought to its current state through the application of basic principles—end-to-end responsibility for service, the systems concept, the common carrier principle itself—that the doctrine of competition for competition's sake puts in jeopardy and could in time destroy. . . .
>
> The time has come for a thinking-through of the future of telecommunications in this country, a thinking-through sufficiently objective as to at least admit the possibility that there may be sectors of our economy—and telecommunications [is] one of them—where the nation is better served by modes of cooperation than by modes of competition, by working together rather than by working at odds.
>
> The time has come, then, for a moratorium on further experiments in economics, a moratorium sufficient to permit a systematic evaluation not merely of whether competition might be feasible in this or that sector of telecommunications but of the more basic question of the long-term impact on the public.[53]

With these words deButts issued a clarion call to regulators to stop the forces of competition. And he made his pitch to a particularly sympathetic audience: state regulators concerned about the impact on local rates of the high-handed actions of the FCC in approving competitive telecommunications services. In the meantime, AT&T faced off with specific competitors.

Defeating Datran

Just two months earlier, the FCC authorized construction of AT&T's response to Datran's digital network. Dubbed "Data-Under-Voice" (DUV), it used digital transmitters on the Bell System's existing analog network. For nearly three years, AT&T had been making plain its ability to serve the digital market. Early announcement of this system naturally hurt potential investor and user interest in Datran, which was a new and untested service provider. Though not as sophisticated or fast as Datran's all-digital service, DUV cost less as only the switches were new.

Need for more construction and operating funding led Datran to lease facilities from SPCC (to which it also sold and from which it leased back its initial Chicago-to-Houston line) for access to West Coast customers (approved by the FCC in April 1972), and from AT&T for East Coast trunk and virtually all of its customer local loop connections. Having begun service in early 1974, for 1975 Datran earned revenues of some $4 million from 140 customers, and projected profitability by 1978 with a minimum of $20 million in revenues.[54]

In the meantime, the FCC permitted AT&T's DUV tariff to go into effect in December 1974, allowing the company to provide commercial service. At the same time, the Commission expressed concern about Datran's accusation that the DUV tariff was predatory (prices to customers being less than the costs of providing the service), and called for an investigation. Only after three years did the FCC finally determine that AT&T's DUV tariff *was* unjust and unlawful. By that time, however, Datran had already declared bankruptcy and ceased service (December 1976), selling its facilities to SPCC. Whether or not AT&T's DUV tariff was predatory would become a central example of the Bell System's anticompetitive behavior in the government's 1974 antitrust case against the company.

DeButts' 1973 speech before NARUC had by no means scared off MCI. Datran's struggle showed what might be risked. But McGowan was not confident of his chances for success at a still somewhat skeptical FCC. Although the Commission had acted favorably (after six years) on MCI's initial petition, he considered the FCC all but captured by AT&T. The relationship between AT&T and the Commission dated back decades, and both sides had had ample opportunity to get to know each other, understand what was acceptable, and know what lines not to cross. In such a cozy setting, McGowan reasoned, MCI had little chance of making further headway. He was wrong, for the FCC was to surprise MCI with a huge break.

FX Service Decision

In early October 1973, representatives of MCI, SPCC, and the FCC met to determine who could offer foreign exchange (FX) service.[55] About a third of AT&T's private-line service revenues derived from FX offerings,[56] and months of often fractious interaction among the parties and AT&T predated the October meeting. The previous year, AT&T and its local companies refused to interconnect an Illinois MCI customer for FX service. Among other reasons, said AT&T, it had long had a policy against allowing other firms to "piece out" part of its interexchange network though a joint service arrangement. AT&T also felt that FX service was outside the scope of what SCC companies could provide. If the FCC gave MCI and SPCC the right to connect their FX lines to AT&T's network, both small companies would gain a huge competitive advantage in their fight against AT&T. However, if the FCC disallowed interconnection, the small competitors would be shut out of an important market, leaving AT&T's dominance intact.

Issuing two letters right after the October 1973 meeting, the FCC authorized MCI (and, by implication, SPCC and other SCCs) to offer FX service.[57] With this FCC support, MCI asked a federal district court to require AT&T to interconnect MCI's FX services. The court did just that in upholding the FCC authorization. AT&T now had no choice but to provide interconnection to MCI customers—which the company began to do in January 1974—while appealing the district court decision. An appellate court ruled in April that the question of which services AT&T had to provide to the SCCs was not yet finally resolved before the FCC, and overturned the district court decision on the grounds that it had acted in advance of the expert agency.

In the face of this opinion (in reality only a ruling on a procedural point, not on the merits of MCI's claim), AT&T made a crucial mistake which would come back to haunt the company later. Flexing its corporate muscles, AT&T began to disconnect the dozen or so MCI customers who had been connected just months before, infuriating those customers and MCI alike. Not only did the customers lose their service but MCI was made to look unreliable—a potentially fatal blow for a new competitor. Even more harmful, AT&T looked high-handed. To make matters worse for AT&T, the FCC reaffirmed just a week later that MCI *could* sell FX service, and AT&T had to reconnect the customers who had been cut off. AT&T had clearly crossed a line, and at a dangerous time. Pulling such a power play only served to underline AT&T's marketplace position and ability to crush small competitors.

If anything, the sorry FX mess convinced McGowan (and at least some regulators) that AT&T had behaved in a manner that defined arrogance. By its overly exacting attention to detail in connecting, then unconnecting, and then reconnecting MCI's customers, AT&T had established, perhaps too clearly, that it had firm control of the public switched telephone network and the interconnections which were vital to the survival of potential competitors.

Going Public

It was past time to turn to Congress, the nation's basic public policy forum. Yet McGowan's attempt to persuade Capitol Hill to see his point of view was initially less than successful. Senator Philip A. Hart (D-

Michigan), chairman of the Senate Subcommittee on Antitrust and Monopoly, had introduced an Industrial Reorganization Act which put before the Senate the idea of dramatically reshaping big businesses—or at least airing some new ideas in that regard. McGowan had wanted the subcommittee to hold hearings so MCI would have a public forum in which to air its grievances against AT&T. Committee staffers resisted, fearing the articulate company CEO would divert any hearing to focus on this particular industry, to the detriment of the bill's broader intent.

A master lobbyist, however, McGowan finally overcame staff objections.[58] When he testified in July 1973, McGowan was able to place on record his position that the Bell System should be restructured to separate the local and long distance parts of the network. Only in this way could true long distance competition develop and flourish. McGowan promised that his company would be a good customer of the local phone companies, but only if AT&T would stop directing its local affiliates to reject all requests for interconnection by competitors.[59]

Part of McGowan's clout came from wider public participation in ownership of the new carriers. The constant and rising demands for construction and operational funding had quickly outstripped MCI's ability to raise sufficient cash in the private investment market. After months of preparation (including hiring an accountant to get the books in recognizable order), an initial public offering of about 40 percent of the company—some 3.3 million shares at $10 each—went on the block on June 22, 1972. The entire offering was an instant sell-out, giving MCI a total valuation of over $120 million. That, in turn, triggered the company's ability to borrow further funds as needed.[60]

In the company's first annual report issued just a year later, McGowan wrote the following in a letter aimed at stockholders:

> Recently MCI restructured its corporate organization. A wholly owned subsidiary, MCI Telecommunications Corporation, has been organized to operate our communications common carrier enterprise. At this writing, 12 of the MCI Carrier subsidiaries have been merged into this new corporate identity. . . .

In sentences which a year later looked a bit disingenuous, McGowan went on to assure his stockholders that "MCI is not in the residential telephone business. . . . MCI's business is business communications of all kinds, including private line transmission services and facilities."

NOTES

1. G. W. A. Dummer, as quoted in Stan Augarten, 1984. *Bit by Bit: An Illustrated History of Computers* (New York: Ticknor and Fields), p. 231.

2. Later, and only after a long and expensive legal battle did they and their employing companies agree to share credit so that development could continue.

3. T. R. Reid, 1984. *The Chip: How Two Americans Invented the Microchip and Launched a Revolution* (New York: Simon & Schuster), p. 96.

4. Editors of *Electronics*, 1981. *An Age of Innovation: The World of Electronics 1930–2000* (New York: McGraw-Hill), pp. 85, 87.

5. One of the best survey histories is Paul E. Ceruzzi, 2003. *A History of Modern Computing* Second ed. (Cambridge, MA: MIT Press).

6. See Ceruzzi, and also Gerald W. Brock, *The Second Information Revolution* (Cambridge, MA: Harvard University Press, 2003), especially chapters 6 and 9.

7. Ceruzzi, p. 145.

8. Brock (2003) offers the best discussion of this—indeed the melding of computer and telecommunications technology is what he means by "the second information revolution." See especially chapter 10.

9. "Extra-Terrestrial Relays: Can Rocket Radio Stations Give World-Wide Radio Coverage?" *Wireless World* (October 1945), pp. 305–308. This seminal paper is widely reprinted.

10. For a good history of this period, see David J. Whalen, 1982. *The Origins of Satellite Communications, 1945–1965* (Washington, DC: Smithsonian Institution Press).

11. Brian Winston, 1986. *Misunderstanding Media* (Cambridge, MA: Harvard University Press), p. 253.

12. Steven Lubar, 1984. *InfoCulture: The Smithsonian Book of Information Age Inventions* (Boston: Houghton Mifflin), p. 136.

13. Bray, op. cit., p. 137.

14. Ibid., p. 164.

15. Robert J. Chapuis, 1982. *100 Years of Telephone Switching (1878–1978)* (Amsterdam: North-Holland), p. 293.

16. Lubar, p. 136.

17. Ibid., p. 147.

18. AT&T based its resistance to the Hush-a-Phone on what it claimed were the device's inherent flaws. For example, AT&T asserted that Hush-a-Phones often worked improperly, so that people on the receiving end of a Hush-a-Phone call often heard only garbled sounds. These customers then called an AT&T operator to complain about poor sound quality. The operator would dispatch a repairman who would find nothing wrong with the AT&T network, but would note the presence of a Hush-a-Phone.

19. *In re: Hush-a-Phone* 20 FCC 391 (1955). That the rules could be carried to ridiculous extremes is evident in the oft-repeated tale of the local telephone company, a Bell affiliate, which managed to halt the distribution of advertising-laden plastic covers for telephone books, made by a non-AT&T company, on the grounds that these covers were "foreign attachments" to telephone company property!

20. Fred Henck and Bernard Strassburg, 1988. *Down the Slippery Slope* (Westport, CT: Greenwood), p. 38.

21. 238 F.2d 266 (1956). The *Hush-A-Phone* decision may have been the first time that a competitor of the Bell System found a more hospitable forum in the federal court than at the FCC. It would not be the last.

22. The principle is known as "exhaustion of administrative remedies" and holds that, for most matters involving a regulated industry, the regulatory agency with jurisdiction over the industry must make the first call.

23. *In re: Carterfone*, 13 FCC 2d 420 (1968).

24. Gerald Brock, 1994. *Telecommunication Policy for the Information Age: from Monopoly to Competition* (Cambridge, MA: Harvard University Press), pp. 105–106; and Peter Temin and Louis Galambos, 1987. *The End of the Bell System: A Study in Prices and Politics* (New York: Cambridge University Press), pp. 30–31.

25. Brock, 1994, p. 107.

26. *Cream skimming* was a term first known in the dairy business. When unhomogenized milk sat on a shelf for a few hours, the cream in the whole milk would rise to the top. In the days before anyone worried about their cholesterol levels, it was common for people to skim the cream off the top of the milk and use it on cereal or in some other fashion. Cream was presumed to be the "best" of the milk.

27. *Allocation of Frequencies in the Bands Above 890mc*, 27 FCC 359 (1959), affirmed on re-hearing, 29 FCC 825 (1960).

28. The exception to this were right-of-way companies that were already subject to rate regulation, such as railroads, power companies, utilities, etc. Henck and Strassburg, pp. 84–86.

29. TELPAK originally had four service categories: 12, 24, 60, and 240 circuits. These were referred to as TELPAK A through TELPAK D. Large telecommunications users preferred to view TELPAK as a bulk pricing plan independent of facilities. AT&T revised its initial TELPAK tariff to satisfy the wishes of these large users. However, this view quickly brought up the question of whether such price discrimination is legitimate for a common carrier. In December 1964, the FCC concluded that the revised TELPAK tariff was indeed discriminatory because it could find no material cost difference between individual or bundled private lines. The FCC ordered AT&T to "unify" the TELPAK A and B rates with ordinary private line rates and to submit data cost justifying TELPAK C and D.

30. U.S. Department of Justice, "Green Book," pp. 149–150.

31. An FCC license was required for use of the microwave frequencies needed to operate the system MCI had in mind.

32. Now dominated by Interstate 55, the 290-mile route was then served primarily by U.S. Highway 66.

33. Note that Goeken was essentially following the same path as some "independents" (notably the farmer lines) of the 1890s.

34. Philip Cantelon, 1993. *The History of MCI: the Early Years, 1968–1988* (Dallas: Heritage Press, 1993), p. 30.

35. Goeken had first learned about microwaves while serving in the Army Signal Corps in the early 1950s.

36. Cantelon, p. 57.

37. As noted both in Larry Kahaner, 1986. *On The Line: The Men of MCI—Who Took on AT&T, Risked Everything, and Won!* (New York: Warner), p. 58, and Cantelon's official history of MCI (pp. 74–76) the regional company idea, though it built on Goeken's conception of a larger firm, was based closely on Theodore Vail's conception of how to build AT&T in the 1880s.

38. More formally called the President's Task Force on Communications Policy, and chaired by government veteran Eugene Rostow, the commission had been created by President Lyndon Johnson in 1967, though its December 1968 report was only published after President Nixon took office in early 1969.

39. *In Re Applications of Microwave Communications Inc.*, 18 FCC 2d 953 (1969), and *Petition for Reconsideration Denied*, 21 FCC 2d 190 (1970).

40. *Telecommunications Reports* 35:34 (August 18, 1969), p. 1.

41. The long delay in this and subsequent negotiations between MCI and AT&T affiliates became another central "episode" in the government's 1974 antitrust case against AT&T.
42. Brock, 1981, p. 213.
43. They are listed, with their founding dates, in Cantelon, p. 75.
44. U.S. Department of Justice, "Green Book," p. 395.
45. Peter D. Shapiro, 1974. *Public Policy as a Determinant of Market Structure: The Case of the Specialized Communications Market* (Cambridge, MA: Harvard University Program on Information Technologies and Public Policy), p. 32.
46. As will become evident, this was the genesis of Sprint.
47. The nation's existing public switched network was analog, designed specifically for voice service. Such lines often introduced unwanted electronic noise into computer connections.
48. *In the Matter of Establishment of Policies and Procedures for Consideration of Application to Provide Specialized Common Carrier Services in the Domestic Public Point-to-Point Microwave Radio Service*, 29 FF 2d 970 (June 9, 1971), at paragraph 103.
49. Comsat, established by the Communications Satellite Act of 1962, was the American entity participating in Intelsat, the international satellite organization created in 1964. Building on its acknowledged expertise, Comsat wanted to operate domestic satellites.
50. U.S. Department of Justice, "Green Book," p. 735.
51. Ibid., p. 750.
52. HBO made the first announcement in 1975 followed by Ted Turner's Atlanta-based WTBS a year later. Both services made provision for cable systems to acquire the needed TVRO (TV receive-only) antennas. Once this infrastructure was in place, it was relatively easy for other networks to begin operation. The domsat (and soon many domsats) effectively ended the television bottleneck of a handful of channels in most markets or cable systems with only a few additional channels.
53. The speech was authored by Alvin von Auw, assistant to the chairman of AT&T from 1969 to 1981. It appears as Appendix B in his *Heritage and Destiny* (New York: Praeger, 1983), pp. 422–432, a book valuable for its insight into AT&T top-level thinking at this time.
54. *Fortune* (February 1976), p. 139.
55. FX refers to a dedicated line between the customer's location and a telephone switching center in a distant city. The effect is to allow local calls to be made or received in a distant local exchange area. ("Gold Book," p. 56, para 54)
56. U.S. Department of Justice, "Green Book," p. 473.
57. Temin, p. 104.
58. Steve Coll, 1986. *The Deal of the Century* (New York: Atheneum), p. 29.
59. Ibid., p. 34.
60. Cantelon, pp. 121–122.

Chapter 6

Breaking Up Bell (1974–84)

This chapter focuses entirely on a decade of legal battles that would redefine the shape of American telecommunications—with ramifications still being worked out early in the 21st century. At the end of this decade (1974–84), the unified and long-dominant Bell System would cease to exist, creating a wholly different industry structure.

Underlying much of what follows is a dramatic change in the perception of government and its proper role in this industry. Driven by chronic budget deficits, Congress and the Federal Communications Commission (FCC) had been forced to pick their battles. No longer did most people believe government could undertake anything. This kind of thinking led to more consideration of the costs and benefits of proposed regulation. Indeed, Congress now required agencies to determine the paperwork requirements of proposed new regulations (which, of course, took up more paper to describe).

A feeling of frustration also underlies this period. It is epitomized by the long and eventually failed attempt in Congress to replace or update the Communications Act of 1934. The process began in the mid-1970s when AT&T decided to take its arguments to freeze or roll back competition to Congress, in hopes that Congressional action might override the deregulatory "experiments" by the FCC and the courts. AT&T worked closely with its own operating companies and the independent telephone companies (largely dependent, it must be remembered, on AT&T for much of their revenue) and its own unions in shaping provisions of what became known as the Consumer Communications Reform Act of 1976 (CCRA) which cynics quickly dubbed the "Bell Bill." Intended as a series of amendments to the Communications Act of 1934, CCRA as introduced in 1976 would have nullified most of the decisions by the FCC and the courts, opening telephony to even limited competition:

- Because of their threat to universal service, long distance competitors would be few in number or severely restricted in what they could provide;
- AT&T's competitive tariff offerings would be legalized; and
- Terminal equipment would be returned to an AT&T-dominated state as it was prior to 1968.

Despite initial support, the bill never reached hearings and faded away when it became exposed for what it was: a carefully drawn series of provisions favoring AT&T over virtually all comers. But the bill left behind an ironic legacy, that is, renewed interest in Congress in *real* reform in the field.

House of Representatives members, especially Lionel Van Deerlin (D-California), announced a "basement to attic" attempt to replace the 1934 Communications Act with something more current that took into account the many new technologies and services that had been developed over the past four decades. In late 1977, the first draft bill, 250-page HR 13015, appeared.[1] The legislation took a dramatic, market-driven approach to regulation, did away with the FCC (to be replaced with a five-member Communication Regulatory Commission), and exchanged a marketplace competition standard for the "public interest, convenience or necessity" rationale for regulation. Hearings, including all segments of the industry and many of its critics, took weeks and produced no legislation. By the spring of 1978, Van Deerlin announced that the subcommittee would try again with "son of rewrite" to resolve the many controversies raised in the hearings. HR 3333, introduced in the fall of 1978, took a more gradual approach, intermixing existing features (the FCC, for example) with some new thoughts. Further lengthy hearings demonstrated that the industry was still not ready to compromise sufficiently to back the Congressional effort.

The Senate joined the rewrite effort in 1979 with bills of its own. Yet more hearings demonstrated, however, that after three years of trying, industry positions were hardening and compromise would be difficult to reach. In 1980, both houses cut their losses and tried once more with bills that focused only on common carrier provisions. (Electronic media, including broadcasting and cable, issues were too emotional and politically hot.) The Senate actually succeeded in passing S. 898 in October 1981, legislation that would address competition in telecommunications by forcing AT&T to create separate subsidiaries for its unregulated businesses (Bell Labs and Western Electric) while maintaining the structural integration of the telephone network. But by now many in Congress felt that the antitrust process (described below) should run its course without Congressional interference.

The final gasp of the rewrite saga came in the House in 1981–82, with HR 5158. Some of its procompetitive features were unacceptable to AT&T, which launched an unusually public campaign against the bill (including buttons reading "HR 5158 is a wrong number"). With the failure of HR 5158 to pass out of committee, the Congressional effort to replace the Communications Act died, in part because members of Congress were exhausted from the efforts of industry lobbying and infighting and in part because of the ongoing antitrust trial.[2]

Thus dominating this decade, even in Congress, was the government antitrust suit filed to break up AT&T. But the process began slowly and in the private sector.

6.1   FILING SUITS

In September 1973 MCI had learned that its basic private line interconnection arrangements forged with AT&T after a year of hard negotiation (*see section 5.4*) were about to expire and would not be renewed. Instead, AT&T planned to file separate tariffs with state utility commissions that would dictate interconnection terms and force MCI to begin all over again, doing battle in 52 different jurisdictions. But it was AT&T's early 1974 disconnection of some of MCI's FX customers (*see section 5.5*) that drove MCI's William McGowan to finally file a private antitrust suit against AT&T. Additionally, McGowan, his chief legal officer, Kenneth Cox, and others on the MCI management team began in September 1973 to meet with attorneys from the Antitrust Division of the Justice Department, seeking to persuade them to focus their long-held concerns about the Bell System into courtroom action. He was far from alone. By late 1973, at least 35 private suits had been filed against AT&T.

MCI Brings Suit

On March 6, 1974, MCI filed a 36-page civil antitrust complaint against AT&T with the U.S. District Court in Chicago.[3] The suit charged AT&T with violating the first two sections of the Sherman antitrust act through its domination of domestic telephone service and demanded financial damages. In summary, the complaint alleged that

> Defendant AT&T has exploited unfairly its dominant position in the tele-
> communications industry and sought to substitute for competition vari-
> ous anticompetitive uses of its dominant economic power for the purpose,
> or with the effect, of maintaining and extending monopoly power over re-
> lated markets.[4]

The complaint went on to list 10 examples of AT&T denying or seriously delaying interconnection with specialized common carrier (SCC)
services, and eight ways that interconnection was defined by AT&T so
as to harm MCI or other SCCs. The complaint listed many examples of
AT&T tariff filings designed to harm MCI service offerings and concluded by arguing that AT&T's illegal activities had cost MCI untold
millions in revenue.

From 1976 to 1979, effort on the case was devoted to the tedious process of discovery. Both sides sought relevant (and, it often seemed, totally
irrelevant) documents with which to bolster their case. At the same time,
MCI developed, with some University of Chicago economists, a "lost
profits" study which contended the company had lost nearly a half billion in potential revenue because of AT&T's interconnection and delaying
tactics. MCI also pioneered in developing an intricately indexed computerized database of documents and facts in the case to give it control over
the huge amount of paper received in discovery from AT&T.[5] Late in the
year, the judge asked both sides to outline their likely trial approaches.
MCI said it would bring 17 witnesses in 26 days. AT&T responded with a
promise of 183 witnesses over 18 months. Taken aback by what he
termed an AT&T team that was "thinking more in terms of a theatrical
production than the trial of a lawsuit," the judge limited AT&T to putting
on its case in the same amount of time that MCI would take.[6]

Stories about the pending case appeared in Chicago newspapers, forcing the presiding judge to postpone beginning to empanel a jury until
February 4, 1980. MCI's attack centered on AT&T tariff and interconnection policies, and on its own "lost profits" study. AT&T's defense countered that the Bell System had taken only those actions any competitive
company would consider and that MCI's problems were of its own making. The trial lasted for 54 days, producing some 12,000 pages of testimony and about a thousand exhibits.[7] Of the 15 remaining charges in
the case, the jury found in favor of MCI on 10, and for AT&T on the remaining 5 and concluded that AT&T should be fined $600 million for its
transgressions (MCI had originally asked for $900 million). Under anti-
trust law, the judgment was automatically tripled to a huge $1.8 billion,
then the largest monetary award in U.S. legal history. Interest alone
would accumulate at the rate of $3 million a week or $162 million a
year.[8] And by now there were some 40 pending private antitrust suits
filed by other AT&T competitors.[9]

Justice Files Its Case

When McGowan and his MCI colleagues first met with representatives of the Antitrust Division in September 1973, the MCI chief's denunciations of AT&T reached attentive ears. Two different sets of lawyers undertook informal investigations of AT&T, neither one known to the other. One reviewed data from MCI and other sources and quickly concluded that AT&T had violated the antitrust laws. The second group, somewhat more political in nature and conducted by other officials, contacted Congressional and other executive branch offices, gathering information and testing the waters about a possible formal suit. Sometime late in 1973, the two separate investigations were combined into one more focused drive to develop the necessary data for the filing of a government suit.[10]

The Justice Department took the next step by issuing a Civil Investigative Demand (CID) to AT&T on November 26, 1973, seeking thousands of documents on Bell System relationships (or lack of same) with the SCCs, especially in the little over two years since the FCC had formally recognized those carriers' right to operate.[11] This CID clarified the real problem AT&T faced:

> The FCC regarded MCI as a *customer* of AT&T and was applying regulatory doctrine of nondiscrimination to the Bell System's responses. The Justice Department regarded MCI as a *competitor* and was applying antitrust standards to these same actions. Both of them, moreover, regarded AT&T's reactions to the new demands on the network as too aggressive and ungenerous. What appeared to the Bell System to be protective of the network appeared to the Justice Department to be protective instead of AT&T's monopoly.[12]

As antitrust officials assessed AT&T's mound of documents, the ranks of frustrated competitors grew, increasing pressure on the department to take some action.

On November 20, 1974, the Justice Department formally filed its antitrust complaint.[13] For the opening salvo in the battle that would end with AT&T's dissolution, this initial document was surprisingly short and simple.[14] In a mere 14 pages (far shorter than the 1949 complaint), filed with the U.S. District Court for the District of Columbia (the federal trial court with jurisdiction over the Justice Department), the complaint alleged that a conspiracy existed, in violation of the Sherman Act,[15] among and between AT&T, Western Electric, Bell Labs, and the Bell Operating Companies (BOCs).

> The defendants, among other things, have done the following: . . . attempted to obstruct and obstructed the interconnection of Specialized Common Carriers . . . [and] Miscellaneous Common Carriers . . . [and] Radio

Common Carriers . . . [and] Domestic Satellite Carriers [and] customer provided terminal equipment with the Bell System . . . [and] caused Western
Electric to manufacture substantially all of the telecommunications equipment requirements of the Bell System; and caused the Bell System to purchase substantially all of its telecommunications equipment requirements
from Western Electric.[16]

To resolve such behavior, the Complaint asked that the court order divestiture by AT&T of Western Electric and some or all of the Bell Operating
Companies.

AT&T Responds

The company made clear both its indignation and intent to fight the
charges in a letter from AT&T chairman John deButts to stockholders
sent just a week after the complaint was filed. In its four pages, the chairman expressed confidence in the outcome:

Three considerations support that confidence. The first is that our business
has been structured as it is for the better part of a century and repeatedly
over that span of time has withstood the most searching examinations.
Over most of that period, national policy has recognized that . . . the public
interest is best assured through regulation rather than competition. . . . The
second consideration that makes a breakup of the Bell System inconceivable
is that it would mean fragmentation of responsibility for the nationwide
telecommunications network. That network, to work efficiently, must be
designed, built and operated as a single entity. It is for this reason and no
other that the Bell System is structured as it is. . . . The third and strongest
argument against breaking up the Bell System is its performance. This organization is a unique national resource.[17]

AT&T's formal 23-page legal response came on February 4, 1975, and
indicated in its 68 paragraphs the direction the defense would take in the
years to come:

• The unitary nature of the Bell System and its integrated structure are
consequences of the unique need for technological and economic planning, implementation and coordination to perform the research, design,
manufacture, installation, operation and maintenance of the nationwide
telecommunications system.

• The highly integrated structure of the telecommunications industry
. . . reflects the considered will of Congress.

• The comprehensive federal and state regulation of the rates charged
and of the profits earned by the Bell System has insured that the efficien-

cies, technological advances and cost savings resulting from the unitary Bell System enterprise have been passed on to the users of telecommunications services.

• The Bell System has opposed a number of applications by proposed new entrants in the telecommunications industry on the grounds that the entry being sought, under the terms and conditions proposed, would be contrary to the public interest. Specifically, the Bell System has expressed the views, among others, that a proliferation of customer-provided equipment without adequate safeguards would not meet the highly demanding technical standards of the nationwide telecommunications network and hence would jeopardize the integrity and performance of that network . . . that a proliferation of so-called specialized intercity carriers would produce a needless and wasteful duplication of equipment and facilities.[18]

AT&T also argued the government's failure to state a legally sufficient claim, the lack of subject matter jurisdiction in this Federal District Court (as the New Jersey court still held jurisdiction under the 1956 consent decree), and the prior litigation in the earlier case of the issues raised in this case.

Though not made clear in this first filing, AT&T was focusing two affirmative arguments to explain and defend its behavior—what came to be known as the "thrust upon" and the "regulatory" defenses. The *thrust-upon approach*, which became a cornerstone of AT&T's position in the case, claimed that AT&T's competitive superiority grew out of its efficiency, skill, and foresight. Its integrated structure allowed Bell Labs to develop, and Western Electric to manufacture, the best equipment possible at reasonable prices. The *regulatory argument* held that AT&T's superior position in the marketplace had been approved and encouraged by federal and state regulatory policies over many decades. To bolster this assertion, AT&T claimed that its behavior as a corporation was circumscribed by government-imposed limitations, citing as an example the 1956 Consent Decree which limited AT&T to the provision of regulated communications common carrier services.

6.2   A DEFINING INTERLUDE—EXECUNET

Before the antitrust case could get substantially off the ground, MCI made another move which would lead to landmark court decisions and, more importantly, would permanently redefine competition in telecommunications.

A New Service

Amid all the antitrust legal actions, MCI faced continued financial pressure. Drawing on loans and investors still did not meet daily operational expenses, let alone network construction costs. The various legal problems with AT&T only made everything worse. It became clear the company could not survive limited to a narrow perception of SCC service, a mere fringe role. (AT&T's private line services, with which MCI then competed, amounted to but 3 percent of Bell System's annual business, indicating their limited scope.) After several months of intense internal development, in September 1974, MCI filed with the FCC to offer a new service which would eventually change the face of telephone service by testing regulatory authority.

MCI's new "Executive Network" (Execunet) service, the forerunner to the company's entry into regular long distance telephone service, played a central role in McGowan's concept of MCI's eventual place in the telecommunications industry. For with Execunet, MCI announced that it was taking on AT&T in a direct, head-to-head competitive battle for the core of the service industry. Both AT&T and the FCC were suspicious when MCI approached AT&T with a large number of orders for local exchange interconnection for use with the new service.[19] Service began in Washington, DC, aimed especially at the hundreds of national trade associations based in the capital,[20] and expanded to Dallas and Houston at the beginning of 1975. MCI needed the revenue, and after a slow start in Washington, results were gratifying. Many of the customers for the new service had never used private lines before, indicating Execunet was broadening MCI's market.

Though described by MCI as simply a "metered-use time service" variation of the private line services it had been offering, Execunet's publicity material made clear on close reading that the new service crossed the line into provision of Message Toll Service (MTS)—regular long distance switched service whereby an MCI customer could call any number in an MCI-served city (the number of which expanded all the time).[21] Perhaps surprisingly, AT&T was slow to discover the true nature of Execunet.[22] Only after letters to and discussions with MCI late in 1974 did the Bell System demonstrate its concern to the FCC (by actually placing Execunet calls with FCC officials watching) in early 1975. The Commission was caught in a bind. While it had actively encouraged MCI's development and expansion, the FCC felt MCI had been deceptive in describing Execunet, as the new service clearly exceeded the boundaries of allowable SCC operation.

In July 1975, eight months after test marketing of the service had begun, the FCC rejected Execunet as exceeding the boundaries of proper SCC service offerings and prohibited MCI from continuing to market the service. The Commission's letter, ordering a stop to the service within 30 days, concluded "that MCI sought authorization to offer only private line services, and that it was granted authority to offer only such services. . . . Your Execunet service is essentially a switched public message telephone service, rather than private line."[23] Clearly the FCC and AT&T agreed on the need to limit SCCs to the role the FCC had originally outlined for them in 1971. The Commission's conception of a telecommunications industry with clearly defined sectors was blurring into a free-for-all. Unsurprised by the FCC finding, MCI immediately appealed to the Court of Appeals for the District of Columbia.

FCC Versus the Court

The court issued a stay on the FCC order, thus allowing MCI to continue marketing Execunet, which by its second full year of operation (1976) contributed 40 percent of MCI's total revenue.[24] After long legal battles, and eventually a ban on new Execunet sales while the case was pending, a three-judge panel of the court overturned the FCC in July 1977, saying the Commission had not shown that limiting SCCs to provision of private line service was in the public interest.[25] In other words, the court held that no reason existed to keep MCI from marketing Execunet even if it was equivalent to Message Toll Service (MTS).[26]

Now both the FCC and AT&T committed an additional error. Appalled at the federal court's decision, which effectively broke down the SCC barrier and allowed any firm to offer MTS, the Commission held in February 1978 that AT&T did not have to interconnect this MCI service to the AT&T network. As MCI was still building its national customer base, such a decision would have stopped the company in its tracks. Again, on appeal, the same court came to MCI's rescue, making clear in an angry April 1978 opinion that the FCC's non-interconnection decision was simply an attempt to circumvent the spirit of the court's first Execunet decision.[27]

Save for one thing, the Execunet fight was over. MCI (and by extension, the other specialized carriers) had earned the right to compete directly with AT&T in the primary telephone market. No longer would competitors be restricted to scrambling for a piece of the ancillary private lines market. The one thing left to resolve, however, were the conditions of interconnection.

ENFIA

Just a month later, AT&T filed its "Exchange Network Facilities for Interstate Access" (ENFIA) tariff to define the basis of MCI interconnection to the local loop. As might be expected, MCI and the other SCCs were appalled at rates which would triple their interconnection costs. The companies finally agreed that the OCCs ("other common carriers" as they were now called, since Execunet had shown that there was nothing "special" about them) would for an initial three-year period pay to Bell's local exchange companies 35 percent of what AT&T paid per minute of connection time. As OCC overall income rose, so would their payments—finally to 55 percent of AT&T payment levels.

These discounted rates allowed the OCCs to continue underpricing AT&T as they sought customers. The OCCs needed that cost advantage, for their customers could not dial a long distance call directly. Instead a customer had to dial 10 extra digits to access an OCC for a long distance call. The ENFIA tariff was considered temporary as the FCC was still investigating the entire long distance telephone market, trying to determine what degree of competition was most beneficial for the public interest. As it turned out, this took years, and the tariff, as well as the controversy, continued until the 1984 breakup of AT&T.

6.3 TRIAL AND SETTLEMENT

While much of the industry was focused on the dramatic Execunet case and its aftermath, behind the scenes the parties in the federal antitrust case were jockeying for position. At first the process moved very slowly. The last half of the 1970s was taken up with the pursuit of documents and jurisdictional issues that only an attorney can fully appreciate.

Paper Chase

To begin with, AT&T claimed that it was immune from federal antitrust prosecution because it was subject to a pervasive regulatory scheme administered by that government—specifically, by the FCC. As AT&T could not meet the conflicting standards of FCC regulation (which, the company claimed, promoted AT&T's monopoly status) and competition (as mandated by the antitrust laws), the corporation felt that if the status and practices of AT&T were to be considered at all, the case should be heard at the FCC. Any resulting decision could then be appealed to a federal court, but the first crack belonged to the regulatory agency.[28] Some two years after the case had been filed, Judge Joseph Waddy concluded in a five-page opinion that the Communications Act did not limit the antitrust laws as they applied to communications common carriers and that issues might be more sharply defined if the parties went through a formal court-supervised discovery process.[29]

As the filings in the case piled up in the fall of 1976, Judge Waddy's health was failing (a sad and ironic mirror of what had happened two years earlier in the MCI antitrust case). He retired in 1978, and his replacement (nominated by President Jimmy Carter) was Judge Harold H. Greene.[30] As is always the case when a new judge joins the court, cases were juggled around to even up the caseload, and Greene drew the AT&T case on his first day of work.

One of Judge Greene's first moves was to impose order on the pretrial process with a device known as "pretrial stipulations," a technique proposed to him by AT&T. Stipulations involved a Justice Department attorney and an AT&T attorney meeting with their researchers and other support personnel and hammering out exactly what issues of fact or law could be agreed to, or "stipulated." The packages removed from further discussion those matters that were undisputed, and thus considerably shortened the time needed to try the case. Greene also required three sets of pretrial statements of contentions and proof, each more detailed and definitive than the last.[31] These were long documents where each side detailed its view of the case and events of recent history. The definitive third set, documents identified by the color of their covers and filed early in 1980, totaled nearly 4,000 printed pages with extensive annotation. The two-volume government "Green book" and three-volume AT&T "Gold book" were encyclopedic collections of historical detail and sharply different views of the same events.[32]

Attempts to Settle

Sporadically through this period came reports of attempts to settle the case prior to any trial. The timing of these settlement talks now became critical. The Clayton Antitrust Act holds that evidence entered in a government antitrust suit that the government wins or settles by a consent decree can be used as *prima facie* (in essence, already proven) evidence in a private antitrust case. Thus, a consent decree entered into after the trial began would expose AT&T to all the private plaintiffs who would use the evidence from the government case to win their own cases.[33] The specter of private antitrust plaintiffs lining up at AT&T's door to collect damage awards added pressure on the company to try to successfully conclude a deal before the trial started.

To the Justice Department, any settlement had to involve opening up AT&T's network to competition; anything less, and this case would take its place with too many other failed attempts by the Justice Department to make AT&T conform to the antitrust laws. To AT&T, any settlement that interfered with its way of doing business was unacceptable. But that settlement talks took place at all was an indication of AT&T's change from bluster and attack to a more analytical approach. By January 1980 the outline of a possible deal—involving access to local exchanges and divestiture of two large local operating companies and a portion of Western Electric—had been worked out. This tentative deal fell victim to politics. The potential advent of a Republican administration, which would presumably take more of a pro-business stance than the Democrats, gave the AT&T forces hope. Perhaps the Republicans would simply make the government case go away, as had happened in 1956.

The 1981 MCI antitrust decision handed AT&T a stinging loss (*see section 6.2*). The jury's findings were worrisome because MCI's arguments had so closely tracked the government's approach in the federal case. Many of the same instances of anticompetitive behavior were alleged. Government lawyers now had a useful blueprint for trying their own case—one provided by the corporate nemesis of AT&T. The *Washington Star* reported that a potential settlement included partial divestiture of the company in exchange for letting AT&T into the computer market. The paper also said that attempts to settle dated back to talks in 1976.[34] The *Washington Post* identified the likely operating company to be divested as Pacific Telephone and Telegraph, the heavily regulated California Bell operation.[35]

Into this growing political and legal stalemate came William F. Baxter, a Stanford University law professor nominated by President Reagan to be antitrust chief in late February.[36] Given the recusal of the two officials above him,[37] Baxter became the senior Justice official with the legal and political authority to agree to any settlement of the AT&T case. Baxter was not inclined to settle, however; he felt that AT&T had to lose its grip on its local bottleneck facilities in order for competition to be a reality.[38] After having had a chance to review the record, Baxter vowed at his first public briefing to "litigate the case to the eyeballs." He went further: "I was not going to sign that settlement. It did not do the one thing that should be done—separate all the regulated components of the enterprise from the unregulated."[39] Thus did the public hear for the first time the beginnings of an economic theory behind the eventual breakup agreement.

Trial

When the trial got under way in March 1981, after two months of delay for settlement talks to continue, the government put on nearly 100 witnesses, at the rate of 1 to 4 per day. Representatives of defunct corporations who had tried and failed to compete with AT&T testified, as did expert witnesses who declared that AT&T's pricing policies were anticompetitive. During the government's case, Judge Greene demonstrated that he was very much in charge, cutting off witnesses who repeated themselves or rambled. As a result, chief litigator for the government Jerry Connell found that he was moving through his witness list much faster than anticipated; having told the judge that he would need a year to put on his case, he concluded his presentation in only four months (61 days of testimony), by the end of June. In that period, his witnesses detailed some 80 "episodes" of AT&T anticompetitive behavior in 11,400 transcript pages.

At this stage AT&T attorneys still believed that they would prevail. In July they filed a motion to dismiss supported by an accompanying 550-page memorandum of law explaining why the government's charges in the case were groundless.[40] Because of the need to keep the trial moving, however, AT&T did not have the luxury of waiting for Greene's response before beginning the presentation of its case. Some six weeks of company testimony was on the record by the time Judge Greene responded to the motion to dismiss. In his September 11, 1981 opinion, the judge concluded that AT&T had violated the antitrust laws "in a number of ways over a lengthy period of time, . . . [that the] . . . defendants have used their local exchange monopolies to foreclose competition in the . . . equipment market, . . . [and that] . . . AT&T has monopolized the intercity services market by frustrating the efforts of other companies to compete with it."[41] As he had put it two months earlier: "The Government has presented a respectable case that the defendants violated the antitrust laws."[42]

Settlement

AT&T faced several options that seemed equally terrible. The company could request that the Justice Department prepare a settlement proposal that was simpler and less burdensome than previous offers (that seemed unlikely). It could wait to see what legislation, if any, came from an increasingly hostile Congress (that, too, seemed unlikely as Congress was increasingly inclined to let the case run its course). Or, AT&T could resign itself to what would surely be a stinging verdict from Judge Greene which would then have to be pursued through the appeal process, consuming several more years, costing additional millions, and providing further fodder for a host of private antitrust actions. While all this went on, AT&T would be severely limited in raising operating capital to improve or expand its facilities.

Confronted with this set of unappealing alternatives, AT&T's chief counsel approached Baxter in mid-December to reopen settlement discussions. Now that AT&T was resigned to losing its operating companies, Baxter's demand for complete divestiture was no longer the stumbling block it had once been. When the trial recessed on December 18, those close to the negotiations (which perhaps ironically did not include the trial staff on either side) began to realize that settlement was at hand. After several weeks of further negotiation, on January 8, 1982, the government and AT&T agreed to end the trial and break up the Bell System. The 22 operating companies would be divested and managed as firms separate from AT&T. AT&T would retain Western Electric's manufacturing capability, most of Bell Labs, and the long distance service, while leaving the local service business.

6.4 BREAKUP: DEFINING THE MFJ

To the surprise of both the Department of Justice and AT&T, rather than simply accepting the settlement news, Judge Greene announced that he would seek public comment on the proposed consent decree. He informed the parties that he planned public hearings on the settlement's terms, making clear his intention to subject the agreement to scrutiny before finally ruling on its validity. To the Justice Department and AT&T, the case was over: the proposed settlement was merely a modification of the 1956 Consent Decree, not a brand-new settlement. To Greene, the rather breathless pace of the last-minute negotiations, and the fact that they had taken place beyond his watchful eye, meant that he had not had time to study the details of the agreement to satisfy himself of its fairness.

Agreement

After extensive filings and two days of oral presentations by interested parties, Greene issued an opinion essentially approving the proposed agreement, now formally making it a Modified Final Judgment (MFJ). Greene made clear in numerous places in his decision that the proposed decree prohibited the divested operating companies from providing long distance and information services and from manufacturing telecommunications equipment.[43] The operating companies were to be required to provide that which AT&T had never done: equal access for all long distance service providers. The BOCs would be responsible for reconfiguring their equipment to provide such access and paying the costs of such changes. The restrictions were softened. In then little-noticed clauses in the opinion and the Consent Decree itself, the local operating companies were permitted to apply for permission to enter other, non-telecommunications lines of business as long as they could make a showing that their entry into their chosen markets would not depress competition in the markets so selected.[44]

Working with input from both AT&T and the Justice Department, Greene also modified the line of demarcation between local and long distance service, assigning some short-haul toll services to the local companies. In another attempt to redress the balance between AT&T and its soon-to-be-independent offspring, Greene modified the proposed MFJ to assign the task of publishing the ultra-profitable Yellow Pages to the local companies rather than AT&T as the proposed decree originally intended.[45] And he gave the rights to the word "Bell" and AT&T's old logo of a bell inside a circle to the operating companies, a move as much symbolic as sentimental.[46] On August 24, 1982, the MFJ was entered, and the court phase of the antitrust case was over. But its substantial ramifications had merely begun.

Defining "Local"

An initial task was determining how to set up the 22 Bell operating companies (which comprised two-thirds of AT&T's assets) into post-divestiture businesses that could deliver service with maximum efficiency. Concerned that the companies could not stand on their own in an increasingly competitive industry, AT&T planners decided that they should be grouped in some fashion under regional holding companies (RHCs) that would be responsible for region-wide coordination and functions. Discussions of *how* to group the operating companies, and into how many regions, proceeded on the idea that each RHC should have roughly $15 to $20 billion in assets[47] and should be large enough to attract investors but not large enough to attract further antitrust attention in the future. The final compromise decision—to group the 22 operating companies into seven regional holding companies or Regional Bell Operating Companies (RBOCs)—was announced on February 19, 1982.[48]

Under the MFJ, each operating company was to establish at least one and usually two or more geographical areas in which local exchange service would be offered. These were ultimately dubbed *Local Access and Transport Areas* (LATAs), the new and somewhat clumsy term being adopted to make sure no one confused existing local exchanges with the new local service areas. The divested operating companies would only offer service *within* LATAs, while AT&T and its competitors would provide the connections *between* and *among* LATAs.[49] Creating LATAs was no simple task. Each had to conform to a variety of criteria specified in the MFJ including the following:

- Any LATA would be made up of one or more contiguous local exchange areas serving common social, economic, and other purposes,

even where such configurations spread across municipal or other local governmental boundaries.

- Every point served by a BOC within any state had to be included
  within a [LATA].
- LATAs could not cross state boundaries.

The Bell Operating Companies' 7,000 local exchanges had developed
over the years as cities and towns developed.[50] Never before had the
whole system had to be reconsidered let alone redesigned all at one time.
If, for example, a LATA was too small or had too few customers, operating companies would not earn sufficient revenue and long distance companies might be disinclined to offer service. However, if a LATA was too
large or had too many customers, the highly lucrative market might
make it difficult for competitive provision of interconnecting service by
long distance carriers. Adding to the difficulty of designing LATAs were
the interests of 50 state public utility commissions. Since these regulators
would set rates on calls within and between LATAs (if the latter fell
within a single state), divestiture planners had to take their concerns into
account.

Amidst all this tension, there was great symbolic meaning to the
LATAs. Design of service areas where the RBOCs would offer local service
served to drive home the point that AT&T was no longer in the local telephone business. A new and independent regional company was going to
provide dial tone, install phones, and send the bills for local service.[51]
LATAs were to make practical the theory and fulfill the goal of divestiture. Eventually, Bell System planners created a system of LATAs to
cover the country, confusingly identified with three-digit codes that
made them look to the unwary like area dialing codes with which they
had nothing in common.[52] The initial plan was submitted to the Department of Justice in May 1982. After extensive comments from 120 parties, of the 161 proposed LATAs, the Justice Department had concerns
about only 9.[53] On April 20, 1983, Judge Greene approved the LATA plan
with only minor further modifications.[54]

Reorganization

If divestiture planning had intensified between January and August, by
the fall of 1982 it took on the urgency of a frantic war room. According
to AT&T's chairman, its "Plan of Reorganization" would have to be a
highly detailed and specific document to achieve these tasks:

> More than one million employees must follow their jobs, and tens of thou-
> sands of them must transfer from one former Bell company to another.

Data processing systems must be reprogrammed, reconfigured, and, in some cases, redeployed—systems that control maintenance, repair, and installation work, that generate employees' paychecks and customers' bills, that keep records and manage the large information systems needed to run a giant business. State and federal tariffs establishing prices must be amended and refiled—in perhaps as many as 50 thousand pages of documents. On Day One after divestiture, the former Bell companies must be prepared to process 600 million telephone calls, dispatch 100 thousand installation and repair technicians, and continue a 17.5 billion-dollar modernization program.[55]

One wag defined the process as disassembling and rebuilding a Boeing 747 while in flight.

The seven designated CEOs of the regional holding companies and the head of the Central Services Organization, although all still AT&T employees, were becoming increasingly vigorous advocates for their to-be-divested operating companies. Pointed discussions ensued regarding the economic implications of the breakup and the ultimate effect on the new companies' balance sheets. Although AT&T's senior management kept referring to the time-honored Bell System tradition of collegiality, the CEO-designates were drawing AT&T paychecks but behaving like adversaries. Nothing could have brought home the reality of divestiture more starkly.

On December 16, 1982, the 471-page paperback *Plan of Reorganization* rolled off the presses. Some 10,000 copies were distributed, though interest in the document was surely in inverse proportion to its contents. In dry language, the *Plan of Reorganization* described financial arrangements for spinning off the operating companies, noting that they would be assigned 75 percent of AT&T's total physical assets. Division of assets under the plan were made on the basis of use: the company that had predominantly used the equipment before divestiture would became its sole owner once divestiture took place.

The plan read a bit like a divorce decree as physical assets from land and buildings to shop supplies were allocated. Non-physical assets such as accounts receivable and patent rights were distributed. Employee pension, stock ownership, and benefit plans had to be accounted for. Personnel were assigned to new jobs. A number of ancillary decisions were also spelled out in the plan, including creation of billing services to be performed by the operating companies for AT&T on a temporary basis, and cancellation of the standard supply and license contracts that had long existed between the BOCs and AT&T. New contracts, different in form and substance, were drawn up to guide the relations of the soon-to-be-separated business entities. The plan also detailed how the breakup would impact stockholders.[56]

On August 5, 1983, the judge released his 70-page opinion on the plan of reorganization.[57] Greene noted several changes he would require before final approval, among them:

- AT&T would reimburse the RBOCs for any remaining costs still outstanding from developing equal access facilities and redoing their network reconfiguration;
- AT&T would drop the "Bell" term from its own name and that of its subsidiaries (noting that the use by both AT&T and the operating companies of the Bell name and trademark would imply an ongoing relationship between the companies, something forbidden under the MFJ, the judge assigned the Bell name and logo to the operating companies with the single exception of what would now be called AT&T Bell Laboratories);[58] and
- AT&T would grant to each operating company a five-year royalty-free license to use any of its patents (which the local companies had, of course, helped to support through the AT&T internal contract and payments system).[59]

Line of Business Restrictions

Finally, the MFJ envisioned an RBOC industry that was limited to providing local and limited long distance service within LATA boundaries. Provisions of the MFJ barred RBOCs from providing a host of services that, from the RBOC perspective, represented both lucrative revenue sources and the future direction of the industry. Specifically, Section II(D) stated that no BOC, either directly or through an affiliate, could do the following:

- Provide interexchange telecommunications services or information services;
- Manufacture or provide telecommunications products or customer premises equipment (except for provision of customer premises equipment for emergency services); or
- Provide any other product or service, except exchange telecommunications and exchange access service, that is not a natural monopoly service actually regulated by tariff.[60]

The draft MFJ also originally assigned the lucrative Yellow Pages to AT&T, rather than to the RBOCs. Greene reversed this, finding that "the prohibition on directory production by the Operating Companies is distinctly anticompetitive in its effects."[61] He concluded that transferring the

Yellow Pages from several smaller entities into one nationwide company (AT&T) would be against the antitrust laws; moreover, control over printed directories would give AT&T a competitive advantage in providing electronic directory advertising.

Greene also reversed the prohibition on the RBOCs' ability to *market* equipment, reasoning that the RBOCs would find it difficult to subsidize equipment prices with local exchange revenues and that the participation of a second company (i.e., the equipment manufacturer) would make any efforts at cross-subsidization easier to detect.[62] He also hoped that the participation of the RBOCs in the marketing of equipment would lead to more vigorous competition in that segment of the industry.

Except for the reversal of the Yellow Page assignment and the removal of the restriction on the marketing of equipment, Judge Greene upheld the other restrictions on RBOC activities. He found the restrictions on RBOC manufacturing of equipment to be justified, fearing that the RBOCs would thwart competition by buying their own equipment, even if it were inferior in quality or higher in price than the equipment available from nonaffiliated manufacturers. The RBOCs, moreover, would be able to build inflated equipment prices into their local rates.[63] The judge found the restriction on interexchange services to be necessary in order for competition to flourish in the interexchange market. The MFJ required the RBOCs to provide equal access to their facilities to all interexchange providers. MCI, AT&T, and others needed access to the RBOCs' local network in order to provide long distance service. Greene feared that allowing the RBOCs into interexchange services would subvert these equal access requirements.[64] The RBOCs would have an incentive to discriminate against MCI and AT&T if they were competing with them for interexchange service.

Greene also upheld the restrictions on RBOC entry into information services, which the MFJ defined as "the offering of a capability for generating, acquiring, storing, transforming, processing, retrieving, utilizing, or making available information which may be conveyed via telecommunications."[65] Echoing the fears he had expressed about the anticompetitive effects of allowing the RBOCs into interexchange services, the judge feared that, should the RBOCs be allowed into information services, they would discriminate against their competitors by denying information service providers access to necessary elements of the local network. Indeed, Greene feared that the RBOCs would have an incentive to design their local networks to discourage competition.[66]

The result of the MFJ, then, was to be an industry structure in which the RBOCs operated as the ultimate common carriers. Rather than developing new equipment and new information services or reaping the profits of interexchange long distance service, the function of the RBOCs was to maintain the local network and to provide access to those network facilities to AT&T, and its competitors, and to the growing numbers of information service providers. According to the MFJ, the RBOCs owned the local bottleneck, facilities to which companies needed access and facilities that would give the RBOCs an unfair advantage were they to be allowed to compete in long distance, equipment manufacturing, or information services.

6.5 BABY BELLS

The final important divestiture documents to be issued by the old unified AT&T was the public blueprint defining the new cast of players. The *Information Statement and Prospectus* marked the official emergence of the RBOCs as separate entities.

Concept—Limited Regional Power

The 267-page *Prospectus* released on November 8, 1983, contained a detailed financial and structural explanation of AT&T's divestiture as well as initial RBOC operating data, including detailed balance sheets.[67] The document was designed primarily to introduce stockholders and the investment community to the new regional firms and to provide information to AT&T shareholders on the conversion of their stock to holdings in the new companies. The *Prospectus* also formalized and explained the background of the new names of the seven regional holding companies, each of which had already been formally incorporated. A measure of their size became plain when it was realized that each would be among the top 10 largest utilities in America (only AT&T and GTE would be larger).

Even though these RBOC names could rightfully incorporate the word "Bell," many chose names that not only shunned use of the traditional name but gave little indication that the company was even in the telephone business. The names suggested differing approaches to that crucial aspect of any business—image.

• *NYNEX* (for New York, New England, and X to symbolize the unknown), the holding company that joined the huge New York and much smaller New England telephone companies, would serve 10 million customers in seven states (though much of Connecticut was served by an independent company, Southern New England Telephone, which would not be subject to MFJ provisions).[68] The NYNEX "twins" were not always comfortable with one another and in the months before divestiture usually met on neutral ground—AT&T headquarters.[69] Just one example of the cultural differences between the companies was in how they priced basic telephone service. Nearly two-thirds of residential (and all business) customers in New York paid for local service on a "metered" or usage basis while in New England 96 percent of residences paid the flat monthly fee common to most Americans.[70]

- *Bell Atlantic* would serve customers from New Jersey, Delaware, Pennsylvania, Maryland, Virginia, West Virginia, and the District of Columbia, and incorporated Bell of Pennsylvania, New Jersey Bell, and the Chesapeake and Potomac company (serving West Virginia, Virginia, and Washington, DC). Bell Atlantic included a wave logo in the first letter of its second name. Headquartered in Philadelphia, Bell Atlantic would include in its service area the major urban areas of its headquarters city, Pittsburgh, Baltimore, Washington, and the sprawling New Jersey suburbs of New York City and Philadelphia.

- *BellSouth*, the largest RBOC in terms of revenue (about $11 billion in 1983) and employees (nearly 100,000), was Atlanta-based and combined the former Southern Bell and South Central Bell operating companies. Its nine-state region covered much of the Sun Belt, an area that had experienced a 20 percent growth in population throughout the 1970s. To find its name, the company had combed some 1,800 computer-generated names that combined its regional identity with communication terms.[71]

- *Ameritech*, or more formally, American Information Technologies, Inc., gathered together the five operating companies in the states of Illinois, Indiana, Michigan, Ohio, and Wisconsin. Its name was seen as an indication of its intent to push into a variety of new technologies (it was the first RBOC to begin to offer cellular service), despite the fact that its region was an economically depressed industrial rust belt. CEO William Weiss made clear that he planned to run the most decentralized RBOC operation.[72]

- *Southwestern Bell*, the largest and fastest-growing of the Bell operating companies, would evolve into an RBOC virtually intact. Based in St. Louis (later moved to San Antonio), it included operating companies in the five states of Texas, Oklahoma, Arkansas, Missouri, and Kansas. It employed just under 75,000, the least of all the RBOCs. Said its chairman some months before divestiture, "We're not a go-go company. We're going to stick to our knitting."[73] Years later, SBC (as it had become), would sound very different (*see section 9.6*).

- *US West* was the holding company that covered the largest land mass (some 14 states making up 43 percent of the continental United States, a million square miles). Though limited in population, US West, based in Denver, encompassed three BOCs: Pacific Northwest Bell in Washington and Oregon, Northwestern Bell serving Minnesota, Nebraska, Iowa, and the two Dakotas, and Mountain Bell which supplied states from the Canadian border south to the Mexican border. US West was the first RBOC to come up with its new corporate name.

• *Pacific Telesis* combined Pacific Telephone and Telegraph serving California and Nevada Bell and would continue to serve nearly 20 million customers (most in California). The Greek word Telesis was said to mean "managed growth." The company's California firm had been locked in rate wars with the state public utility commission for years and had fallen behind other operating companies in investment while its debt soared.[74] Only because of special arrangements with AT&T to assume more of the company's long-term debt would Pactel enter 1984 with a debt ratio of 47 percent: the highest of all the new RBOCs. Years later, it would be the first RBOC to disappear (*see section 9.6*).

❶

PACIFIC◨TELESIS

Pacific Bell
Nevada Bell

❷

USWEST

Mountain Bell
Northwestern Bell
Pacific Northwest Bell

❸

Ⓐ
**Southwestern Bell
Corporation**

Southwestern Bell

❹

AMERITECH

Illinois Bell
Indiana Bell
Michigan Bell
Ohio Bell
Wisconsin Bell

❺

NYNEX

New England Telephone
New York Telephone

❻

Ⓐ **Bell Atlantic**

Bell of Pennsylvania
Diamond State Telephone
The Chesapeake and Potomac Companies
New Jersey Bell

❼

Ⓐ **BELLSOUTH***

South Central Bell
Southern Bell

FIG. 6.1.   The seven original Regional Bell Operating Companies (1984).

With more than 900 million shares outstanding, how to deal most fairly with AT&T's 3.2 million shareholders took up a lot of planning time. Negotiators finally determined that all shareholders would receive a single share in each of the seven RBOCs for each 10 AT&T shares they held (soon termed the "seven for ten" plan). Those holding more than 500 shares of AT&T were mailed an option card to make easier sale of securities or consolidation, depending on what the investor wanted.[75] Only 6 percent of AT&T shareholders had 500 or more shares: 75 percent held between 10 and 499 shares, and 19 percent had fewer than 10 shares about which to worry.[76]

People

More than a million AT&T employees knew where to report to work on January 2, 1984, though in many cases that meant members of the same family would be going to work for what would now be competitors.[77] As one business weekly put it, the companies "must redefine managers' jobs, teach them new skills, and—most important—change their mindsets."[78]

This bootstrap re-training of thousands of mid- and upper-level managers who had long "lived in a regulated environment where it has been hard not to make money for the company" was critical to the long run success of all eight segments of what had been a unified AT&T. But now they were being thrust into leadership positions at "the top of very different sorts of companies that are to become public and deregulated" and that would be involved in new and more competitive fields such as cellular communications.[79] Despite the long hours and extensive travel (to meet investors and analysts, for example), many thrived on being in on the ground floor of a radically reshaped industry.

Many long-time employees (those with just the expertise that this difficult time required), however, took early retirement, unwilling to face the difficult transition and disassembly of what they had spent careers building—an integrated system. "The turmoil of the last few years has divided AT&T's management into those caught in the past and those living for the future."[80]

With a serendipitous and fitting sense of timing, at AT&T's long-time New York headquarters at 195 Broadway, workers made plans to leave tradition behind and move to the company's spanking new headquarters building uptown. Some officers had already left for new positions with the RBOCs while others were being groomed for higher things in the newly streamlined AT&T.[81] Of the 15,500 working in headquarters at the beginning of 1983, some 13,000 "have or soon will be dispersed throughout the system" to the local companies while some 68,000 people then at the operating companies would shift to AT&T to help provide interstate and Inter-LATA intrastate long distance services. (The latter had been under operating company control but would shift to AT&T with divestiture.)[82]

6.6 CHANGING ECONOMICS

Just as the Bell System was designing its own breakup, related and important but complicated economic questions had to be faced: specifically, how (or whether) to continue the subsidy to local telephone infrastructure costs from long distance service. For decades, this cost-sharing process had taken place within AT&T (and between it and local independent telephone companies), subject to some (often loose) supervision by the FCC and state public utility commissions (*see section 4.2* for an explanation of this process, called "separations"). Over the years, an increasing proportion of local infrastructure costs had been "assigned" to long distance service until by the early 1980s, long distance charges to users were meeting over a quarter of local company operations.[83] The future of the toll-to-local subsidy seemed even more important during this period, because of various economic factors that were driving up local rates.

Rising Costs

The impending change in ownership of the local exchange carriers, and the MFJ ban on economic ties across the monopoly/competitive barrier, indicated that some new system had to be derived or local telephone costs would shoot up.[84] But several other factors were driving up costs and would have done so regardless of divestiture. As a result, during the early 1980s, the local operating companies were seeking, and getting, often substantial rate increases from the state public utility commissions.

That the rate increases in the 1980s were unusually high is evident when they are compared to earlier increases. While basic local rates increased by only 70 percent during the entire 30-year period between 1940 and 1970, they increased 50 percent during the 1970s and doubled again during the 1980s.[85] Rates rose most quickly during the period from 1980 to 1982 and then again from 1984 to 1986, but for different reasons.[86] The rate increases during the first part of the 1980s were a reaction to the high inflation and steep interest rates experienced in the United States during the late 1970s and early 1980s. Inflation drove up the local operating companies' labor costs, energy bills, and other expenses; the local operating companies, regulated under rate-of-return regulation, filed with the state commissions to increase their rates in order to recover their growing expenses. The local companies also asked for increases in their allowed rate of return in reaction to continually rising interest rates. The companies argued that they needed a higher allowed return in order to cover the increasing cost of debt caused by escalating interest rates. The companies also argued that they needed a higher return in order to attract investors, who were able to earn increasingly high returns by investing in other industry sectors.[87] As a result of these factors, residential local rates rose from an average of $8.19 in 1979 to an average of $9.73 in 1982, a 19 percent increase in just three years.[88]

While the spate of local rate increases from 1980 through 1982 were largely caused by inflation and higher interest rates, the rate increases from 1984 to 1986 may have been fueled in part by concerns and uncertainties regarding the impact of the MFJ on the local operating companies. In 1984 alone, local rates increased an average of 17.1 percent as state commissions approved $3.9 billion in revenue increases.[89] Divestiture did have an impact on the revenue streams that the local companies had come to rely on. Divestiture provisions required the local companies to transfer customer premise equipment to AT&T; as a result, the local companies lost the revenues they had received from leasing that equipment. In addition, the local companies would no longer provide intrastate Inter-LATA toll service; instead of charging long distance customers toll charges for intrastate Inter-LATA calls, they would be charging substantially lower access charges to the long distance companies who were allowed to provide intrastate Inter-LATA toll service. The local companies sought to raise local rates as one way of compensating for these lost revenue sources.[90]

Another factor driving up local rates was a change in the regulatory treatment of depreciation. Depreciation is an accounting concept that seeks to spread the recognition of the cost of an asset over its useful life. For example, if a company purchases a machine for $1 million and uses the machine for 20 years, how should the company recognize the cost of that machine? If the company treats the full $1 million cost as an expense in the year in which it buys the machine, it will greatly understate its earnings for that year and grossly overstate its earnings for the other 19 years in which it is using the machine but not recognizing any expense for that use. If, however, the company recognizes as an expense a portion of the $1 million cost of the machine over its 20 years of use, the company will be presenting a more accurate picture of its earnings. This multiyear recognition of expense is depreciation. Depreciation has been an important and volatile regulatory topic because of rate-of-return regulation (*see section 4.2*). In rate-of-return regulation, the regulated company targets its rates to recover all of its allowable expenses. Depreciation is one of those allowable expenses. The regulatory question becomes how quickly an asset should be depreciated. In the above example of the $1 million machine, a 20-year life would result in $50,000 of depreciation expense per year going into the regulated rates. A 10-year life would result in $100,000 of depreciation expense per year and, therefore, higher rates. By keeping depreciation lives long and depreciation expense low, regulators could keep rates low as well.

The FCC and the state commissions recognized this aspect of depreciation and had for decades required the use of long depreciation lives, often 30 or more years, for assets like switches and copper wire. However, technological developments were reducing the useful lives and therefore the economic value of these assets, suggesting that they should be depreciated over a shorter time period, with greater depreciation expenses being recovered through higher rates and companies purchasing more technologically advanced equipment. The gap between the amount of depreciation expense companies had been allowed to recover by the 1980s and the amount of depreciation that would have more accurately reflected the economic value of the assets depreciated has been estimated as ranging from $10 billion to $26 billion nationally.[91]

In 1980 the FCC, hoping to encourage industry investment in new, more efficient technologies, accelerated the rate at which interstate investments could be depreciated. In 1983, the FCC extended this accelerated depreciation policy to intrastate investment, preempting the state commissions' ability to determine depreciation policy. This preemption was thrown out by the Supreme Court, which found in 1986 that the FCC could not prescribe depreciation rates for intrastate investment.[92] However, between 1983 and 1986, many of the local operating companies asked for, and received, permission to depreciate their assets more quickly and to raise their local rates in order to recover the resulting increased depreciation expenses. After the 1986 Supreme Court decision, many states returned to slower depreciation schedules and lower rates.

Rising inflation, more aggressive depreciation schedules, and the loss of revenues from customer premise equipment rentals and intrastate Inter-LATA toll service were not the only factors that were raising local service rates. Changes in the allocation of investment and expenses between the intrastate and interstate jurisdictions were shifting costs out of long distance charges and into local service rates. These changes in allocation were driven in part by the fear of bypass.

Bypass Fear

New technologies such as microwave systems were making it possible for private companies to *bypass* the public switched network by taking care of their intracompany long distance communication needs themselves. Indeed, the FCC had fueled these efforts through such decisions as. *Above 890* (*see section 5.3*). Major companies, instead of paying long distance charges to AT&T for calling among their multiple business locations, could save money by building and maintaining their own networks. The economics of building a private network entailed some hefty up-front costs and some ongoing maintenance expenses. But a private network might prove a financially attractive alternative to companies who were paying as much as $2.70 for a five-minute long distance call,[93] especially if the companies generated a large volume of long distance traffic.

The FCC was concerned at the prospect of bypass by large long distance users because these users contributed substantially toward covering the cost of the public network. If they took their long distance business away from the public network, the remaining subscribers—residential customers and smaller business users—would have to shoulder a larger portion of the cost of the public network through higher ser-

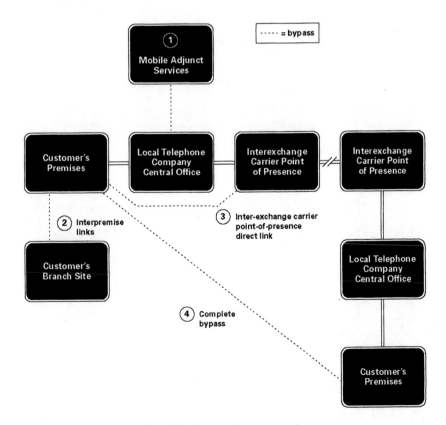

FIG. 6.2. Types of bypass service.

vice rates. The FCC's concerns were based on the underlying economics of the public network, which required a substantial amount of fixed investment, including switches, central office equipment, and interoffice cabling. With more users of the network, there would be more users and more usage over which to spread the costs of the investment. The FCC recognized that long distance was, in effect, a declining unit cost service. In other words, if the fixed costs involved in hauling traffic from City A to City B were $10 million, and the traffic on that route totaled 50 million minutes, the per-unit cost of each minute of traffic would be 20 cents. If however, the traffic declined to 25 million minutes, the per-unit cost would increase to 40 cents. The FCC feared that if large users turned to bypass, the significant fixed costs of the network would have to be recovered through rising prices that might prove unaffordable to smaller users who might then also be driven off the network. The Commission hoped to keep large users on the network by making bypass a less attractive alternative; the best way to do that was to lower long distance prices.

NOTES

1. See "And It Is from the Basement to the Attic," 1978. *Broadcasting* (June 12), pp. 29–41, for details on principal points of the first bill.
2. See *Telecommunications Law Reform*, 1980. (Washington, DC: American Enterprise Institute) for 1979–80 revision attempts plus a comparison of House and Senate bills; and "Major Telecommunications Policy Proposals," U.S. National Telecommunications and Information Administration, May 5, 1982 for a tabular comparison of 1981–82 final attempts.
3. *MCI Communications Corp. v. AT&T*, 462 F. Supp. 1072, U.S. District Court for the Northern District of Illinois.
4. U.S. Department of Justice, Antitrust Division. *Complaint*, United States of America v. American Telephone and Telegraph Co., et al. (1974). Civil Action 74-1698, U.S. District Court for the District of Columbia, para. 23. (Cited below as *Complaint*).
5. Early in the process, District Court Judge William J. Lynch became ill and died (a premonition of what would happen in the later government antitrust case). The case was assigned to Judge John F. Grady.
6. Cantelon, p. 300, quoting Grady's opinion.
7. Ibid., p. 311.
8. Ibid., pp. 313–314.
9. AT&T rapidly appealed the Chicago jury's decision. In January 1983, in a 205-page divided opinion, an appeals court panel dropped 2 of the 10 counts for which AT&T had been found liable. The judges also found that the fine assessed against AT&T was grossly excessive. The penalty portion of the original case would have to be re-tried. On April 8, 1985, some 11 years after the original suit had been filed, a new trial limited to a discussion of the amount of damages due to MCI for the 8 remaining counts began. When the jury came back with a decision that MCI should receive $37.8 million, which, even when tripled to $113 million was but six percent of the original fine five years ear-

lier, it was MCI's turn to be stunned as investors saw the stock drop 20 percent in a day.

10. See Coll, pp. 59–61.

11. A CID seeks formal answers to questions posed by a government agency. Its purpose is two-fold: to give a government agency an opportunity to collect information in support of a presumed wrongdoing, and to give the target of the inquiry a chance to show that no wrong has been committed.

12. Temin, p. 109. Emphasis in the original.

13. See Coll, pp. 63–72 for a dramatic retelling of the events of November 20th, 1974. Coll suggests that Saxbe, worried about the low image of the Justice Department after the Watergate scandal, and only in sporadic touch with a Ford White House displeased with some of his other actions, viewed the potential AT&T suit as one way of showing that the Justice Department had not lost its teeth. Break-up historian Peter Temin feels that Attorney General William Saxbe may have been acting on his own authority as both President Ford (then on a good-will trip to Japan) and Treasury Secretary Simon later recalled no cabinet-level discussion of such a suit. See Temin, 1984, p. 110.

14. Every lawsuit begins with a complaint. Many complaints list elaborate grievances by plaintiff against defendant and ask for exorbitant damages to compensate plaintiff. By contrast, this Complaint was notable for its plain and relatively low-key statement of the case.

15. The Sherman Act (15 USC sec. 1 *et seq.*) was passed by Congress on July 2, 1890, and was named for Senator John Sherman, a late 19th-century member of Congress who was known as a trustbuster. The Sherman Act, the nation's first antitrust law, was passed in response to public outrage at anticompetitive corporate practices. It outlaws monopolization and attempted monopolization that is achieved by unlawful means. It is possible for a company to be a lawful monopoly, that is, to have achieved its status as the sole provider or supplier of a product or service through lawful means, including skill, industry, thrift, adequate capitalization, and so on. What the Sherman Act prohibits is the illegal monopoly: the firm that keeps competitors out of its market through acts of bad faith (refusal to deal, price-cutting, control of an essential facility, and so on). The early history of the Sherman Act is interesting for the mixed results which greeted the Act in court.

16. *Complaint*, para. 29 as reprinted in Sterling et al., eds. 1986. *Decision to Divest: Major Documents in U.S. v. AT&T, 1974–1984* (Washington: Communications Press), p. 15.

17. Letter from John deButts to Share Owners, dated November 26, 1974 (mailed out with monthly bills), p. 2.

18. *Answer* in Civil Action 74-1698 (February 4, 1975), as reprinted in Sterling et al., pp. 20–43.

19. AT&T, "Gold Book," p. 692.

20. Cantelon, p. 186. His chapter 8 provides the best behind-the-scenes view of how MCI developed the idea and the service.

21. "There were no lines dedicated to one customer's use; the bulk of the charges were for the calls themselves, not for access to the system" as in private line service. Temin, p. 133.

22. Cantelon (pp. 192–193) speculates too many other things were going on at the time for the Bell System to careful weigh what the new MCI tariff indicated. Just one example is that the government's antitrust complaint was filed only six weeks after the Execunet tariff entered into force.

23. FCC letter to MCI (July 2, 1975), pp. 3–4.

24. Cantelon, p. 250.

25. *MCI v. FCC*, 561 F.2d 365 (1977). In light of subsequent events, this is often referred to as *Execunet I*.

26. And MCI did just that—by the late 1970s, its "regular" long distance services were responsible for 80 percent of the company's revenue.

27. *MCI v. FCC*, 580 F.2d 590 (1978). Often called *Execunet II*.

28. The FCC could not apply the Sherman Act; only a federal court had jurisdiction to determine if a company had violated that federal law. Thus, if the case had been sent to the FCC, the agency would have evaluated AT&T's behavior under the much more lenient standards of the Communications Act of 1934. This was a point not lost on Trienens and Saunders.

29. *US v. AT&T, et al.*, 427 Fed Sup 58 (1976) at part V.

30. Judge Harold Herman Greene (1923–2000) was born in Frankfurt, Germany, on February 6, 1923. He came to the United States in 1943. In 1952, after serving in military intelligence with the U.S. Army in West Germany, he was graduated at the top of his class from The George Washington University Law School. From 1958 to 1965 he worked in the Civil Rights Division of the Department of Justice. In 1965, he was appointed a judge of the District of Columbia Court of General Sessions, a post he held until 1971 when he became Chief Judge of the Superior Court of the District of Columbia. He was appointed a U.S. District Judge by President Jimmy Carter in 1978. Judge Greene took senior status (largely retiring) in 1996 and died in Washington early in 2000.

31. To some extent, Greene was reacting to a parallel and even older government antitrust case, this one against IBM. Originally brought in 1969, that dispute had bogged down in a huge flow of paper, and lacked the order and deadlines Greene imposed in the AT&T proceeding.

32. "Green" came first—more formally known as *Plaintiff's Third Statement of Contentions and Proof*, January 10, 1980, two volumes totaling 1,842 pages. "Gold" followed: *Defendant's Third Statement of Contentions and Proof*, March 10, 1980, three volumes totaling 2,147 pages. Copies of these are now difficult to find. Their tables of contents and key portions of their conclusions were reproduced in Sterling et al., pp. 189–459.

33. On the other hand, this aspect of the Clayton Act was largely untested at the time "and attorneys disagree over how serious an impediment it is to reaching a settlement during trial." See "AT&T–U.S. Talks on Antitrust Settlement Break Down, Increasing Chance of Trial," *Wall Street Journal* (February 24, 1981), p. 2.

34. "U.S., AT&T Reported Near Accord," 1981. *Washington Star* (January 7), p. A1.

35. "Settlement Reported Near in AT&T Antitrust Suit," 1981. *Washington Post* (January 17), p. B1.

36. William Francis Baxter (1929–98) was born in New York City but grew up in California. He earned a bachelor's and law degree from Stanford University. At the time he was named to the antitrust job, he had taught at Stanford for more than 20 years. He was reported to have studied "differential calculus so he could understand all the economic literature" so essential in antitrust cases. He authored several books. See "Government Watchdog," *New York Times* (January 9, 1982), p. 34.

37. Both William French Smith, the nominated Attorney General, and his deputy, had worked at various times for elements of the Bell System.

38. Baxter was cited by several observers for a 1977 law journal article which stated that "divestiture was often the appropriate remedy in cases involving regulated monopolies." See "Bell Setback Seen in Trial Resumption," *New York Times* (March 2, 1981), p. D1.

39. "Reagan Antitrust Chief Vows to Pursue AT&T Breakup," *Washington Post* (April 10, 1981), p. E1.

40. Such a filing by the defendant at the conclusion of plaintiff's case is a standard legal maneuver in almost any civil case.

41. *Opinion* [on Denial of Motion to Dismiss], 524 F. Supp. 1336 (September 11, 1981).

42. "U.S. Rests Arguments in Bell Suit," *New York Times* (July 2, 1981), p. D1. For the full text of Greene's opinion, see 534 Fed Sup 1342, or Sterling et al., Vol. 2, pp. 860–899.

43. *Opinion* [on Modification of Final Judgment], 552 F. Supp 131 at 165. In the years following divestiture, the leadership of the operating companies never missed an opportunity to bemoan these restrictions; as many pointed out with some asperity, the operating companies were never the target of the government's antitrust suit and had not committed any independent acts in violation of the Sherman Act, since they were never free enough from AT&T to act on their own. See chapter 8.

44. Ibid., at 225. See "VIII: Modifications," paragraph C and paragraph 2.

45. Ibid., at 225. See "VIII: Modifications," paragraph B.

46. Rights to the word "Bell" and the time-honored logo were matters of huge importance to the participants in this case. As one commentator remarked at the time, the true measure of the trauma attending the breakup was the sight of grown men fighting in open court over the ownership of the word "Bell".

47. Temin, pp. 293–294. That level was approximately equal to Pacific Telephone and Southwestern Bell at the time. And, while not considered in 1982, it made the firms sufficiently large to avoid takeover battles that, as Temin puts it, were soon to became a "corporate sport."

48. "AT&T Proposes New Regional Structure," *Washington Post* (February 20, 1982), p. A1.

49. Confusing to many at the time (and somewhat since) is the use of the term "local." Some LATAs are sufficiently large that even though a BOC provides service within them, that service may encompass toll telephone service. Indeed, such *intra*-LATA calls are "long distance" to users who pay a usage-sensitive toll for such calls. They are not so considered under the divestiture scheme, however, for which only *inter*-LATA calls are long distance services forbidden to the RBOCs.

50. As had the approximately 11,000 local exchanges served by independent telephone companies. See *Opinion* [on LATA Plan], 569 F. Supp. 990 at 993 (April 20, 1983), as reprinted in Sterling et al., Vol. 3, p. 1689.

51. Whether this bit of psychology worked fully was questionable when even 15 years later polls often indicated that well over a third of Americans thought AT&T continued to be the provider of their local telephone service.

52. "AT&T Unveils Plan for 161 Service Areas in Nation," *Washington Post* (October 5, 1982), p. D7. The document is partially reproduced in Sterling et al., Vol. 2, beginning on p. 1303.

53. Temin, p. 300.

54. 569 F. Supp. 990 (1983) as reprinted in Sterling et al., Vol. 3, pp. 1687–1753.

55. Tunstall, p. 50.

56. For the table of contents and selected summaries of specific asset assignments of this 475-page document, see Sterling et al., Vol. 3, pp. 1401–1657.

57. 569 F. Supp. 1057 (1983), as reprinted in Sterling et al., Vol. 3, pp. 1756–1826.

58. Especially vexing to AT&T was that the company had just spent upwards of $30 million promoting the name of its new subsidiary, American Bell, created to sell computer equipment and services under the separate subsidiary provisions of the FCC's Computer II decision (*see section 7.4*). Greene's decision on use of "Bell" meant the subsidiary would have to be renamed. "Views Vary on Long-Used Bell Name," *New York Times* (July 9, 1983), p. 35. American Bell became AT&T Information Systems a month after Greene's decision.

59. Op. cit. (note 41) at 1123-1124, as reprinted in Sterling et al., Vol. 3, pp. 1818–1819.
60. *United States v. AT&T*, 1982. 552 F. Supp. 131, at 228 (D.D.C. 1982).
61. Ibid., at 194.
62. Ibid., at 192.
63. Ibid., at 191.
64. Ibid., at 190.
65. Ibid., at 229.
66. Ibid., at 190.
67. The *Information Statement and Prospectus* is excerpted in Sterling et al., Vol. 3, pp. 1827–1852.
68. SNET was one of two local operating companies in which AT&T held minority shares (the other was Cincinnati Bell). Under provisions of the MFJ, AT&T sold its stock in both companies which thus became independents, not otherwise subject to MFJ restrictions.
69. "Ma Bell's Kids Fight for Position," *Fortune* (June 27, 1983), p. 66.
70. "At Nynex, X is Unknown," *New York Times* (October 27, 1983), p. D1.
71. "Ma Bell's Kids . . . ," p. 63.
72. Ibid., p. 64.
73. Ibid., p. 65.
74. Ibid., p. 68.
75. The Internal Revenue Service, recognizing that the divestiture was taking place because of settlement of a government antitrust case and not because of normal investor decisions, decided in October to exempt nearly all of the initial stock transactions from taxes.
76. "Confusion on AT&T Stock," *New York Times* (November 21, 1983), p. D1.
77. Many stories ran in the closing months of 1983 like "The Sense of Loss at Ma Bell: Phone Family Breaking Up," *New York Times* (December 28, 1983), p. D1. Local papers all over the country ran similar features on long-time Bell System workers who were now heading to work in different directions for different employers.
78. "Culture Shock Is Shaking the Bell System," *Business Week* (September 26, 1983), p. 112.
79. "Bell's Breakup Offers Bonanza," op. cit., p. D22.
80. "Culture Shock," p. 114.
81. "Included in the announcement of these moves was also the news that [Robert] Allen, the one-time Chesapeake and Potomac Telephone Cos. chief, has a new title: executive vice president-corporate administration and finance. That is seen as a possible grooming spot for Allen, 48, who insiders saw as could be in line to succeed [CEO Charles] Brown, 62." "Bell's Hour: It's Quiet at the Top," *Washington Post* (December 22, 1983), p. D3.
82. Ibid.
83. U.S. Congress, Congressional Budget Office, 1984. *The Changing Telephone Industry: Access Charges, Universal Service, and Local Rates* (Washington, DC: Government Printing Office), p. 10.
84. To some extent local rates would rise after divestiture anyway. But, ironically, this had little to do with the divestiture process itself. At the same time, and totally unconnected with the AT&T breakup, the FCC had modified the rate at which telephone carriers could depreciate their equipment and facilities. The change had the effect of allowing telephone companies to depreciate at a much faster rate so that newer (often digital) equipment could come on line faster. But the faster rates of depreciation meant that ex-

pensive equipment was now depreciated over roughly 10 years whereas closer to 30 had been the norm in the industry for decades.

85. Roger G. Noll and Susan R. Smart, 1991. "Pricing of Telephone Services," in *After the Breakup: Assessing the New Post-AT&T Divestiture Era*, ed. Barry G. Cole (New York: Columbia University Press), p. 188.

86. Ibid., p. 189.

87. Interest rates during this period were indeed unusually high, with home mortgage rates often exceeding 15 percent. One measure of the magnitude of accepted interest rates during this period is the allowed return that the FCC approved at the beginning of access charges in 1984; it was 12 percent.

88. Noll and Smart, p. 189.

89. U.S. Department of Commerce, National Telecommunications and Information Administration, 1988. *NTIA Telecom 2000: Charting the Course for a New Century* (Washington, DC: Government Printing Office), pp. 209–210.

90. Ibid., p. 210.

91. James C. Bonbright, Albert L. Danielsen, and David R. Kamerschen, 1988. *Principles of Public Utility Rates*, Second Edition (Arlington, VA: Public Utilities Reports), p. 594.

92. *Louisiana Public Service Commission* v. *FCC*, 476 US 355 (1986).

93. A five-minute call placed during weekday daytime hours from New York to California cost $2.70 in 1982 according to the FCC's *Statistics of Communications Common Carriers*, 1995/96 edition, p. 280. At this time, long distance rates were time-of-day and distance sensitive. Calls placed during peak periods (8:00 to 5:00 on weekdays) were charged peak rates, with evening and weekend calls discounted by 40 percent and night-time rates discounted 60 percent. Rates were also mileage sensitive; while a five-minute call from New York to California cost $2.70, one from New York to Chicago cost $2.34.

Chapter 7

Operating Under the MFJ (1984–96)

In this short time span (1984–96), the American telecommunications business underwent much of a difficult transition from the old, ordered, monopoly system to the beginnings of something quite different. Triggered by the implementation of the Modified Final Judgment, the period became especially complex with the rise of two important modes of telecommunications that had been limited in their application earlier—wireless (cellular) systems, and the Internet. Either was sufficient to change the rules; together (and combined with the end of the old Bell System) they created new players, policy concerns, and options for consumers. As will be seen in chapter 9, these and other changes finally forced substantial modifications in the ruling Communications Act.

7.1 "FREE THE RBOC SEVEN"

Implementation of the Modified Final Judgment (MFJ) was widely expected to be complicated but relatively straightforward, with everyone playing their intended part. Bell's 22 local operating companies, organized into seven Regional Bell Operating Companies (RBOCs), would develop their role as monopoly local service providers subject to close regulation by both states and the Federal Communications Commission (FCC). Competitive sectors of telecommunications, somewhat less subject to regulation, would also thrive, multiply, and compete. AT&T, chastened—and exhausted—by the experience of divestiture, would quietly reorganize and issue dignified press releases announcing new services, with perhaps a modest price cut thrown in. Competing long distance companies would enjoy robust economic health as a result of access to markets formerly closed to them by AT&T's anticompetitive practices. The Justice Department would enforce the MFJ with vigor, and Judge Greene could turn his attention to other cases.

But reality quickly differed sharply from expectation. Divestiture was *supposed* to produce a neatly segmented telecommunications world with a regulated monopoly sector (the RBOCs) and a less-regulated competitive sector (AT&T and its many competitors). Instead, thanks to rapidly changing technological options and parallel changes in industry economics and regulation, the new post-divestiture world became far more complicated than expected, with companies behaving in ways that the architects of divestiture never foresaw. In part, the surprises of the post-divestiture period resulted from shifting alliances among companies: those that had formerly slugged it out as adversaries now joined forces while former allies turned into vigorous competitors.[1]

Petitions for Change

To begin with, the RBOCs had no intention of operating strictly within the narrow confines outlined for them by the MFJ. Relying on Section VIII (C) of the MFJ, which allowed local operating companies to file for permission to enter other lines of business "upon a showing by the petitioning BOC that there is no substantial possibility it could use its monopoly power to impede competition in the market it seeks to enter,"[2] the RBOCs began to file what quickly became a flood of requests to enter new business ventures. The process that Judge Greene had established in the MFJ for considering waiver requests was straightforward: if a company, after filing a request, was able to meet the burden of showing that the proposed new venture would not lessen competition in the relevant market, the request was granted. The filings were to be reviewed by the Antitrust Division which would recommend findings to the judge.

Less than a month after the MFJ's implementation, Bell Atlantic filed a petition asking Greene's permission to enter the equipment leasing market. The next day, BellSouth filed a motion seeking to provide certain software programs and related services. The BellSouth motion was typical: a brief two-page pleading supported by a longer memorandum explaining how the separate subsidiary offering the proposed service would be set up.[3] Soon after, on February 8, 1984, Pacific Bell and Nevada Bell (together they formed Pacific Telesis) filed for permission to enter into foreign business ventures.[4]

What began as a trickle soon became a torrent of requests to enter different lines of business.[5] All seven regional holding companies filed for permission to go into the cellular monitoring and consulting business. Two wanted to offer paging services. The companies began to piggyback their requests: all of them wanted to become insurers, offer software, and provide cellular radio services outside of their regions.[6]

By the spring of 1984 Judge Greene had been overwhelmed by the number of filings.[7] Between a quarter and a half of his total time was focused on dealing with RBOC filings which amounted to what one party termed the "second phase restructuring of the American telecommunications industry."[8] In a July 1984 opinion he made clear that the RBOCs were not to turn their attention away from providing basic local service nor treat that function as some sort of "pedestrian sideline"[9] to the potentially more lucrative businesses into which the companies sought entry. The operating companies, "each of which retains in its particular area a monopoly over local telecommunications service and thus has the potential for using its monopoly power to discriminate against others,"[10] would be required to show that entry into new businesses would not impede competition or harm the company's ability to provide local exchange telephone service.

Remembering that AT&T had cross-subsidized competitive services with profits from monopoly customers, Greene required the RBOCs to establish a separate subsidiary for the conduct of each proposed new business,[11] to obtain financing for the new ventures on their own credit,[12] and to limit new revenues of the new lines of business to no more than 10 percent of the companies' total net revenues.[13] He also declared that waiver requests would go first to the Department of Justice. If the Department found that the standards had been met and interested parties had no viable objections, the requests would be granted.[14]

Greene's opinion described an industry that seemed not to have changed much. Despite divestiture, the specter of anticompetitive behavior still lurked. Customers still had to be protected from the monopoly businesses serving them. Markets still had to be kept open and thus some restrictions were still needed. The only change was that in the post-divestiture world, the monopoly business was now in the hands of the RBOCs. Greene concluded that without his constant oversight, old patterns of behavior might reappear in the hands of new managers.[15] To the RBOCs, the opinion was so comprehensive that Greene seemed to be creating a position as *de facto* regulator of the telephone business. As Greene wrote:

> [The former Bell operating companies] are at present the only entities with the license and the capacity to provide local telephone service to the general public. Thus, if they neglect their responsibilities in that regard in order to pursue what they may consider more interesting ventures, or if they are

diverted from those responsibilities by unrelated or speculative business objectives, the public, unable to go elsewhere, will suffer in higher rates, deteriorating service, or both. Such a result would be entirely inconsistent with the basic purposes of the decree and the Court will not approve requests which would help to bring it about.

Beyond these considerations, there is the overriding principle that the Court is obligated under the decree to make certain that the Regional Holding Companies do not impede competition in the non-telecommunications markets they seek to enter. Here again, wide diversification presents a serious threat in terms of cross subsidization and other anticompetitive practices.

The decree assumes, as does the Court, that the Regional Holding Companies may diversify on a significant scale only as they demonstrate the centrality of their corporate life of the responsibilities imposed upon them by the decree, their firm commitment to low-cost, high-quality telephone service, and the improbability of their involvement in anticompetitive conduct based upon their monopoly status.[16]

But the companies continued to resist Greene's blandishments. For one thing, no provision in the MFJ restricted the number of waiver requests a company could file. Having tasted freedom in the form of numerous successful waiver requests, the RBOCs concluded that offering only local telephone service was a sure way for them slip behind more flexible competitors. They argued that offering new services would allow them to increase revenue and thus better serve shareholders and customers.[17] Several RBOCs filed motions in 1985[18] asking for clarification of the kinds of businesses that they could enter *without* filing waiver petitions—specifically permitting the RBOCs to provide cellular radio service outside their geographic regions and shared tenant services[19] inside their regions. US West raised an even more fundamental issue, arguing that the restrictions of the consent decree were not binding on the RBOCs because they were not parties to it (they did not exist when the MFJ was agreed to).[20] Greene disagreed on both counts. In an August 1986 decision, a federal appeals court concluded the regional companies would indeed have to toe the MFJ line and that Judge Greene had erred as well:

> We hold that the RHCs are bound by the consent decree, that the decree does not prevent the RHCs from providing exchange services outside their geographic regions, and that the district court therefore erred in ordering the RHCs to stop providing extraregional exchange services.[21]

The decision was significant because it marked the first successful appeal of one of Judge Greene's orders. But it also underlined the difficulty in predicting how MFJ provisions would define the industry: "We suspect the uncertainty surrounding this issue has a simple explanation: the parties and the district court never considered the possibility that the BOCs might want to provide exchange services outside of their geographic regions."[22]

AT&T executives grew increasingly uneasy as the waiver petitions continued. A disturbing new scenario was beginning to unfold: the operating companies, with their monopoly local facilities and expertise, were competing directly with AT&T for the same customers in the same markets. AT&T protested the increasingly aggressive steps that the operating companies were taking.[23] To allow the RBOCs to expand unfettered by the decree's restrictions would dramatically shift the balance of power in the industry. As AT&T increasingly realized that its sharpest competition could come not from other interexchange carriers but from its former operating companies, the MFJ took on new meaning. While the operating companies chanted "Free the RBOC Seven!" and fought to provide their customers with one-stop shopping, AT&T urged the court to remember the economic theory behind divestiture—to completely separate competitive aspects of the industry from continuing local exchange bottlenecks—and thus to keep the BOCs tightly controlled and restricted to local exchange services.

Triennial Review

As RBOC officials continued to press for change or elimination of MFJ restrictions, they found an increasingly sympathetic ear at the Reagan Administration. Judge Greene's steadfast refusal to budge on MFJ line-of-business restrictions proved too much for a pro-business administration. Beginning late in 1985, a series of meeting between Attorney General Edwin Meese III[24] and RBOC officials set in motion a process that would find the Reagan Administration backpedaling from the MFJ. While the Antitrust Division economists and lawyers most familiar with the case had not changed their minds, senior Justice officials began to suggest that it was time for the department to end its supervision of the telephone business.[25]

The telephone company meetings with Meese were not accidental. As part of divestiture, the MFJ required that three years after the breakup, the court would conduct the first of a planned series of formal reviews to determine if changes were required. In late 1985 the Justice Department contracted with Dr. Peter Huber to carry out a study of the business.[26] Describing himself as "once an engineer but now a lawyer,"[27] Huber sought comments on any aspect of the MFJ—and information flooded in from all quarters.[28] A year after Dr. Huber began his research, the Justice Department finally published his large report which dissected telephone industry structure and operations at length.[29] Unlike the Splawn analysis of 1934 and FCC telephone study five years later (both of which focused on financial/management issues as discussed in section 4.3), however, Huber concentrated on changes being wrought by technology.

Huber described what he foresaw as the "geodesic network" of the not too distant future. The MFJ was criticized in the report's opening paragraph:

> As networks expand horizontally the companies that manage them grow vertically. The central paradox of the information age is that the dispersion of consumption is matched by a consolidation of production. Whatever the Regional Bell Operating Companies are or are not permitted to do, the Modified Final Judgment's basic vision of a horizontally stratified telecommunications marketplace, animated by an obsolescent model of the network as a pyramid, will not survive. AT&T, IBM, and other major U.S. and foreign telecommunications and electronics companies are already gathering for the wake.[30]

Huber argued that network design had been driven by the fact that switching was very slow and expensive, while transmission was, by comparison, faster and cheaper: The resulting bottlenecks were not the result of any conspiracy; they were the product of careful and deliberate engineering design.[31] Huber described the network as a set of nodes connected by wires in a hierarchical pyramid and the new telecommunications network as many nodes connected along a geodesic, or a path of minimum length, in a ring.[32] In this network of the future, restrictions on the RBOCs made little sense because "in each one, the geodesic network can support what the pyramid could not: competition, vertical consolidation, and a profusion of small boutique firms clustered around the network nodes. . . . By all market indications, corporate consolidation is absolutely inevitable, whatever happens under the MFJ."[33] He concluded that "notwithstanding divestiture, the geodesic network will end up managed by a small number of giant, vertically integrated firms, AT&T among them,"[34] and that "in all of this, regulation has been a large part of the problem and a small part of the solution, often at the same time."[35]

Armed with Huber's analysis, in February 1987 Justice filed its recommendations with Judge Greene that the

> prohibitions on BOC provision of information services, BOC provision and manufacturing of telecommunications equipment and manufacturing of CPE, and BOC involvement in non-telecommunications business be removed. . . . We recommend . . . that the interexchange services prohibition be modified to permit each BOC to provide interexchange services that are entirely outside of its region.[36]

Pandemonium greeted the Justice filing, for what had formerly been prohibited was now to be allowed by the same Justice Department that had insisted three years earlier that such restrictions were vital. Reactions from competitors, the trade press, Congress, and other regulators ranged from incredulous to horrified. In the face of near-universal denunciation (except for the RBOCs' understandably strong support), the Department withdrew its recommendations. In late April Justice reappeared with a filing that retained its recommendation to lift the prohibitions against manufacturing and information services (as well as non-telecommunications businesses) but that now, more cautiously, suggested that the long distance restrictions be continued for the time being.

Decision

After further proceedings, Greene's September 1987 decision on the triennial review was a sharp rebuff to the Justice Department and a clear victory for AT&T. Greene found that conditions had not changed substantially in the industry and that the operating companies still controlled local bottleneck facilities vital to interexchange service providers. Accordingly, the court would maintain in force the restrictions that kept the BOCs out of manufacturing and long distance service. In a small change of his own position, the judge modified the information-services prohibition slightly to say that the companies could *transmit* but could not *generate* information. Greene expressed his displeasure with AT&T, the RBOCs, the FCC, and the Justice Department. He noted that, even with the restrictions in place, several operating companies had been treating customers badly[37] and included a long section on various abuses by the operating companies, including "efforts . . . to escape regulatory scrutiny,"[38] inability to provide equal access for mobile calls, and refusal to furnish access to information.[39] He did throw the RBOCs a bone—they would no longer have to petition the court for permission to enter a business having no direct relationship to telecommunications.

In yet another alliance between former adversaries, both the RBOCs and Justice announced their intention to file appeals. The RBOCs renewed their entreaties for help from the Reagan Administration, which in turn directed them to the National Telecommunications and Information Administration (NTIA) which filed a petition with the FCC in November 1987 urging the Commission to issue a declaratory ruling that the provision of information services by the Bell Operating Companies would be in the public interest.[40] As if to further urge the Commission to challenge the judge, the NTIA letter stated that "the Commission, not the district court, should be the paramount authority" in the field of telecommunications regulation.[41] The Commission solicited comments, but in the end took no action, understandably fearing the prospect of taking on a federal judge.

But the triennial review fight was not quite over. While parts of his September opinion were already under appeal, the judge moved to close a definitional loophole. In December 1987, Greene fired another salvo which made clear that he had no intention of loosening his oversight on the MFJ. His opinion was directed to an examination of the meaning of the term "manufacture" as used in the MFJ and, specifically, what kinds of activities were off-limits to the operating companies. The judge found that the term—and thus activities forbidden to the RBOCs—included "design and development of telecommunications products and customer premises equipment as well as their fabrication."[42] Several of the operating companies were forced to cancel a variety of planned and ongoing activities that were now prohibited by this broad definition.

The triennial review had solidified the industry into warring camps. Rhetoric had changed nothing other than to make both sides more wary of one another as well as frustrated with the legal process. The limitations of using antitrust law to reorganize an industry were becoming more apparent. Constrained by the provisions of a legal agreement (in this case the MFJ), no judge or court could move quickly enough or display sufficient flexibility to keep up with the industry changes.

7.2 LONG DISTANCE MARKETPLACE

Evident by the late 1980s was AT&T's growing difficulty selling equipment to companies with which it competed in services. The RBOCs, once captive markets in an integrated Bell System, were increasingly leery of buying equipment from an AT&T that was fighting in various court and policy battles to keep them limited to local telephony. Further, AT&T faced strong and growing competition from overseas manufacturers who had taken ever-larger portions of the American equipment market. And when AT&T tried to sell its equipment overseas, it faced stiff opposition from national carriers resistant to AT&T's service offerings.

Two early 1990s AT&T takeovers brought the issues into sharp relief. After a difficult and very expensive battle, in 1991 AT&T took over the National Cash Register Company (NCR), seeing the move as a way to leap-frog into the computer equipment business. Two years later a less rancorous takeover brought McCaw Cellular, one of the major wireless service providers, into the AT&T fold. The McCaw deal, however, raised concerns on several fronts. Eventually the Justice Department and AT&T

agreed on a Final Judgment, to last a decade, that would allow the merger subject to limits on AT&T equipment sales and service agreements with cellular competitors.[43] Chief among these was a requirement for structural separation between McCaw and the rest of AT&T and banning McCaw from use of the AT&T brand in marketing until its cellular franchises had converted to allow equal access to its customers by competing interexchange companies as well as AT&T.[44] This complex legal process highlighted the conflicts the integrated company faced. Had AT&T no manufacturing arm, there would be no need for any legal judgment.

AT&T Trivestiture

This competitive disadvantage opened the door to an obvious but still hard solution. An independent service-based carrier with no ties to equipment manufacturing would be freer to expand and grow. AT&T services need no longer be concerned with possible conflicts with equipment sales and contracts. Both elements might be better off seeking their peak earnings as separate entities than joined together. Further, the computer manufacturing and servicing arm (the former NCR) of the company looked to continue to drag down overall profits for some years in the future. Financial analysts grew increasingly concerned and urged shedding of the problem-plagued division.

There appeared to be no "fixes" for these conflicts within a unified company. Yet across the communications field, the trend seemed to be in the opposite direction as companies added and merged new elements, rather than trimming down. Should—and could—AT&T go against the tide? Pending legislation played a part in AT&T's planning as it became clear by mid-1995 that major legal changes were coming (see chapter 9) that would further increase competition.

In September 1995 AT&T Chairman Robert Allen announced that his company would break up a second time, this time voluntarily.[45] The firm would divide into three independent parts of very different sizes:

- AT&T, which would retain the long distance, wireless service, online service, and research and consulting functions ($49 billion in annual revenues, 121,000 employees);
- Network Systems, an equipment entity which, along with Bell Labs, comprised the former Western Electric ($20 billion, 137,000 employees; not then named, it would later become Lucent Technologies); and

- Global Information Systems (formerly NCR), making bank and business computer systems ($8 billion, 43,000 employees).

Internally, this second breakup became known as "trivestiture." The AT&T core company would have some 80 percent of the pre-trivestiture profit and 60 percent of pre-trivestiture sales. Global Information Systems, the computer manufacturing unit, announced one immediate result of the breakup (though it likely would have occurred anyway given its financial problems): an immediate layoff of more than 8,500 personnel. Much more trimming of personnel was in the offing.

Lucent

At the time of the trivestiture announcement, AT&T's Network Systems arm was the third largest telecommunications equipment manufacturer in the world, outranked by Alcatel and Motorola, and in turn outranking major foreign suppliers including Siemens, Ericsson, NEC, and Northern Telecom. It held about a 13 percent share of the global $150 billion equipment market.[46] Just a year later, its more than $21 billion in revenue moved it to top place.[47] But Network Systems faced the loss of its largest customer: a guaranteed $2 billion a year in equipment purchases by AT&T. To ease the transition, AT&T agreed to purchase $3 billion in equipment over a three-year period after which Network Systems would be on its own competing with other suppliers. The spin-off was also allowed to sell consumer telephones using the AT&T brand name until the year 2000.[48] Balancing this loss was the likelihood of greater success selling to the RBOCs now that the unit would not be a part of AT&T. But looming over the unit was the fact that three-quarters of its market was in the United States where most carriers had already finished their conversion to digital switches, threatening a market slowdown for Network Systems.

Early in 1996 the unit was renamed Lucent Technologies.[49] In March 1996 AT&T fielded "the largest initial public offering in U.S. history" when it placed 111 million shares of Lucent Technologies on the market. Yet even this was only 17 percent of the business as the remaining equity would be spun off to AT&T shareholders in a process reminiscent of what had happened in the 1984 breakup.[50] The proceeds would be used to pay off part of Lucent's accumulated debt of just over $4 billion. Just six months into independent operation, Lucent seemed to be doing very well. Its share price nearly doubled in that period, from $27 to nearly $51, an indicator of investor confidence in the firm's direction. Revenue and profit increases both topped 10 percent. Freed from its AT&T conflicts, Lucent was able to exploit the growing world demand for equipment which amounted to nearly a quarter of its revenue in 1995.[51]

Bell Labs

In what for many was a wrenching psychological change, Bell Telephone Laboratories was also divided in the trivestiture, as it earlier had been with divestiture. But this time, little stayed with AT&T. Behind the shift was a fundamental change in Bell Labs that had been under way for years. The Labs became a highly respected research operation and pioneered work in the transistor, lasers, microwave radio, coaxial cable, lightwave communications, stereo sound, and many innovations, including a crude early form of television. In the two decades prior to the breakup, Bell Labs received some 400 patents a year and its scientists published nearly 2,000 research papers annually. By the time of the breakup, the Labs employed over 22,000 employees with an annual budget of some $2 billion. At divestiture the unit became AT&T Bell Laboratories, and some 3,100 employees and some facilities were turned over to the RBOCs to form part of the parallel Bellcore research operation.

In the dozen years from divestiture to trivestiture, AT&T Bell Labs continued to change to match the needs of the slimmer and more competitive AT&T. No longer able to rely on income drawn from a huge Bell System, the Labs increasingly focused on applied research, slowly diminishing their once-important fundamental research effort as being less central to AT&T's more immediate and changing needs.[52]

Long Distance Competition

One clear purpose of the MFJ was to encourage development of interexchange carrier (IXC) competition. Yet some argued that action of the FCC and the courts "as implemented, has constrained, not furthered, the development of competition."[53] Statistics for this 12-year period show a dramatic change in the interstate long distance marketplace. In December 1984, only 3 percent of the 112 million access lines in the country were capable of supporting equal access; just two years later[54] access lines had grown to over 158.7 million and virtually all were equipped for equal access.[55] The number of companies prepared to provide service to these equal access lines grew appreciably as well. In December 1987, 223 companies provided 1+ service; however, only 19 had a market share of even a half percent of the country's access lines. A decade later, the number of companies providing 1+ service had grown to 621, with 45 of those companies having at least a half percent market share.[56]

TABLE 7.1
Shares of Total Toll Service Revenues: 1985–2002

Year	AT&T	MCI	Sprint	RBOCs	All Other Companies
1985	67%	4%	2%	17%	10%
1990	51	11	8	16	15
1995	45	21	9	10	16
2000	35	21	8	9	27
2002	33	21	9	17	25

Source: FCC, Industry Analysis and Technology Division, Wireline Competition Bureau. See: http://www.fcc.gov/Bureaus/Common_Carrier/Reports/FCC-State_Link/IAD/trend 504.pdf, table 9-8, pp. 9–13. Data are rounded to nearest full number.

Despite competitive growth, AT&T continued to have a lion's share of pre-subscribed lines during this period. In December 1996, AT&T served 63 percent of the total pre-subscribed lines; MCI served 15 percent; Sprint served 7 percent; and WorldCom (later merged with MCI) served 3 percent.[57] AT&T continued to serve a majority of the residential market— AT&T's share of residential toll revenues was 61 percent in 1997.[58] Yet AT&T was serving a declining percentage of the interstate long distance market, as measured by access charge traffic. In 1987 AT&T carried 72 percent of the originating interstate access minutes and 88 percent of terminating interstate access minutes. A decade later AT&T's traffic represented only 51 and 48 percent of those same markets. Of total toll revenues, including intrastate and intraLATA toll revenues, AT&T's market share was 90 percent in 1984 (MCI's share was 5 percent; Sprint's was 3 percent; and all other carriers totaled 3 percent). By 1997, however, AT&T's market share had dropped to 45 percent while MCI was 19 percent; Sprint had grown to 10 percent; and WorldCom was 7 percent; and all other carriers had a market share totaling 20 percent.[59] Table 7.1 illustrates AT&T's declining market share through 2002.

7.3   CHANGING ECONOMIC POLICIES

Before 1984, and thus prior to the changes wrought by the MFJ, two important service boundaries defined the telephone industry: that between local and toll, and that between *intrastate* and *interstate*. The first determined whether calls would be considered local and billed on a monthly flat rate basis, or toll calls billed on a per-minute basis. The second determined which toll rates would apply: intrastate approved by state commissions, or interstate under the jurisdiction of the FCC. The MFJ created the new Local Access and Transport Area (LATA) boundary which became the defining one in the industry. LATAs partitioned the industry and established parameters of service for the newly created RBOCs. They also made clear the need for new ways of routing traffic, of serving subscribers, and of dividing up revenues among the newly created industry players. Yet, while the need to create new policies and procedures was evident, many firms sought to continue many arrangements of the past, especially those that had maintained low residential rates and assured toll service to sparsely populated and remote locations.

Equal Access

The MFJ assigned the local exchange portion of the former Bell System network to the newly created RBOCs. This meant that the RBOCs owned the facilities closest to the telephone subscriber. The RBOCs controlled the local loop and switch that linked subscribers to the public switched network, as well as the tandem switches and interexchange trunk facilities that linked switches within their LATA boundaries. The RBOCs would use these facilities to provide their own services, but AT&T and its competitors needed access to these same facilities in order to originate and terminate calls. In effect, an inter-LATA long distance call would originate over RBOC facilities, and the RBOC would then hand off the call to an IXC (AT&T or one of its competitors). This hand-off is illustrated in Fig. 7.1. Conversely, at the terminating end of an inter-LATA call, the IXC would hand off the call to the RBOC for delivery to the called party. To make this arrangement possible, each IXC had to negotiate a Point of Presence (POP) with each RBOC; the POP would be the physical point at which the RBOC and IXC facilities would interconnect so that the hand-offs could take place.[60]

AT&T held a huge competitive advantage in POP designations, of course, because it had designed the former integrated network. In most instances, AT&T's POP was in the same building as the RBOC's switch. Therefore, connections to AT&T usually consisted of a short span of cabling and some central office equipment. For Other Common Carriers (OCCs) such as MCI, however, the POP location was more problematic. They had to establish facilities, and many were some distance from the RBOCs' central offices. If the IXCs were to pay for their connection according to the actual facilities used, the OCCs would have paid more than AT&T, thus giving AT&T another competitive advantage. To "level the playing field," Judge Greene determined that OCCs should be billed on the same basis as AT&T, as long as the OCCs' POP locations were within five miles of an AT&T POP.[61] This was but one example of the detailed efforts made by Greene and the FCC to mitigate AT&T's advantages.

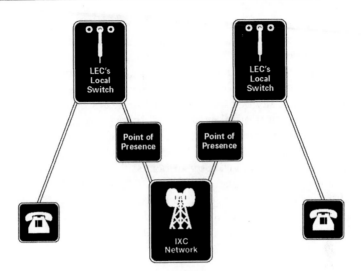

FIG. 7.1. An inter-LATA call—the hand-off from LEC to IXC.

Prior to divestiture, provision of a long distance call was a relatively seamless process. Subscribers dialed "1" to signify that they were placing a long distance call, and then dialed an area code and seven-digit number. The call was routed through the Bell System to the appropriate local switch. Dialing was simple; service was quick; subscribers received one bill for all services. With divestiture and the introduction of long distance competition, this simple arrangement could no longer be the norm. For long distance competition to succeed, there had to be *dialing parity*. In other words, a subscriber who dialed 1+ to place an inter-LATA call had to be handed off to the IXC network of his or her choice.[62]

The MFJ required that within about two years "the Operating Companies must provide access services to interexchange carriers and information service providers which are equal in type, quality, and price" to the access provided to AT&T.[63] Despite divestiture, the physical connection and routing of calls for subscribers dialing 1+ was virtually unchanged for AT&T subscribers; indeed during this period, any customers dialing 1+ for an inter-LATA call were routed to AT&T. Customers using another IXC had to dial additional digits, thus giving AT&T yet another significant competitive advantage. RBOCs were allowed to phase in upgrades to their switches to make equal access possible.[64] When equal access was deployed, subscribers could use any IXC by dialing 1+, just as AT&T subscribers did.

This dialing disparity was caused by the superior form of connection to the RBOC network enjoyed by AT&T. The RBOCs collected AT&T traffic and delivered it over trunk group facilities called a Feature Group C connection, as shown in Fig. 7.2. The call itself plus needed routing information was delivered to the AT&T POP, making calling quick and simple for the customer. MCI and the other competitors were initially connected to the RBOCs with a cruder and more complex Feature Group A link also shown in Fig. 7.2. Only one OCC customer at a time could use it and had to dial an identification number and only then the area code and seven-digit number..This clearly was not "parity." Intermediate Feature Group B provided toll-free access to the OCC network, which helped but still didn't equal what AT&T enjoyed. Subscribers still had to dial more than 20 digits to complete a call. In compensation, OCCs using these cruder Feature Group A or B connections paid 55 percent less than AT&T paid for its premium access. This created a trade-off of sorts, with AT&T having superior service but at a higher cost than the OCCs paid.

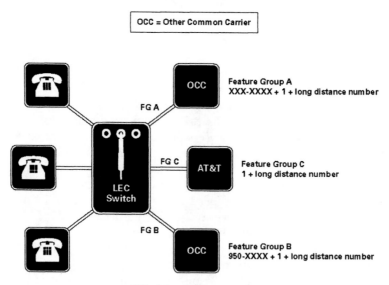

FIG. 7.2. Feature groups.

Equal access called for ending these differences. Once the RBOCs upgraded their switches, all IXCs would receive Feature Group D and would pay the same access charges. Feature Group D was a trunk connection that delivered more information than even AT&T's Feature Group C had provided. In order to achieve this, RBOCs had to route subscribers to the IXC carrier of their choice (see Fig. 7.3). As an added component, subscribers would be able to override their IXC carrier on a call-by-call basis by dialing 10XXX rather than simply dialing 1+. This was referred to as casual or dial-around calling.

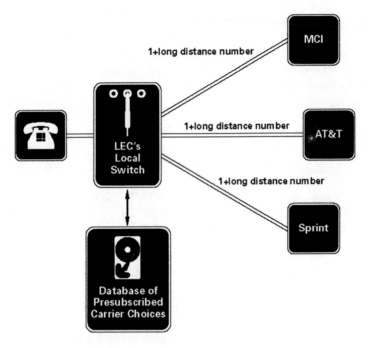

FIG. 7.3. Equal access—Feature Group D.

When an RBOC serving area had made the needed technical modifications, the RBOC was required to survey all existing subscribers to ascertain their presubscribed primary interexchange carrier (PIC) so that the customer would be routed to the appropriate IXC network. RBOCs had to be impartial in this presubscription process and subscribers could change their PIC at will, though changing PICs has remained a difficult process.[65]

Access Charges and NECA

The MFJ required that RBOCs be compensated for these interconnection services by filing tariffs

for the provision of exchange access including the provision of each BOC of exchange access for AT&T's interexchange telecommunications. Such tariffs shall provide unbundled schedules of charges for exchange access and shall not discriminate against any carrier or other customer. Such tariffs shall replace the division of revenues process.[66]

These had to be cost justified, with carriers paying only for services received. Tariffed "access" rates, applied in a nondiscriminatory manner to all IXCs, would form the basis of the new system. Prior to divestiture, the Bell System formed one big toll revenue pot from which all parts of the industry covered their expenses and generated a profit. Through the separations and settlements process, local companies had reported the interstate toll revenues they had billed and then based their claim to a portion of those revenues on the results of annual separations studies (*see section 4.2*). To replace this division of revenues process, the seven RBOCs and 1,400 independent telephone companies would have to file "access charges."

For the FCC, the prospect of dealing with 1,400 individual access tariff filings was daunting. To facilitate matters, the Commission created the National Exchange Carrier Association (NECA) to file access rates on behalf of the hundreds of small telephone companies that lacked the resources to file their own access tariffs and to administer the arrangements that replaced the former Bell System division of revenues process.[67] Soon NECA was managing its own version of a revenue pot. Each company reported to NECA their costs for supplying IXCs with access to loops, switches, and other facilities. NECA filed a set of access rates designed to recover all of the reported costs. Each company billed the access rates and then reported the resulting revenues to NECA. Each company then recovered its share of the revenues by filing cost studies with NECA. NECA, also, as will be discussed further, played a major role in implementing universal service policy in an increasingly competitive telecommunications marketplace.

Switched and Special Access

Access charges fell into two broad categories: switched and special access. Switched access charges covered the use of the local telephone company's facilities in the origination and termination of message toll service. When a subscriber originated a long distance call, the following local telephone company facilities were used:

- The *local loop* from the customer's premise to the telephone company switching office;
- The telephone company's *local switch*; and
- *Transport facilities* to carry the call to the designated IXC's POP; transport facilities could include a tandem switch and extensive interexchange trunk facilities.

Similar facilities were used at the terminating point of the call as well, and were provided by the local telephone company serving the customer receiving the long distance call. As shown in Fig. 7.4, the switched access charges corresponding to these three components of the local telephone company network were the following:

- The carrier common line charge (CCLC): billed per minute of use
- Local switching charge: also billed per minute of use; and
- Local transport: billed per minute of use multiplied by mileage.

Each of these access charges was billed at both the terminating and originating ends of a long distance call.[68]

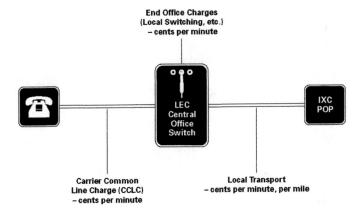

FIG. 7.4. Switched access charges.

Implementation of these access charges had several implications for the IXCs and for long distance competition as the charges were significant. They exceeded 10 cents per minute just for the CCLC, before local switching and transport charges were added. If long distance rates were to decline, access charges would have to decrease. IXCs shared the benefits of increased traffic volumes with local companies because access charges were billed on a per-minute basis. As inter-LATA long distance minutes grew, so did access charge minutes of use billed.

Development of special access charges reflected major developments both in service categories and underlying technologies. Prior to divestiture, dedicated lines provided by the Bell System to subscribers were analog and designated for specific use. The identified use reflected additional equipment required to "condition" the line to provide the desired service. The monthly charges reflected the equipment and preparation used. Initial special access rates reflected the former dedicated lines. With development of digital transmission technologies, however, special access rates and provision of dedicated services changed greatly. Now local companies provided digital lines that carried any service—voice, data, audio, or video. Issues involving digital facilities concerned units of bandwidth that local companies would provide for special access services.

High bandwidth digital circuits carried traffic between local central offices but were not initially available to special access customers. The local carriers enjoyed the benefits of the economies in transmission by continuing to bill customers at lower-bandwidth rates. For example, while the local company might use a high bandwidth circuit, it would charge for 24 lower-bandwidth facilities. Eventual inclusion of high bandwidth facilities for special access services was a significant development for long distance competition. Special access customers could now purchase bandwidth in bulk and then reap the benefits of the scale economies involved. Providing bandwidth in bulk also in effect underlined the wholesale character of access charges. Special access services raised two questions about service that were ultimately decided by arbitrary means. If a special access circuit terminated on a company's private branch exchange (PBX), a customer could connect to the switched network and make calls on the public switched network without incurring access charges, especially the CCLC. Because of this "leaky PBX" phenomenon, the public switched network could be used without making the appropriate contribution to local loop costs represented by the CCLC. Because it was impossible to determine how many minutes and how many CCLC charges were potentially involved, the FCC determined that special access customers leasing dedicated lines that terminated on a PBX would pay $25 per month as a contribution to local loop costs. The special access customer was expected to report circuits that terminated on its PBX equipment.

While jurisdiction over a long distance call was easily determined, the same could not be said for special access services because they were quite often multi-leg circuits involving many locations. Jurisdictional questions posed by such arrangements were difficult to unravel, especially since the charges for special access differed widely for state and interstate special access. The FCC determined that jurisdiction over a multi-leg special access network would be interstate if 10 percent of the usage was interstate. This presented challenges to local companies seeking to bill special access rates accurately as it required information about all of a multi-leg special access circuit arrangement, including the portions outside of any one local area.

Billing and collection functions were also considered an access service after implementation of the MFJ. Initial tariffs included these, and NECA administered a billing and collection pool as AT&T and the OCCs would have to rely on RBOCs and independent companies for billing services. Judge Greene sought to assure that billing would not confuse customers about who provided their IXC services, and required that bills from local companies had to identify billing provided on behalf of the IXC separately.[69] As equal access was rolled out, initial billing and collection tariffs were replaced with contractual arrangements between the IXCs and the local telephone companies. These included separate bill pages for each IXC, and payments to the local carriers to cover the costs of updating local telephone company billing systems to accommodate new billing plans and special offers crafted by the IXC, and for the treatment of billing inquiries and uncollectible accounts.

Impact on Universal Service

Divestiture had a major impact on universal service policy because a major component of that policy had been to keep local service rates low through subsidies from long distance rates. Whether subsidies could continue in an era of long distance competition was by no means clear.

Universal service and monopoly appeared to go hand-in-hand. Theodore Vail, the president of AT&T during the period of disorganized competition early in the 19th century (*see section 3.2*), seems to have been the first to use the term "universal service" in several AT&T annual reports. He focused on the phrase "one system, one policy, universal service" to argue for the end of "inefficient" competition, in which telephone service providers refused to interconnect their networks with one another, and to lobby for the creation of a system of monopoly in which one (Bell) system served all subscribers.[70]

Subsequently, the term came to mean the provision of affordable residential local service for all, although the origins of this interpretation of the term are not clear. Some argue that the term fell out of use until the Bell System, in the 1970s, made it part of a defensive strategy to ward off long distance competition by arguing for the importance of the long distance subsidy provided by the Bell monopoly to the realization of universal service.[71] Observers point to the Communication Act of 1934 as tacitly laying the groundwork for a universal service policy in its expressed desire for "reasonable charges."[72] While the 1934 act never used the term "universal service," it outlined that policy goal in explaining the purpose of the legislation as well as the creation of the FCC:

> For the purpose of regulating interstate and foreign commerce in communication by wire and radio so as to make available, so far as possible, to all the people of the United States a rapid, efficient, Nation-wide, and world-

wide wire and radio communication service with adequate facilities at reasonable charges.[73]

No matter which interpretation is applied to the idea's development, it is
evident that, for decades, policymakers, in regulating the pricing of telephone services, followed a policy that resulted in low residential local service rates and complicated cross-subsidies to keep those rates low, and so
affordable to the vast majority of subscribers.

Universal service policy was based on a complex pattern of telephone
service pricing that included subsidies from business to residential rates,
from urban to rural rates, and from long distance to local service rates.
Although the facilities providing local service to business and residential
customers were identical, business customers paid more than twice as
much. Regulators and service providers justified this pricing disparity by
noting that business customers made more local calls, thus putting more
pressure on the network. The urban-to-rural subsidy could be seen in the
pricing of urban and rural local service rates. Rural rates were often
lower than urban rates for local service, even though, per customer, it
was more expensive to provide service to rural customers.

Long distance services supported universal service through geographic
toll averaging and through subsidies to help defray the cost of local service. Long distance rates were based on distance, not traffic density on
specific routes. A 500-mile call originating from Chicago was rated the
same as a 500-mile call originating from North Platte, Nebraska, even
though the unit costs of providing toll service to high-density locations
like Chicago were lower. The impetus for this geographic averaging was
to assure that remote, low-population areas received affordable toll service; the effect was that those calling on high traffic routes were subsidizing those using lower traffic routes.

Before divestiture, AT&T's long distance rates subsidized local service
through the practice of "separations" (*see section 4.2*), in which the local
companies (the local Bell companies and the independents) allocated their
investments and expenses between the state and interstate jurisdictions.
Through the separations process the local companies allocated an increasing amount of state investment and expense to the interstate jurisdiction. AT&T Long Lines recouped these investments and expenses from
long distance callers instead of the local companies recovering them from
local service customers.

It was difficult to quantify these implicit subsidies, though some estimates place the urban-to-rural subsidy at $5 billion and the long dis-
tance-to-local subsidy at $20 billion annually.[74] The impact of ending
these subsidies was no less difficult to estimate, though one study

claimed that customers of rural telephone companies would see an average increase of nearly $31 per month in their local and toll bills.[75] The obvious message from all these estimates was that the end of subsidies would mark the end of universal service.

The state commissions could continue to require the newly divested RBOCs to continue the business-to-residential and the urban-to-rural subsidies. The future of the toll-to-local subsidy was in doubt, however, because it had been based on the Bell System monopoly. AT&T's Long Lines division had, through the separations process, subsidized residential local rates for the local Bells and the independents. Without that subsidy from AT&T or its long distance competitors, would local rates have to increase? AT&T, moreover, through geographic toll averaging, had provided affordable long distance rates to the remotest areas of the country. Would that continue after divestiture?

Accommodating Competition

The new access charge scheme would be used to maintain at least part of the toll-to-local subsidy and to encourage the continuation of geographic toll averaging. Non-traffic-sensitive (NSF) costs, which included local loops and a portion of the local switch, were the main component of local service. Local companies who had recovered a large percentage of these NSF costs from AT&T's Long Lines division through the separations process would now use that same separations process to allocate those NSF costs to interstate access charges, instead of raising local rates.

Exactly how this would be done proved to be highly controversial. Some policymakers argued that it was well and good to allocate NSF costs to the interstate jurisdiction, but that these costs should be recovered from the subscriber through a flat monthly charge, noting that since NSF costs didn't vary with traffic, they should not be based on usage. Others argued that these NSF costs should be recovered from long distance carriers through per-minute charges. Adopting a middle ground, the FCC devised a plan that would recover half of the local carriers' interstate NTS costs through a flat rate subscriber line charge (SLC) levied on each subscriber and half through a CCLC levied on the IXC.

After considerable debate, the FCC agreed to delay the SLC until 1985 and to cap it at $4 until 1990.[76] Business customers (defined as any with more than one telephone line) were charged $6 per line from the beginning of the plan. SLC charges for residential customers and for single-line businesses were gradually increased to $3.50 by early 1989.[77] With each SLC increase, the CCLC decreased. These decreases were partially passed along by the IXCs in the form of lower long distance charges which made the whole process more palatable to both subscribers and policymakers.

The SLC and CCLC did perpetuate some semblance of the toll-to-local subsidy; local telephone companies continued to recover a significant portion of their NTS costs from interstate services. This allowed them to avoid increasing rates for local service, though to many subscribers the monthly SLC appeared to be a local rate increase. Fearing that the SLC might force existing low-income subscribers off the network, thus impairing universal service, the FCC created yet another charge to be paid by interstate long distance carriers. Under a new "Lifeline" plan, the SLC would be waived and local service rates reduced for low-income subscribers meeting a state-specified means test. In addition, a new "LinkUp" plan would encourage new low-income subscribers by waiving local service installation fees (up to $30) and allowing deferred payments for the balance. Local companies would recover the foregone revenues from the waived SLC and installation charges from a NECA-administered fund. Long distance carriers paid monthly "Lifeline" rates into the fund, based on the number of long distance customers they served.[78]

The FCC sought to maintain geographic toll averaging by requiring all local telephone companies to charge the same nationally averaged CCLC, a charge calculated and filed by NECA. The Commission feared that company-specific CCLC rates would prove to be prohibitive in sparsely populated or remote areas, thus discouraging long distance carriers from serving those areas. If all local carriers charged the same CCLC, long distance carriers would be more likely to serve all areas. Long distance carriers objected to paying the CCLC from the beginning, arguing that NTS costs should be recovered solely from the SLC. The CCLC was a significant charge, with the initial CCLC exceeding five cents, at each end of a call. Even though the CCLC declined with each increase in the SLC, long distance carriers continued to complain about paying the charge.

Local telephone companies also began to chafe at the access charge regime, with the RBOCs and the larger independent companies wanting to set their own CCLC, fearing that long distance companies would seek ways to bypass local networks (especially when serving large businesses and when delivering terminating traffic), in order to avoid paying a high CCLC. Departing from a nationally averaged CCLC would impair geographic toll averaging, however. In 1988, the FCC finally agreed to end the requirement that all local companies charge the same NECA-filed CCLC. Local companies filing their own CCLC would, however, have to provide support payments to those local companies electing to stay with the NECA-filed rate. Because of these support payments, called Long Term Support, the companies who stayed with NECA could afford to charge a lower NECA-filed CCLC. To keep as much cost out of the NECA-filed CCLC rate as possible, while still protecting those local companies with especially high NTS costs, the Commission created yet another access charge to be paid by long distance carriers: high cost local companies would defray some of their NTS costs from a newly created Universal Service Fund (USF) that would be paid for by the long distance companies.[79] NECA administered the USF.

Access charges provide an interesting example of how policymakers attempted to foster competition on the one hand while maintaining universal service on the other. Introduction of competition and the ending of the division of revenue arrangement terminated the toll-to-local subsidies in place during the monopoly era. The FCC, however, extended some of the elements of the division of revenues into the access charge regime. Averaged rates and intercompany subsidies continued. The specific arrangements among companies were changed and the companies participating in the arrangements changed, but the underlying motives and goals remained the same. The policy objective of keeping a flow from toll services to local services in order to keep local rates relatively low continued.

Implementing Price Caps

Competition increased after the MFJ, as AT&T and the RBOCs faced competitive threats to their services, as well as participating in competitive, unregulated services themselves. This increasing complexity put pressure on existing regulation. The FCC and state commissions again faced the challenge of overseeing entities that engaged in both regulated and unregulated activities. Regulators had to assure that AT&T and the RBOCs were not using their regulated profits to subsidize their competitive undertakings.

Although the MFJ encouraged long distance competition, neither the FCC nor the MFJ had deregulated AT&T: the company's non-competitive services were still fully regulated. As long distance competition increased, and AT&T lost market dominance, it would eventually be deregulated, but in the meantime, the FCC had to determine the best way to regulate the company on a transitional basis. FCC economists became increasingly convinced that the traditional rate-of-return (RoR) regulation (*see section 4.2*) was not equal to the task. RoR was a broad-brush system that calculated a carrier's total revenue needs; it was not a method that lent itself very easily to developing service-specific costs. Without the latter it was difficult to determine whether a carrier was engaging in cost shifting.

As competition was appearing in the 1970s, the FCC required AT&T to develop a procedure for allocating costs among private line, WATS, and message toll services and to document those procedures in an *Interim Cost Allocation Manual* (ICAM). The major competitive threat to AT&T was in private line services, so the focus was on assuring that costs were not shifted out of private line and into other services.[80] A decade later AT&T faced competitive threats to all of its services. Something much more refined than the ICAM was needed to protect against cost shifting from services that faced significant competition to services that faced little competitive threat.

Instead of advocating more detailed cost allocation, FCC economists recommended an approach that would focus on regulating price movements. In a 1987 staff study, *Competition Policy in the Post-Equal Access Market*, two of them proposed that AT&T's market power be controlled by maintaining geographic toll averaging (so AT&T could not discriminate against areas with less competition), and by limiting or "capping" AT&T's ability to raise prices for a specific set of services that faced little competition. At the same time a Commission attorney found that such a plan would result in rates that were "just and reasonable" as required by the Communication Act. The Act did not require the use of RoR regulation; moreover, the proposal was a form of regulation, so that its implementation would not be tantamount to deregulating AT&T. The proposal caught the attention of Dennis Patrick (who became FCC chair in 1987), and he decided to make such price caps a top priority (see Appendix A for a more complete explanation of how price caps work).

Rather than presenting the price cap approach as a needed intermediate step on the way to deregulating an AT&T increasingly faced with competition—a strategy bound to meet huge opposition from competitors and others—the FCC decided to make a case that price caps provided greater efficiency and lower prices. Price cap regulation would allow AT&T to retain the fruits of its efficiency and innovation and to share those benefits with consumers. The emphasis on efficiency opened the door for price caps to be considered as an appropriate regulatory approach for the RBOCs and the other local exchange carriers (LECs) as well. LECs held a monopoly, and any regulatory method that encouraged them to be more efficient was seen as a plus. The FCC, recognizing that a price cap plan for the LECs would involve greater complexities, decided to implement price caps for AT&T first, and then address the LECs.

Even this approach delayed adoption of price caps for AT&T, in part because of Congressional resistance. The FCC adjusted its plan to assuage Congressional complaints. To protect competitors, price floors were put in place so that AT&T had to justify any rate decreases greater than 5 percent. To protect subscribers, AT&T's ability to raise residential rates was limited to 1 percent a year. The price cap plan for AT&T became effective in mid-1989; a final price cap plan for the LECs was adopted in September 1990.

There were several differences between the two plans. The plan for AT&T focused on retail rates—those charged to residential and business customers. AT&T's business services were under greater competitive pressure, making it important to assure that no cost shifting would take place between business and residential services. To avoid such shifting, the AT&T price cap plan created separate "baskets" of services, with a price cap calculated separately for each. The three baskets included one for ordinary long distance services used mostly by residential customers; another of "800" services over which AT&T still maintained monopoly power (because competitors did not yet have access to the necessary 800 service databases); and a final basket containing the services used mostly by big business customers and subject to competition. Prices of services within any basket could be lowered or raised so long as the net effect of these price movements did not exceed the assigned cap on that basket. Prices could be changed on services in one basket without having any effect on prices of services in the other baskets. In other words, if AT&T wished to lower prices of services for large business customers, it could not make up the difference by raising residential toll prices. The service baskets in the LEC plan consisted of its different categories of access services.

Perhaps the most difficult aspect of implementing price caps was determining the "productivity offset." FCC analyses suggested that the former Bell System had made productivity gains of 2 to 3 percent each year. The price cap system would encourage continued annual increases in productivity by subtracting an offset from the price cap each year and, by so doing, limiting how much the prices in a service basket could increase. There was a good deal of controversy about what that offset should be. The productivity figures for the post-divestiture industry were inconclusive, and it was difficult to extrapolate post-divestiture figures from pre-divestiture data. The FCC finally decided that AT&T's offset would be 3 percent. The LEC plan differed in its productivity offset, and also in its basic assumptions about limitations on earnings. While there were no limitations on the level of earnings AT&T could realize, the FCC placed limitations on LEC earnings and potential losses.

One danger of a price cap system was that the carrier might not be able to earn a sufficient amount to provide quality service or to even stay in business. That prospect was not an acceptable risk for the LECs, as they were the only providers of local network facilities to most of the country. As a result, the LEC plan could not divorce itself fully from RoR regulation. The LEC plan included a backstop provision: if a LEC's earnings fell below a specified level, the LEC could raise the price cap and so raise prices. The LEC price cap plan contained a higher productivity factor than that for AT&T, the assumption being that competition would provide

AT&T with the incentive to be productive. In the absence of competition, the LEC price cap plan would provide the incentive. LECs who wished to keep a greater percentage of the benefits of greater productivity were given the option of accepting a higher offset (4.3 percent) in return for which they were able to keep a higher RoR. In effect, the LEC price cap plan resulted in an increase in the LECs' allowed return. To earn that greater return, LECs were encouraged to cut costs and increase efficiency.

Price cap regulation for AT&T continued until AT&T was declared a nondominant carrier in 1995 and as such fully deregulated, making even price cap regulation unnecessary. Price cap regulation continued for LEC access charges; however, recent plans to lower access charges to specified levels have made price cap regulation of interstate access charges all but irrelevant. Price cap regulation continues to be used by several states to regulate local and state activities.

7.4 TO REGULATE OR NOT TO REGULATE

The period between the MFJ and the passage of the Telecommunications Act of 1996 saw new regulatory regimes and the continuation of ongoing regulatory controversies. On the one hand, the relationship between the newly formed RBOCs and the growing competitive IXC industry had to be defined, and the efficacy of traditional RoR regulation for the new industry structure had to be examined. On the other, there continued to be ongoing questions about how and where to draw the line between information services and telecommunications, and about the role of the newly formed RBOCs in the evolving information services marketplace. Regulators, facing a reorganized common carrier industry and a developing information services market, struggled with questions about how much and whom to regulate and sought to craft regulatory schemes that would encourage the development of new services while at the same time perpetuating some of the arrangements of the past.

FCC Computer Inquiries

Decades before the MFJ, the FCC began to struggle with the question of computing. With the growth of mainframe computers came the growth of applications combining computing with communication.[81] Telecommunications lines were being used to link terminals to mainframe computers for data entry and information retrieval purposes. As mainframe computers grew in size and capacity, companies shared their computing power with remote users through telecommunications facilities. Computing and telecommunications were being combined, and the FCC was increasingly concerned about its role in overseeing this development.

By 1966, the FCC "faced a fundamental choice: whether to regulate the burgeoning data processing field as an integral part of communications. If not, how was it to draw the line between regulated communications and unregulated data processing? And on what terms would common carriers be permitted to participate in unregulated data processing?"[82] To answer those questions, the FCC began its first Computer Inquiry, which resulted in a 1971 decision (Computer I). The Commission determined that data processing was competitive and therefore not in need of regulation. It also found that telephone companies could participate in data processing, but only under specific conditions. A policy of maximum separation was needed to prevent the telephone companies from using their monopoly in telecommunications to gain an unfair advantage in the competitive computer industry. Telephone companies,[83] except for the small independents, could provide data processing service only through totally separate affiliates. These affiliates would maintain separate books and separate officers and would not be able to use the parent company's name or logo; nor would the parent be able to purchase data processing services from its affiliate.

While Computer I resolved questions about whether data processing should be regulated and established the conditions under which the telephone companies would be able to participate, it did little to clarify exactly where the line fell between communication and data processing. The decision created four categories of services: (1) message switching, (2) data processing, (3) hybrid communication, and (4) hybrid data processing.[84] Message switching would be regulated because it was, for all intents and purposes, telecommunications. Data processing would not be regulated. The situation for the hybrid services was less clear, because they combined elements of both communication and data processing. The FCC determined that those hybrid services that contained only incidental communication elements would not be regulated, while those hybrid services that were primarily communication and only incidentally data processing would be regulated. The Commission would determine on an ad hoc basis which services fell into which categories.

Not surprisingly, the ad hoc approach of Computer I failed, especially because technological developments continually blurred the line between data processing and communication. It was no longer a simple question of telephone lines linking mainframe computers. The telephone network was itself becoming increasingly computerized. Mechanical switches were being replaced by digital switching systems capable of providing a host of services beyond simple message transmission. Drawing boundaries within such a network was increasingly difficult. The question of boundaries spread beyond the network to customer premise equipment (CPE). Instead of dumb terminals linked to mainframe computers, there was now distributed processing with smart terminals capable of performing computing functions and communicating with other computers. The status of computerized CPE became an issue with AT&T's introduction of its Dataspeed 40/4 smart terminal as part of its Dataphone Digital Service.[85] IBM objected, noting that the device was actually a computer, and as such not equipment AT&T should be able to provide. While the FCC approved the tariff, it did so pending further consideration of the rules governing CPE.

The FCC made the consideration of CPE part of its Computer Inquiry II proceeding, which began in 1976 and ended with the issuance of a decision in 1980 (Computer II). Although the original intent was to draw a distinction between telephone-type CPE and data processing equipment, the FCC decided to deregulate all CPE.[86] In making this decision, the FCC found that

> the deregulation of CPE fosters a regulatory scheme which separates the provision of regulated common carrier services from competitive activities that are independent of, but related to, the underlying utility service. In addition, the separation of CPE from common carrier offerings and its resulting deregulation will provide carriers the flexibility to compete in the marketplace on the same basis as any other equipment vendor.[87]

In examining the line between communication and data processing, the FCC decided to go even further in delineating those services that were to be competitively provided and those services that were to remain regulated common carriage.

Computer II also revisited the question of where to draw the line between communication and data processing. Realizing that it was impossible to draw such a clear line, the FCC decided instead to establish a distinction between "basic" and "enhanced" services. Basic services would continue to be regulated and would comprise the switching and transmission of a message. Anything beyond basic services would be enhanced services and thus deregulated. Enhanced services "acted on the format, content, code, protocol or similar aspects of the subscriber's transmitted information, or provided the subscriber additional, different, or restructured information, or involved subscriber interaction with stored information."[88] In other words, basic services would be pure common carriage; the carrier would merely transmit the message, doing nothing to the message itself. Everything else would be enhanced services; such services as voice mail, store-and-forward faxes, interactive voice response and protocol processing would be considered as enhanced services.[89]

As a result of Computer II, CPE and enhanced services were to be deregulated and therefore provided separately from regulated basic services. Computer II continued the maximum separations policy of Computer I, but only for AT&T and its affiliates.[90] The local telephone affiliates of the Bell System would provide only regulated basic services; CPE and enhanced services could only be offered by the Bell System through totally separate subsidiaries. After the divestiture of the Bell System, the FCC extended the separate subsidiary requirements of Computer II to the newly formed Bell Operating Companies.[91] GTE and the other independent telephone companies would be held to a lesser standard. Instead of maximum separation through separate subsidiaries, these companies would practice accounting separation by keeping separate books for their enhanced services.

Fearing that the state commissions would undo the provisions of Computer II by requiring the regulation of enhanced services at the state level, the FCC preempted state jurisdiction by claiming "ancillary jurisdiction" over enhanced services under Title I of the Communications Act. Having claimed jurisdiction over these services, the FCC then declared that it would forebear from regulating them.[92] The FCC's action was upheld on appeal. Despite all this effort, it soon became evident that the Computer II decision was doomed to be short lived. Its underlying assumptions were called in question by the development of new services that relied on a combination of basic and enhanced services, and by AT&T's arguments that the separate subsidiary requirement made it impossible to reap the benefits of scale economies in providing these new services. Beginning in 1981, AT&T began to request waivers of the Computer II provisions.[93]

In 1985, the FCC opened Computer Inquiry III (to the amusement of critics who argued this process would never cease) to once again address where to draw the line between regulated communication and unregulated information services. The number of waiver requests of the Computer II rules, as well as a concern that its structural separations requirements were inhibiting the introduction of new services, caused the FCC to begin to rethink its position. Indeed, as the network became more sophisticated and more dependent upon computerization, it was becoming more and more difficult for AT&T and the RBOCs to introduce new services that did not involve the integration of both communication and what Computer II would categorize as enhanced service. The FCC decided that structural separations had proven ineffective and that accounting separations could be adequate in preventing anticompetitive behaviors, so long as other protections were in place. Those protections were to be an arrangement called open network architecture (ONA).

Computer Inquiry III and ONA

ONA marked a curious chapter in the FCC's efforts to encourage competition in information services. The FCC had no clear roadmap for its ONA plan, relying instead on the industry to provide the format and details. The resulting plan was little used; few information service providers ordered ONA services. At times, it seemed that ONA was a solution looking for a problem. Still, ONA was a breakthrough of sorts. It outlined a methodology for unbundling the network for use by competitors and by those seeking to provide new services. That approach is certainly basic to the objectives of the 1996 Telecommunications Act. The ONA approach also highlighted the profound disagreement between the FCC and Judge Greene regarding the proper role for the RBOCs in information services.

The FCC undertook Computer III because it recognized that the growing information services industry often needed access to the Public Switched Telephone Network (PSTN), especially that portion which was owned and controlled by the RBOCs, in order to do business. The FCC also recognized that the RBOCs wished to provide information services, and could do so at a distinct advantage because of their control of the PSTN. Requiring that AT&T and the RBOCs compete through separate subsidiaries was proving inefficient and unworkable. The trick, then, was to create a mechanism that would allow AT&T and the RBOCs to offer information services without creating separate subsidiaries, while at the same time assuring that the information service providers received access to the PSTN on an equal basis. The Computer II rules had required AT&T and the RBOCs to treat their information services affiliates on an arm's length basis and to give them no preference over their competitors. Computer Inquiry III replaced structural with accounting separation.

Instead of creating a separate subsidiary through which to offer an information service, the RBOCs would now only have to account for that information service separately, as the independent telephone companies had been doing since Computer II. The accounting separation required was a bit more specific, however. Each company would develop a Cost Allocation Manual in which it would very specifically outline how, through its books of account, the company was separating the expenses and investment associated with its unregulated services from its regulated activities.

The other component of Computer III—the ONA provisions—were to assure information service providers nondiscriminatory access to the PSTN resources they needed. The concept behind ONA seemed simple. The RBOCs would break their networks into Basic Service Elements (BSEs), like call forwarding, which they would offer through tariffs to information service providers. To get access to these BSEs, however, information service providers would purchase Basic Serving Arrangements (BSAs) which were basically a standard connection to the PSTN, like a voice grade telephone line. Information service providers argued that, with BSAs, they were being asked to purchase more facilities than they actually needed. The Commission, however, realized that doing away with the BSA requirement would have unbundled the network further and necessitated collocation, or allowing a competitor to connect its equipment in the RBOC office, something the Commission was not yet prepared to require.[94]

Developing ONA plans was a confused process. No one seemed sure what the information service providers wanted, and the RBOCs offered widely different sets of BSAs and BSEs.[95] The FCC envisioned ONA as an ongoing process that would slowly open up the network to competitors. However, few BSAs or BSEs were sold because information service providers did not want or need them.[96]

Since the ONA development process was protracted, the FCC provided the RBOCs with the opportunity to offer information services immediately by filing Comparably Efficient Interconnection (CEI) plans, instead of waiting until they had full-blown ONA plans in place. CEI plans specified how an RBOC would unbundle its network to provide a specific service, providing access to those unbundled elements to its competitors. Once the FCC approved the CEI plan, the RBOC could offer that service without a separate subsidiary. ONA and CEI requirements were not extended to the independents, except for GTE in 1994.[97]

In Computer III, the FCC was also aware that AT&T and the RBOCs not only controlled the physical network but also controlled information about that network and its use. Computer III, therefore, also required disclosure to competitors of any new network services or changes to the network that would have an impact on the connection of enhanced services to the network.

The status of Computer III remained unclear after the passage of the Telecommunications Act of 1996. The Act did not affect these provisions, nor did the FCC rescind the ONA/CEI rules. The FCC argued that its ONA and CEI provisions were the only rules governing access to RBOC facilities by providers of intra-LATA information services. The FCC also has found that, because information providers are not defined as telecommunications carriers, the LECs are not required to provide them with the unbundled network element required by the 1996 Act, suggesting the need for the ONA provisions. Information service providers, however, need only become affiliated with a telecommunications carrier, or begin to offer telecommunications services, to receive the benefits of the unbundling provisions of the 1996 Act. The FCC opened proceedings to determine whether ONA is still necessary after the 1996 Act.[98]

Jurisdiction—Congress, the Courts, or the FCC?

The period following divestiture was often a confused and contention one. The MFJ had totally reconfigured the industry, and it was not always clear who had oversight authority. The FCC had not been a party to the MFJ, and the DOJ seemed to believe that the agreement made the FCC unnecessary.[99] At various times different players (the courts, the FCC, and even Congress) stepped forward claiming to have the authority to set the path for the industry. Nowhere was this confusion more evident than in the matter of the RBOCs' ability to offer information services.

The MFJ stated very clearly that the RBOCs were not to provide information services, which it defined as "the offering of capability for generating, acquiring, storing, transforming, processing, retrieving, utilizing, or making available information which may be conveyed via telecommunications."[100] This definition was similar, though not the same, as the definition of enhanced services crafted by the FCC in the Computer Inquiries. As Brock has pointed out, it was often

> unclear whether the divested BOCs could provide any enhanced services without violating the MFJ prohibition on the provision of information services. . . . The close relationship between restrictions on the BOC provision of enhanced services, administered by the FCC, created many opportunities for conflict between the district court and the FCC. Both claimed primacy and suggested the other should yield to avoid conflict, but neither had clear power to override the other.[101]

The tension between the Court and the FCC was exacerbated by the split in direction between the Greene Court and the FCC. While Judge Greene continued to insist that barring the RBOCs from information services was the only way to protect against anticompetitive behavior, the FCC, in the Computer Inquiries, became increasingly convinced that the development of new services would be enhanced by allowing the RBOCs to provide enhanced services, at first through a separate subsidiary and eventually on an integrated basis. As long as the MFJ's ban on information services was in force, the FCC could not fully implement the provisions of Computer III.

In a very short period, Judge Greene became virtually the only figure seeking to hold on to the ban on information services. The Department of Justice came out against the ban in its first triennial report on the MFJ in 1987.[102] The DOJ expressed support for Computer III and claimed that the FCC was better equipped to oversee RBOC activities in information services. Judge Greene angrily rejected the DOJ's recommendations, issuing a report that criticized the FCC and took issue with the Computer III approach. To Greene, the FCC was ineffective because it was pursuing deregulation, a course with which he did not at all agree. Even the Department of Commerce weighed in against Greene when the NTIA asked the FCC to clear up uncertainties by ruling that the RBOCs' provision of information services was in the public interest. Probably wisely, the Commission declined to challenge Greene's authority.

The RBOCs and the DOJ did challenge Judge Greene by appealing his decision on the information services ban. In 1990, the appeals court decided against Greene, finding that he had substituted his own judgment for that of the parties in the case. The RBOCs and the DOJ were the original parties to the agreement that resulted in the MFJ; if they were willing to change the provisions of the MFJ by lifting the ban, Greene had to accept their decision. The appeals court also noted that Greene, in using the public interest standard, had applied the wrong legal standard in arriving at his decision. The appeals court remanded the case to Greene with instructions that forced him, albeit unwillingly, to lift the information services ban in 1991. After seven years of uncertainty, the FCC would now oversee the RBOCs' involvement in information services. Attitudes toward the information services ban had changed dramatically, a change that has been explained as being "a widespread shift of concern from problems of controlling monopoly power . . . to problems of enhancing productivity and international competitiveness."[103]

The information services ban was not the only example of confusion regarding appropriate regulation and the proper regulator. The MFJ also provided for the creation of access charges to be paid by the IXCs to the RBOCs for the use of RBOC facilities. It was not clear who should be responsible for determining what these access charges should be. The Communication Act granted the FCC rate-setting authority for interstate traffic; however, the RBOCs would be levying these charges and their facilities were intrastate, suggesting that the state commissions should regulate access. Some in Congress felt federal legislation was necessary to establish access charges because they would have income distribution effects analogous to income tax changes.

The FCC was adamant in pressing its case and emerged as the leader in crafting the access charge regime.[104] State commissions, Congress, and Judge Greene were all concerned that the Commission's implementation of access charges would result in increases in local rates. The National Association of Regulatory Utility Commissioners (NARUC) went to court to challenged the FCC's right to establish access charges. Judge Greene attacked the access charge plan as undermining universal service. Members of Congress attempted to pass legislation that would continue the subsidy flow from toll revenues to local rates. NARUC was unsuccessful; Judge Greene lacked authority over rate making; and Congress stopped short of legislation. The FCC was able to formulate an access charge plan, but did so in a manner that cushioned the impact on local rates.

7.5   EXPORTING DEREGULATION

Growing information service availability forced substantial change in international telecommunications during the last 20 years of the 20th century. Until the early 1980s, the United States was one of a handful of countries in which telecommunications was privately owned. Elsewhere telecommunications was usually provided by a government-owned monopoly and was regarded as a utility, much like electricity or water. Two decades later, the situation had changed greatly. More than 100 countries had privatized telecommunications, selling government-owned monopolies to stockholders or to consortia often made up of foreign telecommunications companies. Some nations opened all areas of telecommunications to competition, from CPE to long distance, local service, and mobile telephony. They had come to see telecommunications as a strategic resource, an engine for economic growth and development, and an important element of international trade. The United States played a substantial role in facilitating this change, providing a domestic example and by pushing for privatization and competition through FCC actions, trade negotiations, and lobbying efforts in international organizations.[105]

Privatization and Liberalization

Two themes dominated international telecommunications after 1980: privatization and liberalization. *Privatization* involves ownership and control. Countries that privatize telecommunications shift those resources out of government ownership and control into the hands of private entities. In *liberalization*, countries replace monopoly with some measure of competition. While there has been a tendency to lump privatization and liberalization together, they do not necessarily have to be linked. Even the United States long had a private system without liberalization (the Bell System was privately owned, but functioned as a monopoly). Iceland and Luxembourg both have state-owned telecommunications providers, but also allow for competition in public switched network services; in effect, they have liberalization without privatization. In general countries have moved toward both privatization and liberalization.

The primary motivations for these shifts have been to upgrade a nation's telecommunications infrastructure and to expand the range and quality of available services. The Post Telephone and Telegraph (PTT) system, in which a government agency provided monopoly telecommunications services, failed to produce advanced, efficient telecommunications systems. In a PTT system, the most common way to upgrade telecommunications is to raise taxes, a prospect seldom seen as politically feasible. As one of many government agencies, PTT needs are often secondary to other concerns. Indeed, in many countries, PTT revenues are often used to subsidize other parts of government, instead of being plowed back into telecommunications infrastructure improvements. Increasingly, countries found PTT-controlled infrastructures insufficient to support expanding telecommunication needs.

Advances in computerization and telecommunications fueled growth of multinational corporations, which ran their global enterprises through intracompany networks which pressured countries to provide more advanced telecommunications infrastructures. This was especially evident to British prime minister Margaret Thatcher. Faced with the unattractive prospect of raising taxes in order to upgrade an increasingly deficient telecommunications infrastructure, the Thatcher government turned to both privatization and liberalization. In 1984 just over half of the former PTT, British Telecom, was privatized. At the same time, a new subsidiary of Cable and Wireless, Mercury, was created to compete with British Telecom. Subsequently, the government sold its remaining share of British Telecom and, in 1991, replaced the duopoly of British Telecom and Mercury with a policy supporting full competition.

Britain was the first of many nations that privatized and liberalized telecommunications after 1980. Countries have taken a variety of approaches to the process. Those with strong capital markets have sold all or portions of their former PTTs through stock offerings while many Third World nations, notably those in Latin America, have sold their PTTs to foreign investors, often consortia of telecommunications carriers from several countries. Other countries, especially those in Asia, have stopped short of selling their telecommunications provider, opting instead to sign management contracts with private entities to be responsible for day-to-day operations, or to make BOT (buy-operate-turn over) arrangements in which private entities agree to build a portion of the telecommunications infrastructure (perhaps a digital network), to operate the infrastructure, and to then turn ownership of the infrastructure over to the government after a specified period of time.[106]

Before the turn of the 21st century, 70 percent of the nations in the Americas had privatized their telecommunications providers, and 55 percent of European nations had done the same. Lesser developed nations were less quick to privatize, with only 46 percent of the nations in the Asia-Pacific regions, 33 percent of the Arab states, and 28 percent of African nations engaging in privatization efforts.[107] The extent of privatization by the most developed nations can be gauged by the membership of the Organisation for Economic Co-operation and Development (OECD).[108] By the end of 2002, only one OECD member nation (Turkey) was 100 percent state owned, while most OECD member nations pursued a course of mixed state and private ownership.[109] As shown in Fig. 7.5, 56 percent of 201 nations had privatized all or part of their former PTT.

Countries approached the introduction of competition in stages. Much as the United States had done, countries initially allowed competition only in value-added (information) services and in customer premise equipment. A European Union 1987 Green Paper regarding liberalization of telecommunications recommended that value-added services be provided competitively and that monopoly be limited to basic services like voice telephony. Many countries used mobile services as a way to introduce competition into the telecommunications marketplace, issuing multiple licenses, especially for digital services. Another European Union directive *required* competition in basic services by 1998. Less developed members were granted a delay to prepare for competition. Assessing OECD member states as a gauge toward liberalization, in 2001 only Hungary and Turkey still had monopolies in some of their telecommunications markets.[110] Europe appears to have been more aggressive in introducing competition. The spread of privatization and liberalization provided U.S. multinational companies with more advanced and efficient services and with new markets and new investment opportunities.

Market Flexibility

The United States is a dominant player in international telecommunications, originating and terminating more international traffic than any other nation. American policymakers have used this dominant position to persuade other countries to adopt more open policies toward the provision of telecommunication and information services.

During the 1990s the FCC capitalized on foreign carriers' desires to serve the American market, domestic carriers' desires to serve foreign markets, and the dominant role of the United States in generating international long distance traffic as it sought to encourage other countries to open their telecommunication markets to increased competition. The Commission focused on the accounting and settlement process, the issue of international resale, and the promise of decreased bureaucratic red tape to encourage market entry. The FCC argued its approach would assist "the protection of US customers from potential harm caused by instances of insufficient competition in the global telecommunications market."[111]

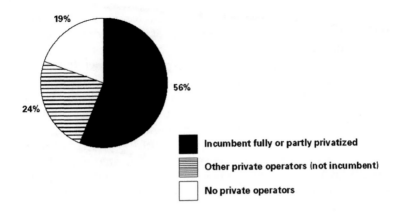

19%

56%

24%

- ■ Incumbent fully or partly privatized
- Other private operators (not incumbent)
- □ No private operators

Source: ITU *Policy and Strategy Trends*, 2002

FIG. 7.5.   Telecom privatization status of 201 countries, 2001.

A primary FCC concern has been the international accounting and settlement regime. This developed when international calling was rare and expensive and all of the service providers involved were monopolies (the Bell System in the United States and government-owned PTTs in other nations). The underlying premise is that there are two major parties involved in any international call: the carrier originating the call and the carrier terminating it. The originating carrier is compensated by the customer who makes the call while the terminating carrier is compensated by the originating carrier.

As an example, if an AT&T customer places a call to France, that customer pays AT&T. AT&T, in turn, is responsible for compensating the carrier (probably France Telecom) for terminating the call in France. AT&T would pay France Telecom a negotiated "accounting rate" that both firms have agreed represents the cost of carrying the call from the United States to France. Since this covers the whole call, the two carriers must decide on how to split the accounting rate (traditionally that split, the "settlement rate," has been 50/50). In this example if the accounting rate were negotiated to be 50 cents per minute and the settlement rate were negotiated to be 50/50, AT&T would pay France Telecom 25 cents per minute for terminating the call. Accounting rates are route specific.

This traditional regime is subject to some interesting problems. If there is symmetrical traffic on a route, accounting and settlement payments are even and there is no outflow of money from one carrier to another. If, however, traffic is asymmetrical (as is usually the case), a payment imbalance may result. The United States generates many calls, as immigrants call their home countries and international companies operating here place many calls. The imbalance in international calling was exacerbated by the introduction in the United States of domestic long distance competition, resulting in lower international charges and even more calling. Because most traffic originates from the United States, significant imbalances occur and substantial payments have been made by American carriers to those in other countries. Indeed, from 1985 through 2000 U.S. carriers paid out nearly $53 billion more than they received.[112] This imbalance reflects both traffic patterns and the accounting rates themselves. Many nations traditionally negotiated high accounting rates and used the settlement dollars generated as a means of funding national needs.[113] Such countries were less enthusiastic about decreasing the traffic imbalance either through competition or lower rates.

Alarmed by the settlement amounts paid by U.S. carriers, the FCC sought to increase competition (and thus lower international charges) in other countries, and to lower accounting rates to better reflect underlying costs. In a 1997 order, the FCC directed U.S. carriers to negotiate accounting and settlement rates at or below specific benchmarks tied to the economic situation of the receiving nation. As a result of the Commission's stepped approach, the average settlement rate fell from 35 cents in 1997 to an average of 14 cents in 2001.[114]

As an added incentive for foreign carriers to agree to all this, the FCC promised greater ease in entering the huge U.S. market. The price for this was increased competition in the foreign carrier's home country. A monopoly PTT could play one U.S. carrier off against another in order to negotiate a better deal, a process referred to as "whipsawing." To protect American carriers, the FCC required the same accounting rate on a route for all U.S. carriers, a settlement rate of 50/50 on all routes, and the rule of proportionate return specifying that the percentage of terminating traffic a carrier received on a route would be determined by the percentage of originating traffic that the carrier generated on that route. Hoping to encourage competition in other countries, the FCC later granted carriers greater flexibility in their negotiations, so long as the U.S. carrier was negotiating with a foreign counterpart that met what the FCC called an "effective competitive opportunities" (ECO) test. This gauged whether a U.S. carrier in a foreign country would have opportunities similar to those accorded foreign carriers in the American market.[115]

The FCC has also encouraged International Simple Resale as a way to reduce the amount of traffic that is subject to the accounting and settlement process. In resale, a service provider leases private lines that are connected to the public switched network and then "resells" traffic on those private lines. Resale has proven to be an effective method for introducing competition, because it does not require a service provider to build facilities. Hoping to encourage resale, the FCC has designated nearly 90 countries to which U.S.-based carriers may provide resale service. The Commission notes that "U.S. carriers can now send 46 percent of all U.S. international traffic outside of the traditional settlements system, thereby benefitting customers by reducing prices, encouraging greater service options, stimulating demand and spurring technological innovation."[116]

Opening Markets

There are several bureaucratic steps and some restrictions involved in entering the U.S. telecommunications market. Those wishing to land a cable must obtain a Cable Landing License. Those wishing to provide either facilities-based or resale services must file for authorization under Section 214 of the Communications Act of 1934. The Communication Act restricts ownership of broadcast or common carrier radio licenses by companies with more than 25 percent foreign ownership. The FCC has expedited the bureaucratic process and lifted the foreign ownership restrictions for carriers whose countries have met the ECO test or have joined the WTO. In the case of carriers whose countries have done neither, and in the case of arrangements between U.S. carriers and foreign carriers having market power in their countries, the FCC continues to require traffic reports, full-blown Section 214 filings, and tariffs as a way to monitor possible abuses of market power.

U.S. policymakers have also encouraged the spread of privatization and liberalization in satellite communication. In the 1960s the United States spearheaded the effort to create a satellite system to be used for commercial communication. This culminated in the 1964 creation of Intelsat, a satellite system owned by its members; these members were a government-designated signatory (Comsat) in the United States and the government-owned PTTs of other nations. Intelsat enjoyed monopoly status in global satellite communication, with only regional satellite systems receiving authorization into the 1980s. In 1984 the Reagan administration argued in favor of competition in satellite systems; in 1985, the FCC, in its separate systems decision, found that separate systems that did not connect to the public switched network and that had limited capacity would not be harmful to Intelsat. Intelsat was, for all practical purposes, a government-owned entity, with Intelsat shares being held by government-owned PTTs and decisions about Intelsat operations being made by representatives of the governments of the member nations. Intelsat's governing Assembly of Parties voted for privatization in 1999, and Intelsat became a private and commercial operation in 2001.

Global Investments, Mergers, and Joint Ventures

It is perhaps not surprising that, as barriers to market entry fell in telecommunications markets around the world, a torrent of mergers, acquisitions, investments, and joint ventures flooded international telecommunications. According to the OECD, the largest 20 mergers and acquisitions involving telecommunications services between 1990 and 2000 totaled $440 billion in value.[117] The increase in global enterprises fueled demand for simplification in acquiring telecommunication services across national borders. Instead of dealing with different, and sometimes conflicting, rules about service provision, varied service pricing, and differing quality-of-service standards, telecommunications managers of global enterprises sought uniformity in services, rates, and rules. Global joint ventures emerged as one way to meet this demand.

Privatization efforts provided opportunities for acquisitions, investments, and joint ventures, as once government-owned enterprises created a profit-oriented organization. In many instances, governments turned to foreign investors to provide the monetary resources and the expertise required. Joint ventures provided a way to spread the risk involved in financing privatization moves or the introduction of new technologies like wireless systems. In many cases, foreign ownership restrictions made joint ventures necessary, as companies native to the country teamed with foreign investors who brought resources and expertise to the enterprise.[118]

U.S. companies played a significant role in all of these activities. U.S. carriers engaged in almost 100 joint ventures in Europe alone.[119] According to the OECD, from 1990 to 2000, U.S. firms were involved in 1,100 cross-border mergers or joint ventures having to do with information, computing, or telecommunications services and equipment, and was the principal nation in these alliances 72 percent of the time.[120] At the same time, foreign companies sought to invest in the U.S. telecommunications market. For example, in 1999 British wireless company Vodaphone purchased Airtouch Communications for $62 million. Two years earlier British Telecom, the dominant British carrier sought to purchase MCI, a deal that appeared likely to meet with FCC approval, but which was stopped by WorldCom's successful takeover bid for MCI (*see section 9.6*).

Global mergers, acquisitions, and joint ventures provided U.S. carriers with investment opportunities and with opportunities to gain experience in markets that were closed to them in the United States. The RBOCs, for example, were able to enter information services and entertainment markets outside of the United States when still precluded from doing so within their operating territories. These global undertakings also provided financing and expertise for countries seeking to upgrade their telecommunications infrastructures, to privatize their networks, and to move toward a competitive market structure. The merger and joint venture mania of the 1990s abated by the early 2000s because of weakening national economies and telecommunications markets (see chapter 10).[121]

The ability of global alliances to survive and prosper was also called into question with the demise of Global One, the venture involving Deutsche Telekom, France Telecom, and Sprint. Begun in 1996 as a way for each of the three carriers to extend their global markets, Global One had lost $2.7 billion by 1999. Failure of the three participants to resolve differences in management strategy was cited as a major cause for the venture's failure.[122] In January 2000, France Telecom bought out the interests of its former partners. Another large global alliance, Concert, a joint venture between British Telecom and AT&T, ended in April 2002. The downturn in the telecommunications market, financial problems facing both carriers, and problems with Concert competing for its parent companies' customers were cited as reasons for the venture's collapse.[123]

Trade Issues and the WTO

Trade negotiations have been a major force for privatization and competition in global telecommunications. Countries that participate in trade negotiations are, almost by definition, agreeing to the principles of open markets. Before the 1990s, trade negotiations focused primarily on goods, instead of services like telecommunications. In the new information-based economy, services like consulting, banking, tourism, and telecommunications were crossing national borders and becoming important elements of international trade. In the mid-1990s, the international trading system recognized this development and brought telecommunications under the trade umbrella.

Twenty-three nations signed the General Agreement on Tariffs and Trade (GATT) in 1947, agreeing to the basic idea that trade barriers should be eliminated and markets opened. GATT agreements resulted from rounds of multilateral negotiations that usually lasted three to five years, and dealt with trade in such goods as agricultural or manufactured products. The eighth (or Uruguay) round of negotiations in 1986–94 formed the World Trade Organization (WTO) to implement the GATT provisions. It was also during this round of negotiations that services were added to the trading system.

The underlying principles of the GATT, and now the WTO, are that the free market should govern transactions, that all countries should be treated in a nondiscriminatory manner (the "Most Favored Nation" principle), that foreign companies be treated the same as domestic firms, and that the rules of doing business be clear and available to all. These same principles were applied to services under the General Agreement on Trade in Services (GATS). The GATS framework does allow specific, time-bound exemptions to the Most Favored Nation principle, and takes a "positive list" approach. Countries list the industries and activities for which they are willing to make trade commitments and only those are then subject to GATS obligations like transparency and the national treatment principle.[124]

The creators of this framework realized that certain services, because they were becoming such important elements of international commerce, merited special attention through the drafting of specific rules, or annexes. Telecommunications was identified as such a service and rules called for nondiscriminatory access to public telecommunications, cost-based pricing, the ability to lease private lines and to connect them to the public network, and other provisions that, at least in principle, obligate countries adhering to the GATS to allow a certain level of access to their telecommunications infrastructure. After a difficult start and extensive negotiations,[125] nearly 70 governments (whose markets represented 91 percent of global telecommunications revenues) agreed to liberalize both basic telecommunications and value-added services.

It is easy to overestimate the effectiveness of this agreement in fully opening telecommunications markets. Not all countries made commitments in all areas of telecommunications services, as many did not reduce their foreign ownership restrictions, and few developing nations have been enthusiastic participants. The agreement has loopholes that make it possible for countries to avoid fully acting on their commitments.[126] Other commentators argue that the agreement provides a useful regulatory code for developing nations.[127] The United States has been an active participant in the WTO process, urging greater opportunities for foreign investment and more significant commitments for opening basic telecommunications services to competition from foreign, as well as domestic, firms. The WTO agreement provides a framework for a global marketplace in which companies can cross national boundaries in order to invest in telecommunications enterprises or services.

NOTES

1. For a useful, albeit biased, view of these changes and how they argued for releasing the RBOCs from the line-of-business restrictions, see Richard S. Higgins and Paul H. Rubin, eds. (1995). *Deregulating Telecommunications: The Baby Bells' Case for Competition* (New York: John Wiley).
2. Modification of Final Judgment, 1982. *United States v. AT&T*, 552 F. Supp. 131, 228, Section VIII(C), (D.C. Circuit). Hereafter cited as *MFJ Opinion*.

3. Christopher H. Sterling and Jill F. Kasle, eds. 1988. *Decision to Divest IV: The First Review* (Washington, DC: Communications Press), used this very filing as an example. *See* pp. 2061–2084. Judge Greene granted this request on May 25, 1984.

4. Ibid., p. I-9. (I = introduction)

5. As Judge Greene noted in his opinion of July 26, 1984, these were requests by the operating companies to enter *new* lines of business. Not included in the list were the requests of the companies for the establishment of "transitional rules with respect to existing network arrangements, the grandfathering of existing services such as time and weather, and the adaptation of exchange area boundaries of the needs of BOC public mobile radio services." *See US v. Western Electric Co.*, 1984. Memorandum of the United States Concerning Removal of Line of Business Restrictions, p. 2, fn. marked with an asterisk. In other words, the BOCs had been busy sending petitions to the judge on routine matters; what was worrisome now was the BOCs' collective move into uncharted territory.

6. Sterling and Kasle, p. I-9.

7. At an April 1994 hearing on several divestiture-related matters, a peevish Judge Greene asked Howard Trienens, AT&T's general counsel, if Trienens had expected the divested companies to diversify so fast. Trienens emphatically replied that he had not. *See US v. Western Electric Co., Opinion* of July 26, 1984, p. 23, fn. 44, as reprinted in Sterling et al., *Decision to Divest* (1985), Vol. 3, p. 1926.

8. Ibid., p. 2, fn. 2; Sterling et al., p. 1005.

9. Ibid., p. 24; Sterling et al., p. 1927.

10. Ibid., pp. 5–6; Sterling et al., pp. 1908–1909.

11. Ibid., pp. 53–55; Sterling et al., pp. 1956–1958.

12. Ibid., p. 55; Sterling et al., p. 1958.

13. Ibid., p. 56; Sterling et al., p. 1959.

14. Ibid., p. 60; Sterling et al., p. 1963.

15. The judge was not shy about expressing his peppery opinions on the integrity of the operating companies. As he wrote (Ibid., p. 38; Sterling et al., p. 1941): "It may confidently be predicted that, even if the Regional Holding Companies could, somehow, reap significant profits from their outside ventures, they would not use them to benefit their regulated telephone affiliates. In fact, the opposite appears to be true."

16. Ibid., pp. 63–64; Sterling et al., pp. 1966–1967.

17. For example, in a brief filed on March 23, 1984, Ameritech noted that provision of local telephone service had become increasingly risky as a result of uncertainties in the regulatory process and advances in technology that made it more economical for customers to provide their own telephone service instead of purchasing the service from the operating companies. Sterling and Kasle, p. I-10, fn. 5.

18. *US v. Western Electric*, 797 F.2d 1082 (D.C. Cir. 1986), as reprinted in Sterling and Kasle, pp. 2085–2095.

19. As described in the appellate opinion, shared tenant services "essentially involve bulk purchase of long distance service from inter-exchange carriers and discount resale of this service to groups of office building tenants." In the case, Ameritech did not propose to sell a complete package of shared tenant services, but rather only certain elements of such services. Ibid., at p. 1082; Sterling and Kasle, p. 2088.

20. Specifically, US West argued that the application of the consent decree's restrictions to the Regional Bell Operating Companies offends due process because the RBOCs were not parties to the underlying antitrust proceeding, did not have independent legal representation, and never actually consented to the decree. Additionally, Ameritech asserted that even if the application of the consent decree to the RBOCs did not offend due process, the

RBOCs were nonetheless free from the decree's restrictions because the terms of the decree were directed only to the operating companies, not to the regional holding companies. Ibid., at 1087; Sterling and Kasle, p. 2089.

21. Ibid., at 1085.

22. Ibid., at 1091.

23. In some cases the company's protests had an effect on the operating company's request. In September 1984, for example, US West sought permission to offer financial services and engage in financing transactions on a nationwide, commercial basis. The Department of Justice approved the petition, but AT&T contested US West's provision of such services to telecommunications equipment manufacturers. Ultimately, US West was permitted to enter the financial services business, but only subject to the limitations proposed by AT&T. Sterling and Kasle, p. I-10.

24. Former presidential advisor Edwin Meese, who had been against pursuing the case when Reagan first came into office in 1981, was named Attorney General in 1985.

25. Certain aspects of the meetings between the Attorney General and officials of the Regional Bell Operating Companies could have had unpleasant consequences for the nation's highest law enforcement official. Like so many others, Meese and his wife had been AT&T stockholders and at the time of the meetings the Attorney General and his wife held 17 shares of stock in each of the seven Regional Bell Operating Companies, an investment worth more than $10,000. Although the stock was physically in the possession of Meese's financial advisers, the Attorney General's actions seemingly violated federal conflict of interest rules; at the very least, the meetings between the Attorney General and RBOC officials to devise a strategy to get the companies out from under the MFJ—and presumably enhance the companies' balance sheet and the stockholders' investment—had the appearance of impropriety. Material on the meetings was turned over to a special prosecutor investigating several of Meese's actions in office, but there was no result.

26. Peter Huber received a B.S., M.S. and Ph.D. in mechanical engineering all from MIT, the doctorate granted in 1976. He served on MIT's faculty from 1976 to 1982, the year he received a J.D. (summa cum laude) from Harvard Law School. He then clerked for Ruth Bader Ginsberg of the U.S. Court of Appeals for the District of Columbia (and later a member of the Supreme Court), and Sandra Day O'Connor of the Supreme Court. He then practiced law and consulting in a variety of fields. See U.S. Department of Justice, 1987. *Report and Recommendations of the United States Concerning the Line of Business Restrictions Imposed on the Bell Operating Companies by the Modification of Final Judgment,* Civil Action No. 82-0192 (February 2), (as reprinted in Sterling and Kasle) at fn. 80 (p. 37). Hereafter cited as "DoJ Report."

27. U.S. Department of Justice, 1987. [Huber, Peter]. *The Geodesic Network—1987 Report on Competition in the Telephone Industry,* p. 1.34. Hereafter cited as "Huber Report."

28. Huber worked with two attorneys, two economists (one from the FCC), and a paralegal. See DoJ Report, p. 38 (February 2).

29. The report was typed (instead of type-set) and bound in a bright yellow cover. And the press of speed prompted a failure of editing here and there. For example, footnote 62 in chapter 2 and footnote 105 in chapter 3 each offer a long, anecdotal account of the trouble visited upon a large corporate customer of AT&T by the local operating company. The text of both is identical.

30. Huber Report, p. 1.2.

31. Ibid., p. 1.3.

32. Ibid.

33. Ibid., p. 1.10.

34. Ibid., p. 1.6.

35. Ibid., p. 1.21.

36. DoJ Report, p. 158.

37. Ibid., pp. 112–113.

38. Ibid., p. 154.

39. Ibid., p. 156.

40. "In Step Designed to Place Conflict Between Greene's Jurisdiction and FCC's Before Appeals Court, NTIA Asks for FCC Declaratory Ruling . . . ," 1987. *Telecommunications Reports* (November 30), p. 3.

41. "Petition for Declaratory Ruling of the National Telecommunications and Information Administration" (November 24, 1987), pp. XX. Reprinted in Sterling and Kasle, eds. *Decision to Divest* IV, pp. 2270–2803.

42. *Opinion* [on Meaning of "Manufacture" and on Enforcement], December 3, 1987, p. 33; reprinted in Sterling and Kasle, p. 2662.

43. *United States v. AT&T Corp. and McCaw Cellular Communications, Inc.* Civil Action 94-01555, U.S. District Ct. for the District of Columbia (July 15, 1994).

44. Ibid., pp. 84–85.

45. That the story was important beyond the United States is evident in the play given to the news in most London newspapers. Coauthor Sterling was in the British capital to speak on changes in American telecommunications and had to scramble for details of the announcement.

46. "AT&T's Big Switch: A Tool to Compete?" 1995. *New York Times* (September 22), p. D1.

47. "How Glowing Is Lucent's Future?" 1996. *Business Week* (March 25), p. 42.

48. "A New Name in Telephone Gear," 1996. *Fortune* (April 29), p. 126.

49. "AT&T Names a Unit It Plans to Spin Off," 1996. *New York Times* (February 6), p. D-6, citing *Bloomberg Business News*.

50. "How Glowing Is Lucent's Future?" 1996. *Business Week* (March 25), p. 42.

51. "Lucent: A Sassy Spinoff Enjoys a Lot of Luck," 1997. *New York Times* (April 13), p. 3:1, 3:6.

52. For useful background, see A. Michael Noll, 1987. "Bell System R&D Activities: The Impact of Divestiture," *Telecommunications Policy* 11:2:161–178 (June).

53. Paul W. MacAvoy, 1996. *The Failure of Antitrust and Regulation to Establish Competition in Long-Distance Telephone Services* (Cambridge, MA and Washington, DC: MIT Press/AEI Press), p. 6.

54. The last time the National Exchange Carrier Association collected that information.

55. Federal Communications Commission, 1996–97. *Statistics of Communications Common Carriers*, p. 320.

56. Ibid., p. 327.

57. James Zolnierek, Katie Rangos, and James Eisner, 1998. "Long Distance Market Shares: Second Quarter 1998," FCC Common Carrier Bureau, Industry Analysis Division (September), p. 9.

58. Ibid., p. 23.

59. Ibid., pp. 10–16.

60. This was also sometimes called a Point of Interface (POI).

61. The MFJ language in effect equated the AT&T point of presence (POP) with its Class 4 switch. See *MFJ Opinion*, IV, F.

62. It is important to remember that "1+" dialing applied to *inter*-LATA calls only at this point. When a customer dialed 1+ for an *intra*-LATA call, the call was handled by the BOC serving that LATA.

63. *MFJ Opinion.*

64. *MFJ Opinion* at 197, and also Appendix I. A.1. Equal access requirements were extended to the independent telephone companies as well by the FCC (see MTS and WATS Market Structure, Phase III, Report and Order, 100 F.C.C.2d 860), so, although this section speaks of the RBOCs in providing equal access, it is important to remember that the independent telephone companies were required to provide the same services to the IXCs.

65. Many subscribers have had their PIC changed without their knowledge, a practice called slamming. The FCC reported slamming was the largest category of complaints involving telephone service received by the Commission; indeed in 1998, the FCC handled over 20,000 slamming complaints and proposed fines resulting from slamming totaling $13 million. The FCC adopted anti-slamming rules that year. "FCC Adopts New Anti-slamming Rules and Unveils Further Measures to Protect Consumers from Phone Fraud; Slammed Consumers Relieved From Paying Phone Charges," 1998. *FCC News* (December 17). The docket which resulted in the anti-slamming rules was CC Docket No. 94-129.

66. *MFJ,* II. B.1.

67. See 47 CFR Section 69.601 ff.

68. These three access charge elements were the major components of switched access. The RBOCs did add other charges, including a charge for the use of the RBOCs' directory assistance operators, and a charge to help the RBOCs recoup their costs of implementing equal access.

69. *MFJ Opinion,* at 232.

70. Milton Mueller, 1997. *Universal Service: Competition, Interconnection, and Monopoly in the Making of the American Telephone System* (Cambridge: MIT Press), pp. 4–5.

71. Ibid., pp. 250ff.

72. See for example, Martin Blizinsky and Jorge Reina Schement, 1999. "Rethinking Universal Service: What's on the Menu?" in *Making Universal Service Policy: Enhancing the Process Through Multidisciplinary Evaluation,* ed. Barbara A. Cherry, Steven S. Wildman, and Allen S. Hammond, IV (Lawrence Erlbaum Associates), pp. 69–83. The authors speak of "the goal of universal service adopted in the 1934 Communication Act," p. 70.

73. 47 U.S.C. Section 151. The Telecommunications Act of 1996 kept this language, only adding the words "without discrimination on the basis of race, color, religion, national origin, or sex."

74. See page 115 in Carol Weinhaus et al., 1999. "Overview of Universal Service" in *Making Universal Service Policy: Enhancing the Process through Multi-disciplinary Evaluation* (Mahwah, NJ: Lawrence Erlbaum Associates), pp. 111–134.

75. *Keeping Rural America Connected: Costs and Rates in the Competitive Era,* 1994. (Washington: Organization for the Protection and Advancement of Small Telephone Companies) p. E5.

76. Brock (1994), pp. 201–203.

77. The charge remained $3.50 until FCC activity to restructure access charges in the wake of the Telecommunications Act of 1996. This is more fully discussed in section 9.5.

78. Presubscribed lines are those local telephone lines for which the subscriber has selected a specific long distance carrier as their preferred long distance provider. In other words, when a subscriber decides that all of his or her 1+ long distance calls should go to AT&T, AT&T is that subscriber's PIC (Presubscribed Interexchange Carrier) and the subscriber's line is presubscribed to AT&T.

79. At the same time, however, the commission also created a new subsidy flow to maintain universal service. Local companies electing to file their own CCLC, would provide subsidy payments to the high cost local companies so they could continue to charge a

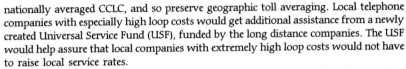

nationally averaged CCLC, and so preserve geographic toll averaging. Local telephone companies with especially high loop costs would get additional assistance from a newly created Universal Service Fund (USF), funded by the long distance companies. The USF would help assure that local companies with extremely high loop costs would not have to raise local service rates.

80. Brock (1994), p. 262. Much of this discussion regarding price caps relies on his excellent and thorough discussion on pp. 257–286—he was an active participant in these developments.

81. Kevin G. Wilson, 2000. *Deregulating Telecommunications: U.S. and Canadian Telecommunications, 1840–1997* (Lanham, MD: Rowan & Littlefield), p. 152. Wilson provides a clear explanation of the technical issues underlying the Computer Inquiries.

82. Huber et al., 1999. *Federal Telecommunications Law*, Second Edition (Gaithersburg, MD: Aspen Law & Business), pp. 1089–90. Huber, Kellogg, and Thorne provide an extensive overview of the FCC's proceedings in the Computer Inquiries and the various legal challenges to the FCC's decisions.

83. As Wilson (p. 155) notes, this policy did not apply to AT&T and its subsidiaries because the 1956 Consent Decree limited AT&T to regulated services only. The maximum separation policy applied, therefore, to GTE and the other larger independents, the so-called Tier I carriers, those telephone companies with annual regulated revenues in excess of $100 million.

84. Huber et al., p. 1090.

85. Brock (1994), p. 95.

86. Ibid., pp. 94–95.

87. 77 FCC 2d 384 (1980) at 446, 447 as quoted in Brock, 1994, p. 95.

88. Computer II, 77 FCC 2d at 387.

89. Huber et al. list more examples of enhanced services. See page 1092, n. 59.

90. Brock (1994, p. 96) notes that this time AT&T's ability to offer non-regulated services was still not clear because of the 1956 Consent Decree. However, in Computer Inquiry II, the FCC interpreted the Decree to say that AT&T could provide non-regulated services through a separate subsidiary.

91. Huber et al., p. 1121.

92. Ibid., p. 1093. As Huber, Kellogg, and Thorne explain, Section 152 of the Communications Act grants the FCC jurisdiction over interstate communication by wire or radio, while Section 153 defines communication by wire as "the transmission of writing, signs, signals, pictures, and sounds of all kinds . . . incidental to such transmission." The FCC found enhanced services to be such incidental transmissions over the interstate network.

93. Ibid., pp. 1126–1129. First, AT&T asked, unsuccessfully, for permission to offer custom calling features without a separate subsidiary, noting that it had begun to install those services on an unseparated basis before Computer II. AT&T was more successful in obtaining a waiver to provide protocol conversion services on an unseparated basis. Protocol conversion was classified as an enhanced service, though some argued that the translation necessary to allow machines on different systems to communicate with one another should have been considered a basic service. Rather than changing the categorization of protocol conversion to a basic service, the FCC granted AT&T a waiver of the separate subsidiary requirement. After divestiture, the newly created Bell Operating Companies sought protocol conversion waivers as well, which the FCC granted.

94. Brock (1994), p. 227.

95. Wilson, p. 164. Bell Atlantic offered five BSAs; Ameritech's ONA plan had four; and NYNEX listed fourteen.

96. Huber et al., p. 433.

97. Ibid., pp. 440–441.

98. Ibid., pp. 446–447.

99. Brock (1994), p. 164.

100. Brock . . .

101. Brock (1994), p. 220.

102. Ibid., p. 228. Brock states that this change in viewpoint on the part of the DOJ was caused in large part by a change in personnel at the DOJ. When William Baxter left his post as Assistant Attorney General and Edwin Meese, who had been opposed to the MFJ, was appointed Attorney General, the DOJ position shifted.

103. Ibid., p. 241.

104. Ibid., pp. 174–175.

105. Al Gore, vice president in the Clinton administration, articulated a vision for a National Information Infrastructure (NII), or "Information Superhighway" that would carry information to and among all citizens. He made it clear that the NII would be built by private interests, not through government financing. A few months later he recast the NII in global terms, calling for all nations to work toward the creation of a Global Information Infrastructure (GII); he called for the GII to be based on private ownership, competition, and regulatory flexibility. The concept of a GII caught the imagination of policymakers in many nations. Al Gore, 1993. "Remarks" (December 21), accessed October 29, 2002. http://www.ibiblio.org/nii/goreremarks.html

106. For a discussion of the pros and cons of privatization, as well as the range of privatization options taken by governments, see Phyllis Bernt and Martin Weiss, 1993. *International Telecommunications* (Indianapolis: Sams), pp. 14–26.

107. International Telecommunication Union, 2000. *America's Telecommunication Indicators 2000*, 2000 edition, Figure 1. (http://itu.int/ITU-D/ict/publications/americas/2000 accessed August 22, 2003.)

108. The OECD was formed in 1961 and is an organization of 30 industrialized countries that serves as a forum for discussion and study regarding matters of economics, political and social concern.

109. OECD, 2003. *Communications Outlook 2003* (Paris: OECD), pp. 41–44.

110. Ibid., p. 25.

111. *In the Matter of International Settlements Policy Reform and International Settlement Rates*, 2002. FCC Docket IB 02-324 and IB Docket No. 96-261, Notice of Proposed Rulemaking, FCC 02-285 (adopted October 10), at 1. Hereafter referred to as *Reform NPRM*.

112. FCC, *U.S. IMTS Net Settlement Payments, 1985–2000*. FCC International Bureau website (accessed November 1, 2002).

113. For example, the accounting rate between the US and Afghanistan in October of 2002 was $5.15. Federal Communication Commission, *Consolidated Accounting Rates of the United States*, October 1, 2002.

114. *Reform NPRM*, at 1.

115. Since the creation of the World Trade Organization (WTO), the FCC has waived proof of the ECO test for countries that joined the WTO, presuming that WTO members are opening their markets. The WTO is explained further later in this chapter.

116. FCC website, http://www.fcc.gov/ib/pd/pf/isr.html (accessed November 3, 2002).

117. OECD, 2002. *Information Technology Outlook: ICT's and the Information Economy* (Paris: OECD), p. 96.

118. E. M. Noam and A. Singhal, 1996. "Supra-national Regulation for Supra-national Telecommunications Carriers?" *Telecommunications Policy*, Vol. 20:769–787.

119. Cliff Wymbs, 2002. "US Firms' Entry into the European Telecommunications Market: A Question of Modality Choice," *Journal of High Technology Management and Research*, 13: 87–105.

120. *OECD Technology Outlook 2002*, p. 101.

121. "Telecom Bottoms out in 2001, but Consolidation Could Spark a Rebound," 2002. *Corporate Financing Week*, 28(23):98–99 (June 10).

122. "France Telecom Gains Full Control of Global One for £2.35bn," 2000. *Financial Times (London edition*, January 27), p. 1.

123. "AT&T & British Telecom Set to End Joint Venture," 2001. *Atlanta Journal & Constitution* (October 13).

124. Transparency requires that all rules and requirements for doing business be open and so transparent to all. The national treatment principle requires that all foreign entities in a country be treated the same as domestic entities.

125. Markus Fredebeul-Krein and Andreas Freytag, 1997. "Telecommunications and WTO Discipline: An Assessment of the WTO Agreement on Telecommunication Services," *Telecommunications Policy*, 21:477–491 at 486. By the end of the Uruguay Round in 1994, countries had only made commitments, or concessions, in value-added services and no concessions had been made in basic telecommunications services like local and long distance services. The U.S. was especially dissatisfied with the commitments being made by Asian and Latin American countries, found the commitments regarding the easing of restrictions on foreign investment to be inadequate, and had reservations about commitments regarding satellite services.

126. For example, Fredebeul-Krein and Freytag, 1999. "The Case for a More Binding WTO Agreement on Regulatory Principles in Telecommunication Markets," *Telecommunications Policy*, 23: 625–644, outline several provisions in the agreement that provide loopholes for the signatories.

127. Peter Cowhey and Mikhail M. Klimenko, 2000. "Telecommunications Reform in Developing Countries After the WTO Agreement on Basic Telecommunications Services," *Journal of International Development*, 12: 265–281.

Chapter 8

Innovating New Services (1980s/1990s)

The pace of technological development picked up steadily in the 1970s and 1980s, making both corporate planning and government policy-making more challenging. Among the most important changes was the widespread introduction of digital devices. These, in turn, helped lay the groundwork for both mobile services and the Internet.

8.1 DIGITAL AGE

Digital technology began its entry in American domestic telecommunications with AT&T's introduction of the T1 Carrier System in 1962.[1] This allowed for both greatly increased system capacity and a far cleaner (less noisy) signal.

Basics

Unlike its analog ancestors, digital transmission allows engineers to design systems that have much lower noise because they use *regenerative repeaters*. A problem with analog amplifiers serving as repeaters had been that background noise was amplified just as much as the desired signal.

The new digital repeaters made it possible to regenerate (or recreate) the original signal, virtually free from noise, at each repeater. Thus, it became possible to engineer systems where the signal quality over a long distance could be the same as that for a short one. Further the manufacturing economies afforded by digital systems would also be greater. Early digital sys-

228

tems required *multiplexers* at each end of the transmission line to convert analog signals to digital and vice versa. This allowed them to interface with analog transmission systems and switches, essentially integrating them into the fabric of the telephone network. As these systems proliferated, and as digital technology became more cost-effective, telephone companies began to develop higher capacity digital systems and to build other system components, notably switches, using digital technology.

Switches

The initial application of digital technology in switches was in the control systems. In these early digital systems, the switching devices were a matrix of relays (called *space division switching*) with computers controlling the process and interacting with other systems. This allowed for ready maintenance and permitted easier development of new services and features. These digital switches interfaced directly with digital transmission systems, saving the capital and operating costs incurred by the digital multiplexers. These switches also supported digital, out-of-band signaling systems, such as Signaling System 7 (SS7), a Bellcore standard.

Out-of-band signaling was a significant advance over in-band-signaling because it permitted faster call setup, was more resistant to fraud, and enabled the ready provision of new network-based services. With in-band signaling, the call and information about the call, such as the destination phone number, all traveled over the same path. This arrangement provided limited space for any information about the call itself, other than its destination point. In some ways in-band signaling was also inefficient. For example, there was no way to check in advance to see whether the called number was busy or did not answer; instead the in-band signal set up the whole call path through the switching system.

In out-of-band signaling, information about the call and call setup information traveled over a different path than the call itself. As shown in Fig. 8.1, while the call traveled over the regular telephone network, signaling information traveled over a separate packet-switched data network. In the call setup process, the signaling system could check ahead to see if the called number was busy or did not answer; if that were the case, no call path would be set up. Out-of-band signaling also provided the opportunity to transmit a wide range of information about the call, including the calling telephone number. This feature proved to be especially important with the advent of long distance competition. In order to do their own billing, AT&T, MCI, and Sprint all needed the calling telephone number (*see section 7.3*). Out-of-band signaling made that possible.

As the signaling systems were installed, RBOCs began to consider ways in which their capabilities could be utilized for new services and

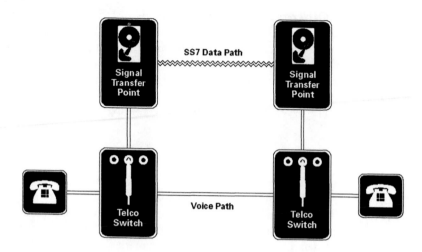

FIG. 8.1. Signaling System 7.

revenue. Since the signaling systems were essentially packet-switched data networks, they could easily use database-enabled features. A simple example of such a feature is "800" toll-free (to the caller) service. This requires a specialized database because an 800 number is not a legitimate network address, so, without additional information, a switch could not complete such a call. When a user dials an 800 number, the central office switch must first send a query to one of the "800" databases, which associates the number with a routable telephone number, the interexchange carrier chosen by the recipient and the payer. The database returns this information to the originating switch, which uses it to complete the call and to route the billing information correctly—all of this in a fraction of a second.

"Caller ID" is another popular feature that requires the use of these signaling systems. In this case, the recipient switch must receive the caller's name and number from the originating switch via the signaling network. The recipient switch then transmits this information (or just the number, if they have different grades of the caller ID service) to the callee's telephone. The RBOCs began to imagine other kinds of information services they could provide using this newly "intelligent" network, only to find that they came up against the information services restrictions of the Modified Final Judgment (MFJ) (*see sections 6.4 and 7.1*).

The deployment of digital switching was accelerated by the equal access requirements of divestiture. The newly divested Regional Bell Operating Companies (RBOCs) were now required to treat all long distance carriers equally, and many of the pre-computer controlled switches could not do this economically, so they had to be replaced. In addition, by the late 1980s, states began entering into "social contracts" with their RBOC in which they made commitments to infrastructure investments (such as digital switching) in exchange for regulatory relief. By the early 1990s, the conversion to digital switching was substantially complete.

8.2   WIRELESS TELEPHONY

As discussed in section 4.1, wireless first became a significant factor in U.S. telecommunications in the 1920s. The original business model for wireless was for point-to-point communications as the Marconi and other companies sold their services to dispatchers, insurers, and ships at sea. With the development of continuous wave transmission (which enabled speech and music to be transmitted), broadcasting came to dominate.

Yet point-to-point applications did not disappear. Even in the 1920s, radio equipment was placed in police cars to improve the effectiveness of law enforcement officers in the field. Postwar FM mobile technology was first used in St. Louis police cars in 1946. The desire (and ability) for commercial subscribers to place and receive telephone calls from an automobile came about in 1949, when a limited MTS (Mobile Telephone System) was introduced. Only 250 subscribers per market could be accommodated, and only 10 percent of them could use their mobile telephones at any one time. While this seemed adequate given the high cost of the service, demand by the early 1950s soon outstripped available service offerings in the largest cities.

Emerging Role of Wireless

Wireless mobile systems have from the beginning included both stationary equipment (consisting of base stations and a switch) and mobile devices. The mobile equipment connects to the public network through the base station and the switch via a radio connection. This radio connection has operated at a variety of frequencies and has used various modulation techniques. The initial systems operated at 450 MHz, while modern systems operate in various frequency bands between 900 MHz and 2 GHz, depending on the country and the system. Traditionally, a user would be assigned to a channel within the frequency band for the duration of the call. Until the introduction of cellular service in the 1980s, wireless mobile services were deployed from a single base station, which covered an entire metropolitan service area. While this was the most technically straightforward way of constructing a system, it resulted in a system with very limited capacity. A caller occupied a channel over the entire service area.

Faced with rising and more insistent demand for more spectrum to serve a growing number of customers, the Federal Communications Commission (FCC) at first resisted and then supported mobile developments. In 1970, the Commission finally yielded to mobile industry pressure and after long debate finally reallocated the upper reaches of the UHF television band (channels 70 through 83) to this new mobile radio technology. This reallocation provided some 450 MHz of bandwidth for mobile uses.

Cellular Systems

To overcome spectrum limitation problems, researchers at Bell Laboratories (who first investigated analog cellular system designs in the late 1940s) and Motorola began experimenting with systems in which a service area would be covered with many smaller, lower power base stations, each of which operated on a subset of the available spectrum. By doing this, radio frequencies could be re-used over the service area (see Fig. 8.2), resulting in greatly increased system capacity. The major pen-

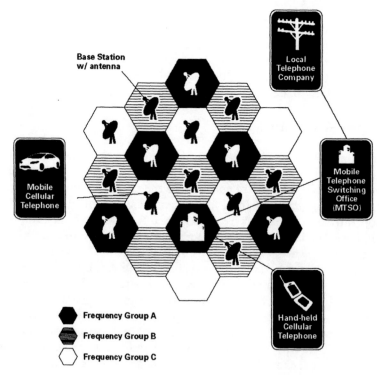

FIG. 8.2.   A typical cellular system architecture.

alty for this was more complex stationary and mobile equipment. By the 1980s, however, advances in semiconductor technology diminished the practical impact of this complexity factor. Systems of this kind were called cellular systems after the strategy of dividing the service area into cells, each with a small transmitter.

The FCC formally approved analog cellular service in 1981. The first offerings were expensive (telephones were the size of bricks and cost $1,000 or more) and customer growth was slow. Initial cellular systems used analog transmission technology, and are now referred to as "first generation" (1G) systems. "Roaming agreements," that is, company contracts to allow subscribers of one system to use the services of another, took some time to complete but were largely in place by the late 1980s and early 1990s, so that users were able to use their mobile equipment in most places that offered service. Such roaming in the United States was a matter of administrative agreement, which largely involved the determination of roaming charges and settlements payments among the carriers. U.S. carriers used the same 1G technology, so that conflicting standards were not at issue. Similar developments took place in Europe, although, because of different analog transmission standards and a lack of roaming agreements, mobile communications were largely a national affair.

Changing Cellular Standards

As the European Community (as it was then called) looked toward integrating and introducing competition into their telecommunications industry, they began to consider a single, new generation mobile communications standard based on digital transmission technology. This system, ultimately called General System for Mobile (GSM), was the first of the so-called second generation (digital, or 2G) mobile communications technologies. GSM-based systems were introduced throughout the member states of the European Union during the early to mid-1990s, usually with multiple competing operators. GSM proved to be a widely successful standard in Europe, and led to pan-European roaming, new and innovative services, and rapid adoption for business and personal use.

Moving in the opposite direction in a less-regulated marketplace, U.S. carriers considered a variety of 2G technologies. Unlike their counterparts in Europe, the FCC declined to set a single nationwide standard, instead determining that the industry was free to choose the transmission standard that they deemed best. The Time Domain Multiple Access (TDMA) technology (on which GSM was also based) was the most developed and was adopted by some carriers, most notably McCaw Cellular (later purchased by AT&T). But an alternative spectrum-sharing technology, called

Code Division Multiple Access (CDMA), was rapidly developing. CDMA promised carriers more users per MHz of radio spectrum, so it was more efficient. A number of carriers, notably Bell Atlantic Mobile (later Verizon Wireless) and Sprint Wireless adopted this technology. GSM was also adopted by some carriers (e.g., VoiceStream), and still another standard was adopted by NexTel. The result was a national system with a variety of incompatible digital transmission technologies.

In turn, such incompatibilities led to carriers trying to establish a large digital "footprint," with the gaps filled in by roaming agreements using analog service. Use of incompatible systems also resulted in handsets useful only with a given service provider and not transferable to a new carrier. Some handsets supported multiple transmission modes (i.e., analog as well as the carrier's chosen spectrum-sharing technology) so that the subscriber had nationwide access. In many cases, the user's calling plan charged higher rates for non-network ("roaming") calls, although, after AT&T Wireless's pioneering "Single Rate" plan, this began to change industry-wide.

These initial digital (or 2G) systems promoted spectrum efficiency in that more users could be supported with the same amount of bandwidth. In addition, the clarity of spoken communications was higher and the handsets were lighter because the amount of power needed to transmit a digital signal was lower, which in turn allowed for a smaller battery. Finally, since the handsets and the transmission were both digital, these systems allowed for easier integration with data services than did analog systems. Data services provided on 2G systems included low speed circuit-switched data and text messaging. Packet-switched data services began to emerge on mobile systems in late 2000 in some countries, by 2002 in the United States. Depending on the carrier, systems supporting these were dubbed either "2.5G" or 3G systems. They all required substantial amounts of spectrum, adding pressure on the management of that resource.

8.3 MANAGING SPECTRUM

One purpose of the Communication Act of 1934 was to "maintain the control of the United States over all the channels of radio transmission; and to provide for the use of such channels, but not the ownership thereof, by persons for limited periods of time, under licenses granted by Federal authority."[2] By virtue of the Communication Act, the FCC is the Federal authority that determines how the radio spectrum will be used and that issues licenses for spectrum use. While the National Telecommunications and Information Administration (NTIA) oversees spectrum use by the government, the FCC has extensive authority over the domestic use of the radio spectrum.

Spectrum management, or allocation, is needed to avoid interference. Because generally only one user can use a frequency at a time, the orderly allocation of spectrum helps assure that a range of service can be provided without interference from other users or services. Government oversight of spectrum use is intended to assure that this resource is used in a manner serving the public good. The FCC has the authority to specify which services will use specific ranges of spectrum, and to issue licenses that restrict the licensees to providing specific services in specific geographic regions. In its early spectrum management efforts, the FCC drew sharp distinctions among services (paging vs. voice service for example) and among providers (common carriers vs. private carriers), and did not allow licensees to subdivide their spectrum for lease to other providers. In recent years, the FCC has loosened these restrictions, allowing more flexibility in spectrum use.

With the development of cellular services and personal communications services (PCS), the radio spectrum has become an increasingly valuable resource. Recognizing this, the FCC has sought efficient allocation methods, with varying levels of success.

Implementing Lotteries

Rather than allocate the spectrum through administrative hearings (so-called beauty contests), the FCC in 1981 decided that the available spectrum in each service area would be divided in half. One half would be set aside for the incumbent landline carrier in the area (typically the local Bell Operating Company), and the other half was placed in a pool that would be determined by comparative license proceedings. Facing a huge number of such proceedings, however, the FCC sought Congressional approval to institute a lottery instead. After some difficulty Congress agreed and amended the Communications Act accordingly by the mid-1980s.

An unintended side effect of lotteries was that they presented a speculative opportunity for groups of investors. These investors would invest in proposal preparation, and would then sell any licenses they won to operators who actually intended to construct and operate systems. As potential applicants began to realize the lottery element of pure chance to win, countless players flocked to be included in the lotteries. From an average of 6 per market in top 30 cities (processed before lotteries were instituted), the number of applicants rose to 19 per city in markets 61 to 90; hundreds per market for markets 100 and greater; and finally more than 500 per market for the smallest markets, numbers 166 to 180.

Auctioning Spectrum

As the demand for wireless communications increased, carriers continued to press for more spectrum to deploy 2G (and forthcoming 3G) systems. Recognizing the inefficiencies of "beauty contests" and lotteries, the FCC began studying the use of auctions as a way of allocating spectrum. While New Zealand had already conducted a spectrum auction, there were questions about how this would scale to a large country with many more carriers. In addition, there were many issues that were left unspecified, some of which would come back to haunt the auction-based spectrum allocation process. Nonetheless, in the 1993 Omnibus Budget Reconciliation Act, Congress authorized the FCC to conduct auctions of the spectrum that these new digital systems were to use.

The FCC reallocated the spectrum from 1850 to 1990 MHz to PCS from private microwave services. With a focus on creating a diverse competitive market for wireless services, the FCC broke the auction into a number of separate "blocks." These included broadband, narrowband, and unlicensed blocks which covered either MTAs (major trading areas) or BTAs (basic trading areas),[3] with set-asides for minority–owned businesses and small businesses. Broadband PCS, which was designed with mobile telephones in mind, was allocated in three blocks of 30 MHz each (blocks A, B, and C) and three blocks of 10 Mhz each (blocks D, E, and F). The A and B blocks were auctioned by MTA and the C, D, E, and F blocks by BTA. The C block contained 30MHz that was set aside for minority-owned and small businesses, and was subsequently subdivided into five sublicenses (C1–C5). In addition, the FCC designated spectrum to be auctioned for narrowband PCS, which was designated for services such as advanced paging.

After consultations with experts in auction design, the FCC set the auction schedule to begin with the national narrowband PCS[4] license auctions in July 1994 (Table 8.1 summarizes PCS auction details). By many measures, these spectrum auctions were considered a great success, especially as compared with the alternative methods of spectrum allocation (comparative hearings and lotteries). The FCC noted that their cost of all auctions (including the PCS auctions) up to 1997 was $74 million, or less than 1 percent of the total money raised. By comparison, the FCC estimated that their cost for processing the applications for the cellular lotteries alone had been $300 million.[5] The FCC further has argued that allocation efficiency can be measured by the number of licenses that are resold or transferred after the initial allocation takes place. If allocation is perfectly efficient, one would expect no transfers or sales of licenses to take place soon after allocation, since it would not be possible to improve on the allocation through sales in the secondary market. After the

TABLE 8.1
PCS Spectrum Auctions, 1994–97[7]

PCS License	Dates	No. of Licenses Auctioned	No. of Bidders	Total Amount Bid (in Millions of $)
National Narrowband	July 1994	10	29	617
Regional Narrowband	Oct–Nov 1994	30 (five regions, six licenses each)	28	392
Broadband Blocks A, B	Dec 94–Mar 95	102 (51 MTAs)	30	7,000
Broadband Block C	Dec 95–May 96	493 (BTAs)	255	10,000
Broadband Blocks D, E, F	Aug 96–Jan 97	1,472	153	2,500

narrowband PCS auctions, only one license was resold (worth 5 percent of the auction revenues), and only 12 licenses were resold after the broadband PCS auctions (worth 6.5 percent of the total revenues). By comparison, 75 resales took place in 1991 after the cellular lotteries, when far fewer licenses were granted.[6]

Auction Troubles

Despite these successes, there were still problems with the auctions, most notably with those for Block C that had been reserved for small and minority-owned businesses. In the wake of the auctions, despite the FCC's introduction of installment plans to help companies pay the amounts they bid, there were a number of defaults. This led to a re-auction of some of the Block C licenses in July 1996. The largest bidder for Block C licenses, NextWave Personal Communications, had bid approximately $4.2 billion for licenses in cities such as New York, Los Angeles, Washington DC, Boston, and others. It was among several firms that would eventually default on their license payment obligations and file for bankruptcy.

Instead of returning the licenses to the FCC for re-auction, however, NextWave's creditors argued that the licenses were an asset that was protected by the U.S. bankruptcy code. The commission argued that because NextWave failed to make timely payments, they effectively ceded their rights to the license. The bankruptcy court supervising the NextWave case sided with NextWave, but this decision was reversed by a federal district court. In the wake of this ruling, the FCC scheduled and conducted a re-auction of the NextWave spectrum from Dec 2000 to Jan 2001, receiving bids totaling $16.8 billion (for spectrum for which NextWave had bid $5 billion). NextWave appealed the district court decision to the federal appeals court, who reversed the lower court and remanded the case to the FCC for reconsideration in light of its arguments,[8] forcing the Commission to put license reallocation on hold pending clarification. The FCC appealed to the Supreme Court, which early in 2003 ruled in favor of NextWave,[9] effectively ending this long legal contest and forcing the FCC to refund revenues from the second auction.

The NextWave case is an unintended consequence of this method of spectrum allocation, because it demands clarification of the extent to which a spectrum license, purchased at auction, is property. Under current law, the bidders were bidding not for ownership rights to the spectrum, but for the license to use the spectrum. The federal government continues to assert ownership rights for all spectrum. On the other hand, the Supreme Court held that bankruptcy laws trump telecommunications statutes and that in this case, NextWave was entitled to retain its property (the partially paid-for frequencies) while under Chapter 11 bankruptcy law protection.

Spectrum Task Force

Amidst these legal maneuvers, the Commission faced larger spectrum issues. Just as the move toward digital technologies has broken down the barriers between voice and data in the wired world, so too has digital wireless technology broken down the barriers between services like paging and voice. All matter of messages—voice or data or paging—when digitized become a string of bits and bytes of data. Moreover, new technologies reduce problems with interference between frequency bands. These technical developments have put pressure on traditional spectrum management, in which service providers are constrained to offer specified services within specified spectrum bands. In addition to technical developments, uncertainty about the property rights of license holders (as demonstrated by the NextWave situation) have inhibited the most efficient use of spectrum. License holders have been precluded from "selling" any unused spectrum to other carriers, and thus creating a secondary market for spectrum. As a result, spectrum has gone unused, even though there are carriers capable of utilizing the spectrum to provide service to willing customers.

The FCC responded to these developments by launching an investigation and proposed rulemaking in 2000.[10] The Commission's Notice of Proposed Rulemaking focused primarily on promoting secondary markets for Commercial Mobile Radio Systems (CMRSs, such as mobile telephone service providers) and would not allow for spectrum transfers between services (e.g., UHF television to mobile telephone services). The FCC also realized that, besides the question of secondary markets and property rights of licensees, there were other significant concerns regarding spectrum management, including the proper treatment of spectrum use for public safety purposes such as police, fire, and ambulance. These agencies would not be able to compete financially with commercial providers; however, they could profit from the sale of any unused spectrum. The FCC was also concerned about the effect of secondary markets and property rights on broadcast licenses, fearing that public interest goals, such as content diversity, would be more difficult to achieve since firms promoting less popular viewpoints might not be financially competitive with those promoting more popular ones.

The Powell-led FCC decided to undertake a comprehensive review of spectrum policy, and to do so, formed the Spectrum Policy Task Force (SPTF). The SPTF recommended that the FCC move toward a more flexible, market-oriented spectrum policy that would allow users to migrate toward new technologies and to more efficiently use the spectrum. To do so, the SPTF recommended that the FCC create flexible rules, facilitate secondary markets, and define users' rights and responsibilities clearly and exhaustively; adopt a standard regarding acceptable interference that allowed for greater access to unused spectrum; consider allowing time sharing of spectrum among multiple users; and shift its spectrum allocation model away from a command and control approach to a more flexible allocation model.

Finding 3G Spectrum

The United States has lagged behind the rest of the world in the deployment of the third generation of wireless service, often called 3-G. The International Telecommunication Union (ITU)'s World Administrative Radio Conference in 1992 began to define 3G and to identify spectrum for its use; further refinements were undertaken at the ITU's World Radio-communication Conference in 2000. The third generation of wireless service promises the greater bandwidth that will enable the development of wireless broadband. 3G promises circuit and packet data of 144 kilobits/second for vehicular traffic, 384 kilobits/second for pedestrian traffic, and 2 megabits/second for indoor use. By the end of 2002, European nations had allocated 155 megahertz of spectrum to 3G providers, while worldwide, licenses had been issued in nearly 25 countries for 3G service.

The United States had not issued 3G licenses by the end of 2004, though Verizon began to offer a form of 3G in early 2002.[11] Because the frequencies recommended for 3G by the ITU were already allocated for other purposes, including military uses, the FCC faced difficulties in freeing up 3G spectrum. The NTIA (responsible for government agency spectrum use), working with the FCC and the Department of Defense, identified 45 megahertz of spectrum in the 1710–1755 MHz band which could be transferred from exclusive government use to mixed use. Some government use of this band would continue to be used for federal power agencies, aviation-related safety communications, as well as Department of Defense fixed microwave, tactical radio relay, and aeronautical mobile stations.[12] Government entities relinquishing use of this band would be reimbursed for their relocation costs. The FCC identified another 45 megahertz in the 2110–2155 MHz band, which was being used by private and common carrier fixed microwave services, and Multipoint Distribution Service (MDS). These incumbents would also be entitled to compensation for relocation to another band.

The FCC determined that these 90 megahertz of spectrum would be used for what it calls Advanced Wireless Services (AWS), which includes 3G services and any other advanced fixed or mobile wireless services. The Commission decided that use of this spectrum would be governed by its Part 27 rules, which allow for flexible spectrum use for a variety of services, not just for 3G. According to the FCC, "We find that permitting licensees to use this spectrum for any use permitted by the spectrum's allocation will not deter investment in communications services and systems, or technology development."[13] The FCC planned to issue geographically based licenses for the spectrum in symmetrically paired spectrum blocks. The geographic areas for the licenses include MSAs and RSAs (units that had been used for earlier cellular licenses) and also Economic Areas (EAs) and Regional Economics Area Groups (REAGs), geographic units that had been used for other wireless licenses.[14]

While the usual term for spectrum licenses is 10 years, with expectations of renewal, the FCC determined that, because of the time required to relocate incumbent spectrum users, the initial license term for this 3G spectrum should run 15 years, with subsequent renewals being 10 years.[15] Continuing their general trend to marketplace spectrum decisions, the Commission also decided that these licenses would be assigned through competitive bidding auctions.

8.4 CABLE CONVERGENCE

In the face of continuing technological development, regulators faced increasingly intense debates about which services telephone carriers should be able to provide. Behind these debates lay continuing discussion about "convergence" or the coming together of formerly different businesses and products and services. One early vehicle for this process was coaxial cable. This technology, developed by AT&T for television network interconnection, helped to create a whole new business that arose in the late 1940s as isolated communities began seeking television service. Many towns were too far from the handful of television stations then on the air, or were blocked from television reception by intervening terrain.

Rise of Cable Television

Communities desiring television service (more often retail stores seeking to sell receivers) would construct an antenna on a nearby hilltop from which a distant TV signal could be received. That signal would then be carried to subscribing homes and businesses by coaxial cable connections. The first of these "community antenna" television (CATV) systems appeared in Oregon and Pennsylvania in 1947–48, and the idea expanded slowly in mountainous areas of those two states, Colorado, and other states into the 1950s. Soon some local telephone companies took up the option. While rarely involved in programming decisions, the Bell companies as well as GTE and United Telecom often built CATV systems and then leased them to independent cable operators.

These combined operations raised increasing questions of cross-subsidy and unfair competition by the late 1960s. In 1968, the FCC began to require carriers to seek its permission before any telephone company could provide any aspect of CATV service in the same region.[16] Further, there was considerable feeling in the regulatory community that the 1956 Consent Decree (*see section 4.4*) disqualified AT&T from participation in CATV as it was clearly not telephone related. Until the 1968 action, however, and thus for the first two decades of CATV's slow development, there was no restriction aside from the Consent Decree preventing any Bell System affiliate from building or buying a CATV television system—and even programming at least some of its channels. But that freedom was soon to disappear.

Cable/Telephone Ownership

In 1970 the FCC issued rules banning ownership of local telephone and CATV television providers in the same coverage area. The rule did continue to allow a telephone carrier to build, though not to operate, a CATV system within its service area. In such a case the facilities would have to be leased to an independent CATV operator. There were many such operations in place at the time the 1970 rule was adopted. Given AT&T's control of 80 percent of the nation's phones (covering half the nation's landmass), the rule effectively banned any continuing Bell company participation in CATV television.

The FCC expressed concern that such co-located systems might be used to hold back development of CATV (by then increasingly called simply "cable") or any non-video services cable might also provide. Additionally, the new rules were designed to promote competition in provision of information by not allowing one monopoly (telephone service) to take over another (cable). There was also more than a little worry about allowing a common carrier to exercise decision-making authority in electronic media programming.[17] The rules were appealed and upheld in 1971.[18]

The FCC soon developed a waiver policy for its cross-ownership rule for rural areas with only limited population. As modified during the 1970s, the waiver allowed cable-telephone co-located ownership and operation in those areas where sparse and spread-out population otherwise threatened the offering of any cable service at all.[19] The cable industry disagreed with this position, fearing *any* relaxation allowing telephone control. In fact, few waivers were requested or granted in the 1970s.[20] In 1980–84 the Commission conducted an intensive analysis of cable television ownership, including the waiver for rural telephone companies. The Commission's staff recommended dropping all ownership limits on cable *except* for the telephone ban.[21]

At the same time, Congress established a largely deregulatory legislative framework over cable television in the Cable Communications Policy Act of 1984. The first legislation to deal with cable issues, one of its few regulatory provisions was to codify the FCC cross-ownership ban as part of the Communications Act.[22] The FCC rules—and the legislative codification—spoke entirely to *co-located* telephone and cable systems. There was far less concern about operation of cable systems outside of telephone service areas. Indeed, that same year, the FCC adopted a proposal eliminating the need for telephone companies to file for advance FCC approval for providing cable service outside their telephone service area.[23]

The RBOCs finally took concerted action to overthrow the hated 1984 congressional codification of the FCC's cross-ownership ban. Supported by an FCC now clearly in favor of dropping the ban entirely (as it had made clear in several reports and statements in other dockets) as it was hindering competition to the now-powerful cable industry, RBOCs took a new tack—that the ban was in violation of their First Amendment rights. Their argument was fairly simple: "Common carriers may be required to carry freight for others, but that does not mean they may constitutionally be barred from carrying expressive freight for themselves."[24] Neither, the RBOCs argued, should telephone (or cable) carriers be limited in the rights to free expression because of holding a government-issued franchise, or because of their use of public facilities. "No rational based on real estate justifies any suspension of First Amendment protections."[25]

The first decision to agree that the 1984 ban was unconstitutional came in 1993 when two Bell Atlantic subsidiaries won a judgment in Federal District Court—a decision later upheld on appeal.[26] The District Court in Virginia initially issued an injunction barring enforcement of the 1984 cross-ownership provision nationwide,[27] but later drew in its judicial horns, and applied it only to Bell Atlantic territory. Closely related cases were brought by all the other RBOCs, and several were soon decided in similar fashion.[28] All of the decisions, including those of two appeals courts who agreed, noted that while the FCC and congressional concerns about wired communication concentration and potential illicit cross-subsidy were valid and serious, a total ban went beyond what was needed to meet those concerns.[29]

Seemingly on a roll, three telephone trade associations and three RBOCs brought suit to end the requirement that telephone carriers obtain FCC Section 214 authorization prior to buying or building a cable system within their telephone coverage area.[30] The Commission had been allowing cable operation by telephone carriers only on a common carrier basis (i.e., the telephone owner could not provide program content) under what they dubbed "video dial tone."

All the ownership concerns appeared to come to a head when the Supreme Court granted *certiori* of the several appeals court decisions in 1995. Oral argument was held in December, the justices appearing skeptical of government and cable industry arguments to retain the statutory ban, but still not clearly resolved among themselves as to what stance to take.[31] Perhaps ironically, Congress came to the Court's rescue with the Telecommunications Act of 1996 (*see section 9.2*). The 1996 law repealed the 1984 ownership codification, replacing it with a somewhat complex set of conditions allowing a greater degree of telephone entry into the video market through so-called open video systems, a variation of the FCC video dial tone notion requiring a telephone carrier to provide co-located cable video only on a common carrier basis. Still disallowed was outright purchase of a co-located cable system by a telephone carrier. All the previous court decisions, being based on a no-longer extant 1984 law, were made moot.

First Merger Moves

In October 1993 Bell Atlantic and Tele-Communications Inc. (TCI), then the largest owner of cable systems in the country, announced a multibillion dollar merger.[32] While other RBOCs had nosed about cable (SBC had purchased two cable systems in the Washington, DC, area earlier in 1993), the proposed Bell Atlantic/TCI merger was by far the largest deal proposed and squarely placed the issue of cable/teleco co-located ownership front and center. Several political cartoons suggested that Americans would soon have but one choice for all their voice, data, and television service and that government would be hard-pressed to keep up any kind of effective regulation.

But despite considerable media hype, all was not well, not least because many aspects of the deal remained unresolved when it was publicly announced. After stirring up a substantial policy debate and considerable newsprint, for a variety of reasons the deal unraveled about six months later (*see section 9.6*). Theoretically because of new FCC rules rolling back cable service prices, the collapse was certainly due in large part to the inability of the principals to resolve issues of control, not to say executive ego.[33]

Five years later TCI would figure in an even larger telephone company takeover deal after passage of the 1996 telecommunications act opened further options as it encouraged competition to the monopoly local cable and telephone carriers.

Broadband Gambles

In November 1997 Michael Armstrong (then the chief executive officer of Hughes Electronics) replaced retiring AT&T Chairman Robert Allen. AT&T was reeling from years of poor management, bad deals, and the wrenching 1995–96 trivestiture which spun off all manufacturing functions (*see section 7.2*). One of Armstrong's first moves was to halt AT&T's then-sputtering attempts to sell competitive local telephone service pending an overall change in approach.[34] Early in 1998 Armstrong announced AT&T's new long-range corporate strategy. Having experienced little success in its attempts to break into local telephone markets, AT&T determined to leap-frog the process and *buy* its way into broadband links to homes over which it might offer voice, data (Internet), and possibly video entertainment services. By purchasing one or more large existing alternative local carriers, AT&T would strive to become a larger integrated firm, a strong demonstration of convergence that others merely talked about. To initiate the strategy, Armstrong launched a soon successful $11.3 billion takeover of Teleport Communications, one of the larger non-Bell local operating companies with a substantial business clientele.[35]

Several months later AT&T made a far more substantial move in the same direction when it bid $48 billion to buy TCI, the nation's second-largest multiple cable system operator, plus another $20 billion deal for Liberty Media (a major cable program source) both controlled by cable entrepreneur John Malone.[36] Buying the pipeline into TCI's 14.4 million subscriber homes (and another 24 million homes "passed") for voice, cable, and Internet service, however, quickly scared investors, causing AT&T stock to decline after the deal was announced. In part this was due to the additional costs AT&T would incur because TCI systems would require considerable expensive rebuilding to bring them up to current standards, let alone interactive services. Regulatory approvals for the huge deals finally came in December 1998 from the Justice Department, and two months later from the FCC.

A few months later, Armstrong added to AT&T's growing cable holdings and the company's debt load. In May 1999 AT&T moved to take over the MediaOne Group, the fourth-largest cable multiple system operator, in a complex $58 billion deal that involved trade-offs with Comcast, another large cable operator. Despite strong media advertising and legal protests from local telephone companies, the deal was eventually approved and AT&T became the nation's largest owner of cable systems. At the same time it established agreements with Comcast and Time Warner Cable to carry AT&T telephone services while Comcast purchased some cable subscribers from AT&T.

The cost of all this activity over just 18 months was massive—variously estimated to total well over $100 billion—and that was before the hugely expensive cable system upgrades necessary to establish AT&T's vision of widespread interactive telecommunication services to a substantial portion (60 million) of the nation's homes. The flurry of deals had grown to become literally a "bet the company" gamble that the pieces would come together, could be integrated, and would at least begin to generate substantial revenue before investors lost their patience or nerve. By the beginning of 2000, AT&T management was projecting an annual return of 25 to 30 percent on their cable holdings by 2004.[37] But this was not to be—the industry's financial collapse after 2000 doomed the colossus and led to yet more divestiture (*see section 10.2*).

One even larger convergence-driven merger announced in January 2000 was predicated upon the potential content-sharing and distribution synergies to be gained. Internet giant America Online (AOL) announced a $185 billion merger (really a takeover and by far the largest merger of any kind to date) of Time Warner, the world's largest entertainment and news empire.[38] The all-stock merger was predicted to be only the first of many Internet and media company mergers. As the two firms were in largely different market sectors, little regulatory delay was expected. That is not to say, however, that negative voices were not heard. Many firms and citizen groups spoke up and suggested limits on the deal.[39] At the end of the year the merger was approved by the Federal Trade Commission, though with strict conditions limiting the firm's freedom of action in competing with other companies.[40] An outside monitor was appointed to assure that the conditions were adequately fulfilled.[41]

The merged colossus shoved AT&T into something of a secondary role. Nearly half of Internet homes linked to the Web used AOL, while Time Warner owned 18 percent of the country's cable subscribers compared to nearly 21 percent for AT&T. The merged AOL Time Warner was a leader in magazines, movies, and television programming and was far larger than AT&T. Both entities demonstrated the trend toward combining older and newer services in an attempt to find the right balance for the converging communications marketplace. Unfortunately, as will become clear (*see section 10.3*), their timing could not have been worse. Yet the stakes were huge, as AT&T, AOL Time Warner, and their competitors were all seeking to become major players in the most important new technology (and promising market): the Internet.

8.5   INTERNET AND INFORMATION

Paralleling the development and rapid expansion of digital, mobile, and converged services, the continued development of computer hardware and software, and the emergence of data network technology lay the groundwork for more new services. These developments were taking place in both commercial enterprises and academic research labs. Their approaches differed.

In general, most commercial systems were "closed" in the sense that content and access were bundled together and sold by a single provider. Systems emerging from academic research labs, in contrast, were considered "open" in the sense that content and access were unbundled (i.e., provided separately). The promise of these computer-based information services was so compelling, even in the early 1980s, that Judge Harold Greene chose to treat them separately in his divestiture order (*see section 6.4*).

Early Information Services

An early form of information services was videotext or teletext. Both of these systems involved transmitting information over a wired or wireless medium to be displayed on either a television or computer screen. Teletext was a broadcast system, in which pages of information were sent from a television transmitter to a television screen; videotext, or videotex, involved sending information over telephone lines from a central computer to individual computer terminals.[42] Videotext was first developed in the late 1960s by the British Post Office, which was seeking ways to encourage use of the telephone, and by the British Broadcasting Company in the early 1970s as a way to provide captioning for the deaf.[43] The

French deployed their own videotext system, called Minitel, in 1979. In an effort to encourage technological development and expertise in what it called "telematics," the government-owned French PTT provided free terminals to thousands of customers in the hopes they would access information from a government-provided central computer. The system remained in service into the early 2000s. The Canadians and the Japanese developed their own systems, while in the United States, Cox Cable, Times Mirror, Dow Jones, Knight-Ridder, CBS, and others began to offer videotext services.[44] Many newspapers regarded videotext and teletext as a new form of publishing (or disseminating) information, and began to experiment with this new "electronic publishing."

Transmission speeds were slow, the resolution was poor, and users could do little more than view the data on the screen. Videotext was, however, hailed as an important innovation. It was a new way of transmitting and accessing information. It demonstrated that information could be transmitted to users who could retrieve and view it at will, and, as such, it set the stage for technological developments that would eventually result in highly sophisticated systems.

What was envisioned as information services in the early 1980s by commercial providers included electronic white and yellow (directory) pages, messaging capability, and the provision of information such as radio, television, and movie schedules. It was not entirely clear how big the proposed market for these services might be. The personal computer as a commercial product was only two or three years old at the time, and modems were quite slow (300 up to 1200 bps was common), so few envisioned the kind of graphics-oriented PC-based information service that is today's Internet. The networking technology of choice for these providers was often X.25, a standard approved in revised form in 1980 by the International Telecommunication Union. This was designed to support public and private data networks, and required a fairly significant amount of technological homogeneity among all components in the system from an architectural standpoint. Nonetheless, many information service providers adapted this technology to provide dial-up access for users.

Users included commercial users, who would access the information services via a permanent connection between computers, as well as private individuals, who would dial in via telephone modems. A number of firms entered the market for these information services, including Prodigy, which was originally a joint venture between Sears, IBM, and CBS television; GEnie, which was operated by GE's information services division; Compuserve; AOL; and others. Despite the promise, these firms struggled to be profitable, and many retrenched or left the business. As PC penetration grew, and as data communications speed increased, such services became somewhat more compelling.

But neither teletext nor videotex were destined for long-range success. Initial interest and investment faded by the late 1980s in the United States.[45]

Information services were also provided via telephone to reach the vast majority of consumers who did not yet have access to a PC modem. Blocked by the MFJ from offering information services until the early 1990s, RBOCs developed portals where consumers could access the resources of various information service providers via telephone.

ARPANET

Academic research labs took a different direction. With initial support from the Defense Department's Advanced Research Projects Agency (ARPA) and then from the National Science Foundation (NSF), an alternative approach to networking began to emerge. This recognized the autonomy of a user's local network, and concentrated instead on constructing a backbone network (it would become known as the Internet) to interconnect these networks. A set of rules, or protocols, governing network interconnection was needed. The Transmission Control Protocol/Internet Protocol (TCP/IP) protocols of the ARPA project, while limited in features and capabilities, were reliable by the late 1980s, so many research institutions adopted them while they waited for improved protocols to come to market.

The TCP/IP protocols stabilized as personal computers interconnected by local area networks (LANs) began to emerge as both a viable and desirable model of computing. The original Ethernet LAN was a product of the prolific Xerox Palo Alto Research Center (PARC). Xerox, together with Digital Equipment Corporation and the Intel Corporation collaborated on a product standard (DIX Ethernet) that they took to the standards committee of the Institute of Electrical and Electronic Engineers (IEEE). The original IEEE goal was to develop a single LAN standard. Competing technologies (such as IBM's Token Ring) were introduced, leaving the IEEE effort in a quagmire. So the group decided to develop several standards including Ethernet, token ring, token bus, and others. The Ethernet standard quickly found its way into products. Subsequent improvements on the original standard found even more market appeal, so that by the early 1990s, it had become the dominant LAN standard.

Emergence of PCs interconnected by LANs was a move away from the centralized model of computing, and it fit well with the model that the fledgling Internet was offering. Availability of this equipment with a stable TCP/IP protocol made this an attractive configuration, particularly for networks covering wide areas. However, TCP/IP was a fairly limited system and few vendors offered system solutions to users and administrators. Stepping into that void were software firms which made services like access control, file sharing, and printer sharing seamless for users and administrators. Examples of firms that delivered these LAN operating systems include Novell, Netware, and Banyan's Vines products.

Commercialization

The developing Internet backbone was funded by the federal National Science Foundation's (NSF) NSFNet and the Department of Defense's MilNet. As a result of this federal funding, commercial use of the network was barred, so application of the Internet was confined to academic research uses, although private sector firms were free to apply these technologies for their networks. As Internet links expanded, pressure to commercialize the system grew. A Commercial Internet Exchange (CIX) was founded to interconnect nascent commercial backbones, notably PSInet. The prime contractor for NSFNet, Advanced Network Services (ANS), created a private Internet backbone that ran in parallel with NSFNet and raised conflict of interest questions in the Internet community. In 1993 the NSF decided that it would cease funding NSFNet operations. This was finally achieved in 1995, after publicly funded network access points (NAPs) were created to maintain connectivity. At this point, the Internet was fully privatized and open to commercial exploitation.

In addition to privatization, commercial exploitation was fueled by the development of graphical user interfaces (GUIs). Early PC users interfaced with their PC operating systems through command line interfaces, which required the user to have some facility with programming commands. A GUI, on the other hand, allows users to manipulate their operating systems by clicking on easily understood graphic icons. By the early 1990s both Apple computers and PCs had GUI interfaces. The next step was a GUI to the Internet. In 1993, Mosaic, the first graphical browser, was developed by the National Center for Supercomputing Applications, and the concept of a World Wide Web, accessible with the click of a mouse, became a reality. Instead of a network used by engineers and computer scientists, the Internet had become a network of information easily accessed with a modicum of technical knowledge. As a result, commercialization of the Internet began with gusto and continued until the "Internet bubble" finally burst after sending many related companies into bankruptcy (*see section 10.3*).

DSL Versus Cable Modems

In order for there to be widespread Internet use, there has to be user-friendly, effective ways to access the Internet. The earliest widespread means of Internet access was dial-up service, in which customers, through the use of a modem, used a regular telephone line to dial up an Information Service Provider (ISP). Dial-up access was slow, with sophisticated modems only reaching speeds of 56 kilobits per second. Such speeds were too slow for the graphics-filled web pages that were part of the World Wide Web. As Internet users sought the ability to quickly download graphics, audio, and video, the use of dial-up became increasingly limiting, and a market developed for broadband connections to the Internet.

Two new broadband access technologies were developed in response to this market demand: digital subscriber line service (DSL) that utilized the public switched network and cable modem service that used the cable television delivery system. Both of these technologies provided subscribers with the ability to download information in speeds measurable in megabits per second. By early 2004 there were 9.5 million DSL connections and 16.4 million cable modem service connections to the Internet.[46] While the technologies underlying these services were relatively straightforward, the regulatory considerations surrounding both of these technologies proved to be complex and often contentious.

The basic idea underlying DSL service is that a regular twisted-pair copper telephone line has the capacity to carry much more than just voice traffic. While voice is carried in the frequency range of 400 to 3400 Hz, a telephone line can handle much higher frequencies. In DSL service, both voice and data are carried over the telephone line into the telephone switching office; the voice traffic occupies the 400 to 3400 Hz frequencies on the line, while the data traffic is carried over the higher frequencies. There is a DSL modem at the customer's location that sends the data from the customer's computer onto the DSL telephone line. In the telephone switching office, the DSL line terminates in a digital subscriber line access multiplexer (DSLAM) which separates voice and data traffic, sending voice traffic to the telephone switch and on over the public switched network, and data traffic over a circuit to the customer's ISP which connects to the Internet.

The most common form of DSL service is asymmetric DSL (ADSL), which providers a much faster download speed than upload speed. This asymmetry is not seen as a problem because, for the most part, Internet users download much more information than they upload. There are several limitations involved in DSL service, however. To be DSL capable, telephone lines cannot have loading coils, which amplify voice signals but interfere with data. Nor will telephone lines with bridge taps support DSL service (bridge taps are extensions to other subscribers that appear between the customer location and the central office). And another major limitation on DSL service is distance; the maximum distance for service is 18,000 feet between the customer and the DSLAM. The farther the distance, the slower the speed.

Provision of DSL service has also been surrounded with a good deal of regulatory controversy. While ILECs were anxious to provide DSL retail service, they were loath to provide the elements of DSL to their competitors on a wholesale level. The FCC, however, expected the local telephone companies to do just that when they added the components of DSL to the list of unbundled network elements (UNEs) that they had to provide to competitors.[47] The local companies argued that the low cost of UNEs, especially the rates they were allowed to charge for the high frequency portion of the loop, sent the wrong signals to the marketplace, discouraging the local companies themselves from investing in broadband technologies like DSL and also providing no incentives for the competitive local service providers to invest in their own facilities. Indeed, the competitive providers were quick to avail themselves of line sharing and other DSL-related UNEs, and were slow to build their own facilities. Finally, fearing that line sharing was discouraging the development of facilities-based competition, the FCC, in its Triennial Review Order in 2003, established a three-year time frame for phasing out line sharing (*see section 9.5*).

Internet users have other means of Internet access besides DSL. Instead of turning to the telephone companies and the public switched network, customers can avail themselves of another line coming into their homes: their cable television service. Congress, in drafting the Telecommunications Act may have envisioned a scenario in which the telephone and cable companies competed with one another to offer voice telephone service. While the development of competition for voice service was exceedingly slow in getting off the ground, competition for Internet access between telephone companies and cable providers was much quicker to develop.

Just as DSL uses existing infrastructure, with some additions and adaptations, so too does the cable modem service use the existing cable television network, with some enhancements. In cable modem service, data ride over the same physical facility as do cable channels; indeed data ride over their own unused cable channel. Each cable television signal is allotted a 6-MHz channel of the cable's total bandwidth. In cable modem service, a 6-MHz segment of bandwidth is allocated to the data that flow downstream from the Internet to the customer, and an even smaller segment of bandwidth is allocated to the data that flow upstream from the customer. Again, the assumption is that Internet users download more information than they upload. The 6-MHz downstream channel and the 2-MHz upstream channel are shared by several cable modem service customers. In effect, the cable company creates a LAN for its cable modem customers.[48]

Regulatory issues surrounding cable modem service have been no less contentious than those surrounding DSL. The major question surrounding cable modem service is how it should be categorized. If it is a cable service, it falls under Title VI of the Communication Act and could be subject to requirements from local authorities from which cable companies receive the franchise to provide service in a specific location. If it is a telecommunications service, it is subject to Title II of the Communication Act and, as a result, the cable companies may face network unbundling requirements or the requirement that the cable provider allow access to multiple ISPs, not to just one ISP affiliated with the cable provider. If it is an information service, it is not subject to regulatory requirements because the FCC has decided that information services are not regulated. Telephone companies argue that, because cable modem service is functionally the same as DSL, it is not fair for the two services to face different regulatory requirements, and, that, in fact, the lack of regulation creates a competitive advantage for the cable companies.

A test case of cable modem's regulatory status occurred in Portland, Oregon. Arguing that cable modem service constituted a cable service, Portland and Multnomah County required AT&T, the cable provider in the area, to provide cable modem users with access to their choice of ISP, rather than automatically routing them to AT&T's affiliated ISP. AT&T refused and went to court. The court found for the city but an appeals court reversed on appeal, finding that the service in question was not cable service but rather telecommunications.[49]

The FCC disagreed with the appeals court decision, noting that the court "did not have the benefit of briefing by the parties or the Commission on this issue and the developing law in this area," and also noting that the "Communication Act does not clearly indicate how cable modem service should be classified or regulated; the relevant statutory provisions do not yield easy or obvious answers to the questions at hand; and the case law interpreting these provisions is extensive and complex."[50] The FCC agreed that cable modem service did not fit the Communication Act definition of cable service, which involved one-way transmission of video programming or other programming services, and subscriber interaction required for the selection of programming services.[51] The FCC further found that cable modem service was not telecommunications because the cable provider "is not offering telecommunications service to the end user, but rather is merely using telecommunications to provide end users with cable modem services."[52]

The FCC determined that cable modem service is information service, noting that Internet access provides the user with the opportunity to create web pages, do e-mail, or engage in newsgroups, activities that all fall under the definition of information service. The FCC, moreover, determined that cable modem service is interstate in jurisdiction, arguing that the "jurisdictional analysis rests on an end-to-end analysis, in this case on an examination of the location of the points among which cable modem service communications travel. These points are often in different states and countries. Accordingly, cable modem service is an interstate information service."[53] By asserting its authority over cable modem service, the FCC assures that cable modem service will not be regulated by any state authorities. And because the FCC has argued that information services should not be regulated, it is doubtful that cable modem service will face regulation by the FCC.

Voice Over IP

The Internet is based on a network that is unlike the PSTN which has been based on a hierarchy of switches connected by circuits. A telephone call that traverses the PSTN has traditionally been provided on a circuit-switched basis. A path is set up for the call through the PSTN, and that path is maintained for the duration of the call. This means that a substantial amount of switching and transmission resources are tied up for the whole call.

The Internet is configured differently. Instead of a hierarchy of switches, there are a host of computers and routers that send information among the computers on the network, which is a packet-switched data network. Instead of creating a pathway for a message, as is the case with circuit switching, in a packet network, the message is broken down into packets; the packets are routed through the network in the most efficient manner, even if that means that packets are routed across different routes; and, at the destination point, the packets are reassembled into the message. Packet networks offer greater efficiencies in routing and also the possibility of integrating many different services because all types of messages can be packetized, including data, audio, video, and voice.

The Internet Protocol (IP) offers service providers the ability to send voice over a data network, including the biggest data network of all: the Internet. This capability, often called Voice over IP (VoIP), or IP enabled services, has revolutionized service provision. Established carriers, like AT&T and SBC, have recognized the efficiencies of IP and have steadily converted their networks from circuit to packet switching. Cable companies and new service providers have begun to use VoIP as a way to provide an alternative to traditional telephone service. VoIP was at first primarily a replacement for more expensive long distance services. By 2004, cable companies were offering VoIP as an alternative to local telephone service, and the FCC had come to regard VoIP as so significant, it began to disseminate information about VoIP through its website.

From a regulatory standpoint, VoIP and IP enabled services present a significant challenge. In its Computer Inquiry II proceeding, the FCC determined that information services would not be regulated, though they fell under the jurisdiction of the FCC (*see section 7.4*). The FCC has regarded the Internet as an information service that is not regulated while telecommunications (i.e., transmission) on the other hand, is regulated. The question then is how to treat VoIP. Is it an information service, or is it telecommunications? The FCC opened a formal proceeding to consider that question, noting that it

> must necessarily examine what its role should be in this new environment . . . and ask whether it can best meet its role of safeguarding the public interest by continuing its established policy of minimal regulation of the Internet and the services provided over it.[54]

In related proceedings, the FCC began to draw some important distinctions. AT&T, noting that it had converted its long distance backbone network into an IP format, argued that it should not be regarded as providing circuit-switched long distance and so should be exempt from paying access charges to local telephone companies (*see section 7.3*). The FCC rejected AT&T's argument, finding that AT&T was providing telecommunications because customers were using regular telephone equipment to place and receive the calls, the calls originated and terminated on the PSTN, and the calls underwent no protocol conversions.[55] In effect, AT&T had merely changed the tools it used to provide telephone service. The case of pulver.com's Free World Dialup service was a different matter. According to the FCC, Free World Dialup service was basically a "directory or translation service, informing its members when other members are online or 'present,' thus available to receive a call, as well as informing them of the Internet address necessary to reach other members during their online presence."[56] Free World Dialup customers provided their own Internet access; and the Free World Dialup system did not transmit the calls or even know the geographical location of its customers. Telecommunications service was not provided because of the lack of transmission and the lack of knowledge regarding geographical location.

In its AT&T and Free World Dialup decisions, the FCC addressed the issue of VoIP in a piecemeal manner, and in each case was able to draw distinctions between information and telecommunications. The FCC may not find it as easy to draw such distinctions in considering how to deal with a cable company that uses VoIP to provide local telephone service. The FCC has found cable modem Internet access to be an information service—not telecommunications. If the cable companies use a variation of that same sort of cable modem access to deliver local phone service, will that cross the line into telecommunications? As the use of VoIP and IP enabled technologies continues to grow, the FCC will have to refine the definition of these services, and their appropriate regulatory treatment, and will also have to decide whether, at some point, VoIP should be regulated as a telecommunications service.

NOTES

1. R. F. Rey, ed. 1983. *Engineering and Operations in the Bell System* (Murray Hill, NJ: AT&T Bell Laboratories), p. 374.
2. 47 U.S.C., Section 301.
3. NMTAs and BTAs are defined by Rand-McNally as follows: A BTA is a geographic area, based on counties, in which residents do most of their shopping for goods. An MTA is a combination of two or more BTAs. Rand McNally defines 47 MTAs covering the 50 states and DC and 487 BTAs. The FCC added three MTA-like areas to this list, in addition to separating Alaska from Seattle, WA, and six BTA-like areas to Rand-McNally's. Thus the FCC's auctions covered 51 MTAs and 493 BTAs. See *Annual Report and Analysis of Competitive Market Conditions with Respect to Commercial Mobile Services*, FCC 02-179, June 13, 2002, p. 10.
4. The national narrowband PCS spectrum consisted of five 100-KHz blocks, three blocks of 62.5 KHz, and three blocks of 50 KHz. This spectrum is too small to be efficiently usable for voice service and was intended for services such as advanced paging.
5. FCC Report to Congress on Spectrum Auctions, FCC 97-353, September 30, 1997, p. 22.
6. FCC Report to Congress on Spectrum Auctions, FCC 97-353, September 30, 1997, p. 23.
7. The data in this table were summarized from results published by the FCC.
8. U.S. Court of Appeals for the District of Columbia Circuit, Docket 00-1402 and 00-1403, October 19, 2001.
9. U.S. Supreme Court, Docket No. 01-653, January 27, 2003.
10. *In the Matter of Principles for Promoting the Efficient Use of Spectrum by Encouraging the Development of Secondary Markets*, FCC 00-401, Nov 9, 2000.
11. Ben Charny, "Verizon Wireless Unleashes 3G Service," CNET News.com, January 28, 2002. Accessed August 15, 2004, at http://news.com.com/2100-1033-824289.html.
12. *Service Rules for Advanced Wireless Services in the 1.7 Ghz and 2.1 Ghz Bands*, Report and Order, WT Docket No. 020353, FCC 03-251 (November 25, 2003), at 6. (Hereafter referred to as the *AWS Order*).
13. NAWS Order, at 14.
14. FCC plans called for a total of 946 licenses allocated as follows:

Blocks	Pairings	Amount	Area	Licenses
A	1710–1720 & 2110–2120	2 × 10	EA	176
B	1720–1730 & 2120–2130	2 × 10	REAG	12
C	1730–1735 & 2130–2135	2 × 5	REAG	12
D	1735–1740 & 2135–2140	2 × 5	RSA/MSA	734
E	1740–1755 & 2140–2155	2 × 15	REAG	12

AWS Order, at 28.

15. AWS Order, at 70.

16. *General Telephone Co. of California, the Associated Bell Companies, the General Telephone System and United Utilities, Inc.*, 13 FCC 2d 448 (1968).

17. *Final Report and Order*, 21 FCC 2d 307 (1970). See, generally, Ferris, Lloyd, and Casey, 1983. *Cable Television Law: A Video Communications Practice Guide* (Albany: Matthew Bender), section 9.10.

18. *General Telephone Co. of the Southwest v. United States*, 449 F.2d 846 (D.C. Cir. 1971).

19. "Rural" was eventually defined by the FCC using Census Bureau classifications of U.S. counties. "D" counties, the most sparsely populated (city downtowns were "A" counties), were to count as "rural" for purposes of this rule.

20. C. H. Sterling, "Cable and Pay Television," 1982. Chapter 7 in B. Compaine et al., *Who Owns the Media? Concentration of Ownership in the Mass Communications Industry* (White Plains, NY: Knowledge Industry Publications), p. 392.

21. Gordon et al., op. cit., p. 175.

22. Section 613, Communications Act of 1934 as amended.

23. 49 Federal Register 21333 (1984).

24. Thorne et al., *Federal Broadband Law* (1995), p. 503.

25. Ibid., p. 505.

26. *Chesapeake & Potomac Tel. Co. v. U.S.*, 830 F. Supp. 909 (E.D. Va, 1993); affirmed 42 F.3d 191 (4th Cir. 1994).

27. *C&P Telephone Company of Virginia v. United States et al.*, Memorandum Opinion, 24 August 1993. "Sec. 533 (b) [of the 1984 Cable Act] is facially unconstitutional as a violation of plaintiff's First Amendment right to free expression. . . ." at pp. 55–56.

28. See *U.S. West v. U.S.*, Case No. C93-1523R (W.D. Wash., 1993); *BellSouth Corp. v. U.S.*, 868 F. Supp. 1335 (N.D. Ala., 1994); *Ameritech Corp. v. U.S.*, 867 F. Supp. 721 N.D. Ill, 1994).

29. Ferris et al., *Cable Television Law* (1983–date), pp. 9–48.14(21–23).

30. Ibid., p. 236.

31. Ferris et al., *Cable Television Law* (1983–date), pp. 9–48.14(25–27).

32. "Wired Worlds Tie the Knot," *Broadcasting & Cable* (October 18, 1993), p. 6. Dollar figures on the value of the deal varied depending on how they were figured, ranging from $12 billion to as much as $30 billion.

33. "Collapse on the Info Highway," *Business Week* (March 7, 1994), p. 42.

34. Mike Mills, "AT&T Corp halts Effort to Sell Local Residential Phone Service," *Washington Post* (December 19, 1997), p. C1.

35. Seth Schiesel, "AT&T to Pay $11.3 Billion for Teleport," *New York Times* (January 9, 1998), p. C1.

36. "At Last, Telecom Unbound," *Business Week* (July 6, 1998), p. 24.

37. Janet Guyon, "AT&T's Big Bet Keeps Getting Dicier," *Fortune* (January 10, 2000), p. 128.

38. Saul Hansell, "American Online Agrees to Buy Time Warner for $165 Billion; Media Deal is Richest Merger," *New York Times* (January 11, 2000), p. 1.

39. Bill McConnell, "Taking on Goliath," *Broadcasting & Cable* (June 19, 2000), pp. 14–16.

40. Stephen Labaton, "AOL and Time Warner Gain Approval for Huge Merger, But with Strict Conditions," *New York Times* (December 15, 2000), p. A1.

41. "Monitor Trustee to Make Sure AOL Time Warner Plays by the Rules," *Washington Post* (April 12, 2001), p. E4.

42. Efrem Sigel, 1982. *The Future of Videotext: Worldwide Prospects for Home/Office Electronic Information Services* (White Plains, NY: Knowledge Industry Publications, Inc.), p. 1.

43. Ibid., p. 2.

44. See Richard M. Neustadt, 1982. *The Birth of Electronic Publishing: Legal and Economic Issues in Telephone, Cable and Over-the Air Teletext and Videotext* (White Plains, NY: Knowledge Industry Publications), pp. 19–24, for a list of videotext trials.

45. For a fuller discussion of the demise of these early systems, see Christopher H. Sterling, 2006. "Pioneering Risk: Learning from the U.S. Teletext/Videotex Failure," *IEEE Annals of the History of Computing*, in press.

46. "Federal Communications Commission Releases Data on High-Speed Services for Internet Access," Press Release, Federal Communications Commission, June 8, 2004, p. 2.

47. The Commission determined that conditioned loops, capable of supporting DSL service (i.e., without bridge taps or loading coils) were a UNE. They further determined that the high frequency portion of a local loop over which the ILEC was providing regular voice service was a UNE. In other words, the ILEC had to share the line with its competitor, allowing its competitor's DSL service to ride over the local loop over which the ILEC was providing basic telephone service. The FCC also required ILECs to provide line splitting, in which two competitive local exchange carriers (CLECs) provided service over the same telephone line, with one offering basic service and the other offering DSL service.

48. Cable modem service requires a cable modem at the customer's location and a Cable Modem Termination System (CMTS) at the cable provider's headend location. The cable modem at the customer's location sends data from the customer's computer to the cable channel or, conversely, moves data from the cable channel to the customer's computer. At the cable provider's headend or transmitter location, the CMTS takes the data traffic from the customers on the LAN and delivers it to an ISP. The CMTS also sends data from the ISP to the customers. Customers share the upstream channel to the CMTS on a time division basis; each customer is allotted a segment of time to use the channel to send information to the Internet. On the upstream channel, customers do not see one another's information. On the downstream side, the CMTS sends all information to all users on the LAN; each customer's network connection determines whether the information is intended for it or not. The CMTS uses Dynamic Host Configuration Protocol (DHCP) to administer Internet addresses for its customers, and also uses the CableLabs Certified Cable Modems standard.

49. *AT&T v. City of Portland*, 216 F.3d 871 (9th Cir. 2000) *reversing* 43 F. Supp. 2d 1146 (D. Ore. 1999).

50. *Inquiry Concerning High Speed Access to the Internet Over Cable and Other Facilities, Internet Over Cable Declaratory Ruling, Appropriate Regulatory Treatment for Broadband Access to the Internet Over Cable Facilities*, Declaratory Ruling and Notice of Proposed Rulemaking, GN Docket No. 00-185 and CS Docket No. 02-52, FCC 02-77 (March 15, 2002), at 54 and 32. (Hereafter referred to as *Cable Modem Order*).

51. 47 U.S.C. Section 522 (6).

52. *Cable Modem Order*, at 41.

53. *Cable Modem Order*, at 59.

54. *IP Enabled Services*, WC Docket No. 04-36, Notice of Proposed Rulemaking, FCC 04-28 (Mar. 10, 2004), at 2.

55. *Petition for Declaratory Ruling that AT&T's Phone-to-Phone IP Telephony Services are Exempt from Access Charges*, WC Docket No. 02-361, Order, FCC 04-97 (April 21, 2004), at 1.

56. *Petition for Declaratory Ruling that pulver.com's Free World Dialup is Neither Telecommunications Nor a Telecommunications Service*, WC Docket 03-45, Memorandum Opinion and Order, FCC 04-27 (February 19, 2004), at 6.

Chapter 9

1996 Act and Aftermath (1996–2000)[1]

Driven by the industry's experience with divestiture as well as years of Congressional failure to create substantial change, passage of the Telecommunications Act of 1996,[2] and the many resulting FCC (Federal Communications Commission) rulemakings that followed, created the most important policy watershed in decades. The 1996 legislation substantially amended portions of the 1934 Communications Act which remained the baseline law.[3] After so many failed attempts, Congress defined new rules of marketplace competition while at the same time seeking—not always successfully—to limit government's long-term regulatory role. For a variety of reasons explored here and in the following chapter, however, the law would not turn out to be quite the landmark its proponents had hoped.

Prompting Congress to take action was its slow recognition that technology was radically changing the industry's operational context. As more and faster transmission capacity became available, thanks both to satellites and especially fiber optic networks, prices declined. Likewise switching and information processing time and cost also dropped thanks to expanding computer capabilities. While the combined potential of these transmission and switching developments was huge, little of it was being realized in the telecommunications sector, save by a few large carriers. Critics argued a key barrier was the 1934 Communications Act, written decades earlier to regulate a dramatically smaller industry.

To a considerable degree, the tension embedded in telecommunications policy proceedings in the 1990s resulted from a growing gulf between technical and economic realities of a digital industry versus the restrictive requirements of a law written in and for an analog age. On several occasions when the FCC had tried to adopt a more flexible approach to meet the requirements of a changing industry, the agency was thwarted by the old act's requirements.

As early as 1979 for example, the Commission had tried to develop less onerous tariff regulation for smaller companies entering the field so as to encourage such competition. Over several years of proceedings, its competitive common carrier proceeding (Docket 79-252) developed just such a reasoned response, deciding to fully regulate only the still-unified AT&T. Smaller companies like MCI would be subject to "streamlined" rules while smaller carriers lacking their own facilities would not be regulated at all. But on appeal (from the larger firms understandably worried about how such asymmetric regulation would impact their business), the courts overturned the FCC action as exceeding the boundaries of what the 1934 statute would allow.[4] An act long praised for how much regulatory discretion it gave to the FCC turned out not to be flexible enough to allow such a dramatic change a half century later. Of course the drafters of the 1934 law could not foresee the rise of multiple service providers in a field they knew as characterized by (and thus regulated as) a monopoly. Demise of the FCC competitive common carrier effort was merely one of the clearer examples of why the 1934 act needed substantial updating or total replacement (attempts to undertake the latter also failed; *see section 9.1*). The breakup of AT&T and subsequent restructuring of the industry under provisions of the antitrust laws (*see section 6.4*) was an even more stark indication that Congress was not doing its job of defining national policy in this sector.

9.1   WHY CHANGE TOOK SO LONG

The Communications Act of 1934 had long been the focus of updating or even replacement attempts, though it had never been a static document. The baseline law had been amended countless times in the six decades following its passage (*see section 4.3*), though usually in minor ways. The most substantial updates included a regulatory scheme for Comsat (1962), creation of a national system of public broadcasting (1967),[5] the first (and largely deregulatory) statutory policy for cable television (1984),[6] and the subsequent re-regulation of cable (1992).[7] But such changes affected only parts of the 1934 legislation (primarily Titles III and VI)[8] and had little or no impact on larger telecommunications questions controlled by Title II which would form the heart of the 1996 amendments.

Earlier Attempts

Congress had grappled with different approaches to broad-ranging tele-communications legislation for years. It even tried to comprehensively address multiple issues by completely replacing the 1934 law. This multiyear process had begun nearly two decades earlier. From 1976 to 1982, House (and later, Senate) committees considered, and in a few cases actually passed, draft legislation that would have updated government policies to match dramatic telecommunications industry changes (*see beginning of chapter 6*). Each subsequent rewrite version introduced, however, moved closer toward the status quo as Congress struggled to placate countless conflicting industry concerns. For while most parties claimed to want change, all could continue to live with the existing law everyone understood (better the devil you know . . .). Indeed, as the process dragged on, competing lobbying groups became more entrenched, increasingly concluding that they were happier with the obsolete 1934 law (perhaps with minor tweaking) than they would be with dramatic change. The old axiom that it is easier to slow or even halt prospective legislation than pass it certainly proved true in this frustrating period.

The failure of this intensive effort after so much work (recorded in dusty shelves of multivolume hearing transcripts[9]) served to erode most House and Senate interest in dealing with communications issues for nearly a decade. Frustrated legislators found the issues arcane, the terminology jargon-filled, and the conflicting industry positions unreconcilable. The 1984 AT&T divestiture further discouraged action as Congress was persuaded to let that legal process work itself through, and let industry players and issues sort themselves out in a marketplace fashion rather than by legislative fiat. But growing resistance to the regulatory barriers inherent in the 1934 law eventually made clear that another try at overhaul was necessary.

1986 Dole Bill

Constantly seeking a means of breaking away from the Modified Final Judgment (MFJ) line-of-business restrictions (*see section 6.4*), Regional Bell Operating Company (RBOC) officials increasingly sought legislative relief. Though no bill passed either house, one attempt introduced by Senate Majority Leader Robert Dole (R–KS) in 1986 got considerable attention. As with most other bills, the Dole bill would have shifted responsibility for MFJ administration from Judge Harold Greene to the (presumably more malleable) FCC.

But Dole's bill went farther than merely shifting the furniture. Sections mandated equal access for all interexchange carriers to local telephone companies, and continued the line-of-business restrictions while opening options for the FCC to abolish them, but only when "there was no substantial possibility" that a local company "could use its monopoly power to impede competition in the market it seeks to enter." AT&T (not named but still the dominant interexchange carrier) was forbidden to "acquire the stock or assets of any Class I local telephone company." But most importantly, the FCC was empowered to consider all options, including doing away with the line-of-business or any other restrictions it felt harmed the provision of effective and efficient service to end users.[10]

Had it passed, the Dole bill would have dumped a huge workload on the FCC as it adopted rules and evaluated the companies' efforts to comply with them. The Commission would certainly have regained its designated place as the primary regulator of the telephone business.[11] But legislators still seemed reluctant to deal with the issues because they were too complicated, too frustrating, and far too politically dangerous to take up in an election year, and the Dole bill faded from view.[12]

Discussion of specific potential language was renewed in the early 1990s when the Democrats still controlled both houses of Congress—and industry need for a legislative overhaul became even more acute. Industry and FCC (and even individual state public utility commission) decisions continued to expand the competitive telecommunications marketplace—at least as far as seemed possible under the 1934 law. While this regulatory progress and the industry's continued rapid pace of change added urgency to the need for the House and Senate to act, many factors conspired to drag out discussions and delay final action until early 1996.

Industry Complexity

First, the issues were (and remain) genuinely complex—nothing less than redefining what various telecommunications players could do and under what conditions. Many economic and technical matters involved were arcane to all but those immersed in the issues. While traditional regulatory regimes for both electronic media and telephone communications had long been in place, the 1934 law on which they were based was increasingly perceived as an antiquated block to the growing convergence of telecommunications services brought about by continuing technological advancement.

Congress therefore needed to understand the implications of increasing digitalization of telecommunication technology as it forced substantial economic and political change. Most U.S. laws and related FCC rules and regulations had been written when broadcasting and telephone carriers utilized distinct and separate analog technologies. Attempts by telephone, broadcast, and cable entities to enter the others' businesses bumped up against obsolete regulatory barriers that prevented offering efficient competitive services to end users. The old law clearly needed updating, though just how—and how much—was a matter of substantial disagreement.

Largely defensive industry positions as well as considerable distrust were deeply entrenched and had changed little since the congressional "rewrite" attempts of decades before. Divisions between interexchange and local telephone carriers, and between the cable and telephone industries, for example, had long divided those players, making it difficult to develop cooperative approaches. While many in industry spoke out in favor of deregulation and change in general terms, when it came down to specific legislative provisions, protection of status quo "turf" was usually preferred to real change.

Political Factors

The shift in control of Congress as the Republican Party (GOP[13]) took over in 1995 for the first time in 40 years, slowed debate while a different set of Congressional members and staff learned the issues, as well as how to manage the legislative process. New committee chairs and changing legislative priorities had to be accommodated while formerly powerful Democrats (and their experienced staffs) retreated to background roles or left entirely. For much of the first year of GOP control, telecommunications issues were not a high priority.

Becoming more evident during this transition was a change in ideological priorities. Even before the GOP takeover, Congress had increasingly reflected the country's growing conservative conclusion that less government was better and, other things being equal, competition within a marketplace was to be encouraged over government rules and regulations. The widely touted Republican 1994 "Contract with America" theme focused on this feeling though it did not specifically mention telecommunications. Other more pressing concerns again postponed any potential telecommunications legislative action.

Most immediately, political debate became caught up in an unprecedented 1995–96 budget confrontation between Congress and the Clinton Administration. Battles over conflicting budget priorities raged while hundreds of thousands of federal workers were kept off the job in a weeks-long shutdown (for lack of operating funds) of most parts of the federal establishment. That battle and the displacements it created—focused more sharply amidst 1996 presidential election year tensions—further delayed action on telecommunications.

Less evident to many was that despite all these divergent concerns, Congress *had* to act if it were to retain a relevant role in communications policy. Growing accustomed to legislative inaction over the years, companies and regulators alike sought to develop viable policies within the obsolete law. Companies continued to make strategic market-driven decisions (as with AT&T's 1995 trivestiture [*discussed in section 7.2*], announced in expectation of passage of what became the 1996 law) and the FCC continued to streamline its regulatory approaches. But both government and industry could go only so far without fundamental Congressional redefining of the statutory arena so further change could take place. Passage of the 1996 amendments placed Congress back in its proper Constitutional role of defining overall national policy.

Finally, of course, the legislative process can be both counterintuitive and counterproductive.[14] Derailing a bill is easy because there are so many places where further amendment or delay can prove fatal to its passage. What became the 1996 legislation nearly derailed several times, including once just days before final passage when Senator (and by then presidential contender) Robert Dole came out strongly against a spectrum feature in the bill. Both local exchange and interexchange companies pressed hard for Congress to develop legislation that would help, or at least not badly hurt, their position. AT&T was one of the most active companies lobbying Congress in the final weeks of debate and mark-up. Such lobbying both helped and hindered the legislative process as education was melded with persuasion in support of a specific point of view.

Both houses finally passed different but sweeping versions of a potential telecommunications act in the fall of 1995. They were consigned to a Senate–House conference committee which reported its compromise results late in the year. Both houses then passed the lengthy compromise on the same day in early February 1996, and the president signed the 100-page bill into law a week later. The changes the new act required were extensive and far-reaching, though some would become evident only with the passage of time.

9.2 LEGISLATING COMPETITION AND CONVERGENCE

The most far-reaching and divisive question dealt with in the legislation was how best to encourage more telephone market entry, and thus greater competition. The 1996 amendments added an important new subpart to the 1934 Act's Title II, entitled "Development of Competitive Markets." This turned the traditional regulatory regime on its head. Most of these provisions were new in both scope and language—yet they grew out of debates in the 1980s and 1990s on how best to fit competitive options into a one-time monopoly industry. And more specifically, they reflected a growing conviction that the 1984 redefining of AT&T (and thus telephony) in an MFJ that had to meet antitrust concerns had been too narrowly drawn for such a dramatic industry restructuring.

(Re)Introducing Local Competition

By the mid-1990s, agreement was widespread (save by some incumbent local exchange carriers or ILECs—meaning primarily the regional Bell firms established after AT&T's breakup) that encouraging new market entry by competitive local exchange carriers (CLECs) would lower customer rates.[15] Such competition would, of course, restore what had once existed in local telephony earlier in the century. The 1996 Act superseded the MFJ, and so promised the RBOCs entry into long distance and other services, but at a price. But while the RBOCs *seemed* to accept that local competition was a *quid pro quo* for entering formerly forbidden (by the MFJ) competitive lines of business, they sharply resisted *real* attempts to introduce such competition. Potential CLECs correctly argued that their provision of local service was a far more difficult, costly, and risky undertaking than was RBOC entry into long distance or manufacturing. Thus, the CLECs continued, RBOCs should be restrained from entering long distance or manufacturing markets until *after* real local exchange competition took root. The law seemingly accepted that CLEC viewpoint, but defining "real" became the central legislative issue.

The 1996 law, while requiring interconnection between all telecommunications carriers, placed a greater onus on ILECs (both RBOCs and Independents), requiring that they provide full interconnection to any CLEC requesting same. This was defined to include such things as equitable resale,[16] number portability,[17] dialing parity,[18] and access to both rights of way,[19] as well as any support and repair services needed. Remembering AT&T's extended battle against interconnection with its long-distance rivals in the 1970s (*see section 5.5*), Congress further mandated quite specific procedures for negotiation, arbitration, and approval of intercompany agreements to assure that such interconnection took place. Any state and local laws restricting competition were to be eliminated or they would face FCC preemption. The Commission was given a year to specifically identify and eliminate market entry barriers for entrepreneurs and small businesses—and conduct an updating review every three years thereafter.

Utilizing the impatience of the Regional Bell Operating Companies (RBOCs) to enter the lucrative inter-LATA (Local Access and Transport Area) long distance market as something of a regulatory club, Congress defined local exchange competition with a 14-point "checklist" by which the FCC was to determine whether any RBOC could offer long distance ser-

Telecommunications Act of 1996 Competitive Checklist
(Section 251, Interconnection)

1. Interconnection at technically feasible points.
2. Nondiscriminatory access to network elements.
3. Nondiscriminatory access to the poles, ducts, conduits and rights of way.
4. Local loop transmission from the central office to the customer's premises, unbundled from local switching or other service.
5. Local transport from the trunk side of a wireline local exchange carrier switch unbundled from switching or other services.
6. Local switching unbundled from transport, local loop transmission or other services.
7. Nondiscriminatory access to (a) 911, (b) directory assistance, and (c) operator call completion services.
8. White pages directory listings for customers of the other carriers' telephone exchange service.
9. Nondiscriminatory access to telephone numbers.
10. Nondiscriminatory access to databases and associated signaling necessary for call routing and completion.
11. Number portability (details to come from FCC).
12. Dialing parity (equal access to numbers, operator services, directory assistance, and directory listing, "with no unreasonable dialing delays").
13. Reciprocal compensation arrangements (for carrying or terminating telecoms).
14. Resale (at wholesale rates of any telecom service provided at retail to subscribers who are not telecommunication carriers).

FIG. 9.1.

vice to its own local customers (see Fig. 9.1). The list's requirements centered on existence of at least one facilities-based[20] CLEC in a given RBOC's region (*or* a total lack of apparent interest by competitors in providing such a service); consultation with the Department of Justice; and RBOC agreement to provide nondiscriminatory CLEC access, interconnection, and all other needed services as noted above. And the FCC was given one more regulatory tool to supplement the list—a requirement that it determine whether provision of interexchange service by an RBOC would be in "the public interest, convenience, or necessity," those classic words repeatedly used in the 1934 act to provide the FCC with regulatory discretion.

Likewise, pay telephone operation (already offered by all ILECs and many CLECs as well) was further encouraged with nondiscrimination requirements similar to those in the FCC's *Computer III* decision.[21] (Unforeseen at the time was how quickly the spread of cell phones would make static pay telephones increasingly obsolete—within five years, pay phones were rapidly declining in use and number nationwide.) The intent of the new legislation was clear: the local telephone carriers would have to provide access to developing competitors; and the RBOCs would face stringent compliance standards if they wanted the freedom to enter other sectors of the telecommunication business.

The "War of All Against All"

While RBOCs argued that their entry into interexchange service would increase competition and thus lower end-user prices (ignoring that the same logic extended to local service), existing interexchange carriers[22] countered that the real competitive bottleneck lay with the local loop, naturally trying to deflect change in their own marketplace. After divestiture, AT&T vigorously opposed RBOC entry into interexchange markets, arguing that as AT&T had been forced to divide into separate monopoly and competitive sectors, surely the monopoly RBOCs should not be allowed to recombine them. In addition to the "checklist" noted above, as a transitional safeguard, Congress mandated that any RBOC provision of interexchange service or manufacturing could be undertaken only through a separate subsidiary, a structural device designed to inhibit illegal cross-subsidy from monopoly to competitive services.[23] But in yet another example of mixing new requirements with deregulation, Congress limited this requirement to a three-year period[24] unless the FCC found that conditions required extending it (and when the time came, the commissioners did not).

Relieved at final passage of what had been expected for months, several RBOCs announced moves to enter new markets within days of the amendments' passage. Pacific Telesis said it was establishing an interexchange subsidiary to offer long distance services outside of its local service area, and Bell Atlantic made public plans to do so as well and to quickly move to ask for FCC permission to offer similar services to its local service customers.

In addition to their quest for approval to enter the interexchange long distance market in all of their territories, which all of the RBOCs finally attained in 2003 (*see section 9.5*), the RBOCs proved eager to enter the wireless, Internet, and even satellite TV markets in the years following the 1996 amendments. In addition to entering new markets, the RBOCs also engaged in significant merger activity that reduced the number of RBOCs from seven to four (*see section 9.6*).

Three RBOCs either entered into, or solidified their positions in, the wireless market during the early years of the new century. In 2000 Southwestern Bell Communications (SBC) solidified its wireless position by combining its holdings with BellSouth's wireless properties to form Cingular Wireless. Verizon (formed from a merger of Bell Atlantic and NYNEX—*see section 9.6*) combined with Vodaphone to form Verizon Wireless, which became the largest cellular provider in the United States.

In various ways the RBOCs also entered the Internet market by offering digital subscriber line (DSL) services. There was also RBOC activity in the CLEC market. As a requirement of its purchase of Ameritech, for example, SBC agreed to become a CLEC in 30 cities outside of its service area by 2002. SBC also sought to enter the satellite TV market, first by unsuccessfully attempting to purchase Hughes Electronics, and then by marketing the EchoStar satellite TV service.

Not only RBOCs sought to enter new markets. The long distance companies, seeking new sources of revenue and a better position in the telecommunications marketplace, also looked to new markets. Before its take over by WorldCom, MCI sought entry into the local exchange market. Its losses in that effort made its merger with British Telecom less likely, thus opening the door to WorldCom's takeover bid (*see section 9.6*). Meanwhile AT&T reconfigured itself into a company with interests in broadband, cable television, and wireless (*see section 8.4*). Passage of the 1996 amendments had encouraged companies to strike out in new directions, and to try to strategically position themselves to be able to offer their customers a bundle of services, from local to long distance to wireless to Internet access.

Convergence?

The 1996 amendments also reflected a telecommunications industry trend already evident: the growing convergence of media, telecommunications, and information services across what used to be sharp regulatory divisions.

A primary factor limiting some convergence efforts had been ownership controls. As one prime example discussed earlier (*see section 8.4*), the FCC for 25 years banned telephone company ownership of cable television facilities operating in the same area, with the exception of rural counties that might otherwise not obtain cable service at all. Telephone carriers *had* been allowed to *construct*, though not operate, cable systems in several urban areas. Despite intense RBOC lobbying to allow local cross-ownership, lawmakers in 1996 were persuaded that simply allowing one monopoly carrier to take over another did little for competition. After considerable debate, Congress decided not to allow simple buy-outs by telephone carriers of existing cable systems, limiting holdings to no more than 10 percent in most cases.

There were exceptions to this restriction. Such buy-outs *were* allowed in smaller markets (under 35,000 people) that otherwise might not receive cable service. Further, under certain competitive conditions, buy-outs might even be allowed in much larger markets (outside the largest 25 urban areas). But while the FCC was given some regulatory discretion here, most telephone carriers seeking to provide cable service would have to develop their own facilities-based system.

A related provision of the 1996 law took this concept a step further and defined a new kind of video common carrier. Such an "open video system" (OVS), owned by a local exchange telephone carrier, was designed to fall between traditional cable and telephone carriers in terms of regulation, subject to final FCC regulations. An OVS operator would be required to make all of its video channels available to unaffiliated programmers without discrimination, and if demand exceeded system channel capacity, the OVS could select programmers for no more than a third of that capacity. OVS systems had to comply with existing restrictions on cable television programming.[25] On the other hand, OVS operators were not required to meet other FCC cable rules or local cable franchise requirements. While it seemed an interesting idea at first, critics on all sides soon agreed the OVS concept would be a failure in competition with an entrenched cable industry, as indeed proved to be the case—none had appeared nearly a decade after the law's passage.

Consistently the 1996 amendments allowed RBOCs to enter lines of business prohibited to them under the MFJ that had broken up AT&T a dozen years earlier (*see section 6.4*). In most cases, however, any such entry required use of a separate subsidiary firm for a transition period of several years, as with interexchange functions. An RBOC would only be allowed to begin research, development, and manufacturing of telephone equipment after establishment of competitive local exchange services within its own region.

9.3   STREAMLINING GOVERNMENT

The 1996 amendments package ended or considerably changed many telecommunication regulations. The 1982 MFJ (*see section 6.4*) was effectively terminated and in a brief order issued April 12, 1996, Judge Greene complied, vacating the document. Swept away was the convoluted regulatory process that for a dozen years underpinned his unique regulatory role (*see section 7.1*). The end of the MFJ paved the way for the FCC to resume its primary regulatory role in telecommunication, subject, as always, to Congressional oversight.

Easing FCC Rules and Regulations

The FCC's renewed central role grew in part from the 1996 amendments themselves, which required the Commission to undertake some 80 rule-making proceedings in 1996–97, many requiring completion within six months of the bill's passage. Some were combined because they were closely related and because of time pressures. Several inquiries would establish basic definitions for who could do what within the new Congressional mandates. Thus these were hotly contested proceedings. Despite these added responsibilities, however, continuing budget pressures and conservative Congressional ideology made certain the FCC would have to undertake all this effort with no more, and possibly fewer, people and resources.

As it often had before, Congress concluded that the FCC's lengthy and paper-laden regulatory procedures had gotten out of hand and needed streamlining. The 1996 amendments required establishment of shorter time limits on handling petitions, applications, and complaints. Additionally, the amendments actively encouraged the FCC to use private sector facilities for equipment testing and certification.[26] Every two years (beginning in 1998) the Commission was mandated to undertake a complete review of its rules and regulations and to eliminate those found no longer useful. This was the first time that the law required what had been an informal FCC practice for many years.

Further, the FCC for the first time received specific legal authority to totally *forbear* from some common carrier regulation. As noted above, the Commission's "Competitive Common Carrier" proceeding had attempted to do just this but was rebuffed in the courts. With the 1996 amendments, Congress remedied that problem and now required the FCC to *at least* consider forbearance in sufficiently competitive situations. This was an open door for further deregulation.

Redefining State and Local Roles

In several instances the 1996 amendments modified the relationship between state regulatory agencies and the FCC, long an arena of delicate political sensibilities. Some amendments appeared to encourage the FCC to do something already common in environmental regulation, that is, share more decisions with the state public utility commissions. The 1996 amendments gave the state commissions the primary role in arbitrating negotiations between ILECs and CLECs.

On the other hand, state public utility commissions were preempted from maintaining regulation in instances where the FCC determined that forbearance was appropriate national policy. Such preemption of state action is usually designed to avoid a patchwork quilt of varied regulations for similar services in different regions. One amendment required state public utility commissions to adopt such regulatory approaches as price cap regulation, forbearance from regulation, or other means to encourage new investment in telecommunications. Another amendment removed a once-contentious issue between the FCC and state agencies in repealing the requirement that the FCC set depreciation rates for common carriers.[27] Instead, the Commission would apply the depreciation rates used in the federal tax code and could use outside auditors to reduce its own personnel requirements.

From the beginning, however, it seemed evident that the eventual impact of the 1996 law on the states might have an ironic twist. For some considerable time they will continue to play an important role in implementing all the mandated changes. In the long run, however, the regulatory pendulum in telecommunication seems to be swinging toward centralized federal policies developed in and administered from Washington as opposed to the variation of state-originated policy initiatives. Indeed, the FCC, in its Triennial Review (*see section 9.5*) was unsuccessful in delegating decision-making authority about local competition to the states; the courts found that such authority rested with the Commission.

This swing in policy initiative was also evident in the constraints placed on local authorities. Cable system operators (telephone carrier-controlled or otherwise) were now allowed to compete without prior permission of the local franchise authority. Nor could that authority play any role in cable equipment selection or standards. Local zoning authorities were also prevented from totally blocking location of cell phone and other antenna towers. Instead, the FCC was to establish national guidelines. Likewise, local authorities (even down to individual condominium boards) were constrained in their ability to control or eliminate use of satellite receiving antennas, or to place taxes on such equipment.

9.4 REDEFINING UNIVERSAL SERVICE

The 1996 amendments envisioned an industry in which the marketplace would govern all decisions. Such a procompetitive stance would seem to leave little room for universal service considerations, which are based on social policy rather than economic efficiency (*see section 7.3*). The Telecommunications Act, however, expressed a surprisingly strong commitment to universal service, codifying the term for the first time, providing a definition, listing universal service principles, and, some would argue, providing a means for expanding the term to encompass much more than residential local telephone service. The Act also required an end to implicit subsidies and the development of competitively neutral support mechanisms. Formulating policies to realize these goals has proven to be a difficult task.

Creating a National Policy

The 1996 Act articulated a strong commitment to a national concept of universal service and provided specific guidelines for the implementation of such a policy. Section 254 contains a definition of universal service—the first time the term appears in national legislation:

> (1) IN GENERAL.—Universal service is an evolving level of telecommunications services that the Commission shall establish periodically under this section, taking into account advances in telecommunications and information technologies and services. The Joint Board in recommending, and the Commission in establishing, the definition of the services that are supported by Federal universal service support mechanisms shall consider the extent to which such telecommunications services
>
> (A) are essential to education, public health, or public safety;
> (B) have, through the operation of market choices by customers, been subscribed to by a substantial majority of residential customers;
> (C) are being deployed in public telecommunications networks by telecommunications carriers; and
> (D) are consistent with the public interest, convenience, and necessity.[28]

Congress envisioned the possibility that the scope of universal service might actually be expanded as new technologies are developed and widely adopted by subscribers. Section 254 goes on to list principles that are to guide the provision of universal service.

> (1) QUALITY AND RATES.—Quality services should be available at just, reasonable, and affordable rates.
> (2) ACCESS TO ADVANCED SERVICES.—Access to advanced telecommunications and information services should be provided in all regions of the Nation.
> (3) ACCESS IN RURAL AND HIGH COST AREAS.—Consumers in all regions of the Nation, including low-income consumers and those in rural, insular, and high cost areas, should have access to telecommunications and information services, including interexchange services and advanced telecommunications and information services, that are reasonably comparable to those services provided in urban areas and that are available at rates that are reasonably comparable to rates charged for similar services in urban areas.
> (4) EQUITABLE AND NONDISCRIMINATORY CONTRIBUTIONS.—All providers of telecommunications services should make an equitable and nondiscriminatory contribution to the preservation and advancement of universal service.
> (5) SPECIFIC AND PREDICTABLE SUPPORT MECHANISMS.—There should be specific, predictable and sufficient Federal and State mechanisms to preserve and advance universal service.[29]

These principles lay out a fairly comprehensive national vision of universal service. Rather than long distance carriers paying for universal service through access charges, all telecommunications providers are to contribute in a nondiscriminatory manner to the "preservation and advancement of universal service." Instead of implicit subsidies, all support for universal service is to be explicit and predictable. Further, instead of universal service being provided exclusively by the incumbent local telephone companies, universal service is to be provided by a new category of provider: the Eligible Telecommunications Carrier (ETC).[30] ETCs are to be designated by the state commissions to provide the services deemed to comprise universal service to all subscribers in a service area defined by the state commission. In return for providing these services, the carrier is eligible to receive federal universal service support.

After passage of the Telecommunications Act, the ILECs were designated as ETCs in their respective states. The language of the Act suggests that universal service is to be provided on a competitive basis, with multiple ETCs designated in nonrural service areas: "the State commission may, in the case of an area served by a rural telephone company, and shall, in the case of all other areas, designate more than one common carrier as an eligible telecommunications carrier for a service area designated by the State commission."[31] Some CLECs have received the ETC designation and have received universal service support.[32] The Act also sanctioned continuation of existing state Lifeline (universal service) programs.[33]

The 1996 amendments did not specify the services for which ETC could get support. Those were determined by an FCC-PUC Joint Board[34] that met after the passage of the Act. The Joint Board was conservative in its approach, creating a fairly short list of service elements to be eligible for universal service support:

- Single-party voice service;
- Voice grade access to the public switched network;
- Touch tone service;
- Access to interexchange services;
- Access to emergency (911) services;
- Access to operator service and directory assistance; and
- Limited toll service for low-income customers.[35]

This list comprises plain old telephone service; no advanced services, such as Internet access, are included. Though Section 254(c) allows for periodic review of this definition, subsequent meetings of the Joint Board have thus far failed to extend this list, considering and rejecting the addition of broadband access, so-called soft dial tone, intrastate or interstate toll, and prepaid calling plans. They also did not include equal access for fear that such a requirement might discourage wireless service providers from seeking ETC status.[36] The Joint Board balanced concerns about universal service with those about maintaining and encouraging competition. As noted in the Joint Board recommendation:

> Requirements that do not unduly prevent new entities from achieving ETC status may serve the public interest and be competitively neutral because they may increase competition, which may lead to innovative new services and lower prices. Moreover, changes that eliminate all potential ETCs in a given area would undermine the goal of providing universal service in all areas. Accordingly, we conclude that it is appropriate to consider the impact of adding a service to carriers' eligibility for ETC status under the public interest criteria and the principle of competitive neutrality when determining whether to modify the existing definition of universal service.[37]

The ETC and the definition of universal service provided in Section 254 (c) apply to traditional universal service: affordable local voice service provided to residential subscribers. The 1996 amendments added a new category of recipient: the institutional subscriber. The Act specifies that rural health care providers, elementary and secondary schools, and libraries shall also receive service at discount rates determined by the FCC. Service providers recoup these discounts from universal support mechanisms.[38] Dubbed the "E-Rate," the discounts for schools and libraries cover Internet access and internal network connections, marking a departure from the move conservative approach to ETC services. While the latter has stayed firmly mired in terms of plain old telephone service, the E-Rate moved support for institutional universal service into advanced information services.

. . . and Paying for It

The 1996 amendments specified an end to implicit subsidies, requiring that all universal support mechanisms be "specific, predictable, and sufficient." The Commission responded by creating a new Universal Service Fund into which all telecommunications providers pay their equitable share and from which all appropriate recipients receive explicit and specific payments. The Commission selected the Universal Service Administrative Company (USAC), a subsidiary of National Exchange Carrier Association (NECA), to calculate, collect, and disseminate payments from the fund. Governed by a board of directors made up of representatives from local telephone companies, long distance providers, and representatives from schools, libraries, and rural health care, USAC calculates a quarterly contribution factor, or percentage, that is applied to the interstate and international revenues of wireless providers, long distance companies, providers of interstate access services, resellers, telegraph and telex providers, payphone aggregators, and other telecommunications providers.[39]

The fund has become a hodgepodge of funds, payments, and recipients. Rather than eliminating the former subsidies that had been in place (*see section 7.3*), the Commission rolled them into the new fund, so that local companies who had been receiving support for high-cost loops through the old Universal Service Fund and Long Term Support mechanisms, would continue to do so. Local companies providing Lifeline and LinkUp services would continue to get their foregone revenues.[40] However, instead of these monies coming from access–related charges levied on long distance carriers, they would come from a new Universal Service Fund paid into by all interstate service providers. And as the Commission has added new subsidies, they have also been added to the fund.

The E-Rate mentioned above added $2.25 billion annually to the fund; these monies, collected from all telecommunications providers, are distributed to service providers who supply eligible schools and libraries with all commercially available telecommunications services, Internet access, and internal connection (including routers, network file services, local area networks) at a discount. The size of the discount ranges from 20 up to 90 percent depending on the level of economic disadvantage of the school or library and whether the school or library is in an urban or rural area. Also added to the fund were monies for service providers offering rural health care institutions with discounted telecommunications services; this support is capped at $400 million annually.

In addition to the former subsidies for high cost companies (like the old USF and LTS mentioned earlier), the Commission has added even more subsidy payments to be distributed through the new Universal Service Fund. These new subsidy payments have resulted from continuing efforts by long distance carriers to reduce the access charges they pay local telephone companies and to avoid paying the Carrier Common Line Charge (CCLC) altogether (*see section 7.3*).

Long distance carriers and some policy makers had long argued that the CCLC charge should be eliminated and that interstate local loop costs should be recovered totally from the subscriber through the monthly Subscriber Line Charge (SLC). The Commission, in its order reforming access charges after the passage of the Telecommunications Act of 1996,[41] proceeded slowly in shifting local loop costs to the SLC, but continued to look for other ways for local carriers to reduce the CCLC and still recover local loop costs.[42] These efforts were not enough for long distance and local companies who wanted to eliminate the CCLC when faced with the prospect of increasing competition. As one commentator noted, local telephone companies feared that AT&T, because of its merger with a large cable television giant, would be able to provide its own last-mile, therefore no longer needing the local telephone companies' facilities, especially if they came at high access rates. At the same time, long distance companies worried that the local companies could use excess revenues from above-cost access charges to subsidize their entry into the inter-LATA long distance market.[43]

A consortium of long distance and local companies formed a Coalition for Affordable Local and Long Distance Service (CALLS) and offered a plan to reduce access charges and eliminate the CCLC. The FCC adopted the CALLS proposal, with some modifications in mid-2000.[44] In addition to reducing switched access charges, the plan phased up the SLC to a capped rate of $6.50 by mid-2003, and phased out the CCLC. For those local companies for whom a $6.50 SLC was not sufficient to cover their loop costs, the plan created a new $650 million subsidy which would be administered through the Universal Service Fund.

The CALLS plan applied to the larger telephone companies. The smaller local companies, most of which were defined as "rural telephone companies" in the Telecommunications Act,[45] also feared bypass. Four groups representing the smaller companies,[46] and calling themselves the Multi-Association Group or MAG, filed a petition with the FCC asking for a plan similar to the CALLS proposal to be developed for the small rate-of-return carriers.[47] The Commission allowed them to also increase their SLC to $6.50 by 2003 and to eliminate their CCLC. The CCLC was replaced by a new universal service subsidy, Interstate Common Line Support, that would be available to all universal service providers serving the small telephone companies' service area.[48] This new subsidy would also be administered through the Universal Service Fund.

In addition to former subsidies like the LTS and new subsidy payments like the E-Rate and the subsidies created by the CALLS and MAGS plans, the Universal Service Fund includes support payments for the ETCs, those local companies who provide the list of services defined as constituting basic universal service. Providing support for ETCs, while at the same time encouraging the development of competition, proved to be a contentious and complex matter for the Commission.

The 1996 amendments envisioned multiple ETCs; the FCC had to develop support payments that would send the appropriate signals to potential ETCs. If support payments were too low, they would discourage ETC entry; if they were too high, they might encourage too many ETCs to enter the market and reward inefficiency. Nor was it clear whose costs to use in calculating subsidy payments. If a CLEC were more efficient than an ILEC, why reward that CLEC with subsidy payments based on the ILEC's costs? If subsidy payments were based on the more efficient

CLEC's costs, would that provide sufficient support for the ILEC? If subsidy payments were based on each carrier's actual costs, would that constitute nondiscriminatory, competitively neutral support, as required by the 1996 amendments?

Traditional subsidy payments had been based on fully allocated historical costs. The FCC decided to abandon this approach, opting instead for an approach based on the "forward-looking" cost of providing service, concluding that "forward-looking costs are the basis of economic decisions in a competitive market, and, therefore send the correct signals for entry and investment."[49] The Commission would no longer worry about whether a carrier recovered every dollar spent on providing service, but rather would worry about whether the carrier received the right signals about whether or not to provide service. Adopting forward-looking costs still did not determine *whose* costs would be considered. The Commission decided that, instead of looking at anyone's actual costs, it would develop theoretical costs. Basing support payments on the outcome of a theoretical model of what it *should* cost to serve a specific area, rather than on the actual costs of a specific provider, would provide identical treatment to both the incumbent and the new entrants, thus giving neither a competitive advantage. Creating a model based on the most efficient way of providing service would also encourage efficiency and keep the size of universal support payments low.

The development of such a model, termed a Cost Proxy Model, proved to be an especially contentious process. Various groups proposed different models.[50] In late 1999 the FCC finally adopted a proxy model that combined many suggested elements.[51] The inputs to the Cost Proxy Model use current costs, assume the use of new technologies and efficient engineering assumptions, and make provisions for such factors as terrain in determining the most efficient cost of providing service to the wire centers in all the areas served by nonrural local exchange carriers.[52] Since 2000, the Cost Proxy Model has been used to calculate high-cost support for nonrural local telephone companies.

In formulating this new high cost support mechanism, the FCC took a new approach. Instead of targeting support to *carriers*, the new system targets support to *states*.[53] Only those states for which the model calculates an above-average cost per line receive support. The Commission's goal was to assure comparable rates across states. The states would have to address the issue of comparability within their borders. In theory at least, those states with an average forward-looking cost below the national average would have fewer high-cost wire centers requiring support and so would not need federal assistance.

The new system created very different support patterns than had prevailed. Initial estimates of the impact of the new approach suggested that nonrural carriers in only eight states would receive support, while under the old approach, nonrural carriers in 19 states had been supported. Nonrural carriers in Maine, who had received no support before, would be eligible for over $10 million under the new system, while nonrural carriers in California, who had received more than $6 million, would be eligible for no support at all under the new system.[54]

The FCC proceeded cautiously in changing the manner in which high-cost support is calculated for rural companies. Recognizing that it would be difficult to create a forward-looking model that would accurately predict the cost characteristics of some 1,300 diverse small telephone carriers, the Federal–State Joint Board appointed a Rural Task Force (RTF) to examine the unique needs of such firms. The RTF recommended that actual costs be the basis for support, that traditional universal service subsidies be continued (with some minor changes), and that $1.3 billion in additional support be provided to rural telephone companies to encourage new investment and to assist those rural companies experiencing significant growth in number of lines. The FCC adopted virtually all of the RTF's subsequent recommendations to be effective for five years from mid-2001.[55]

The varied and complicated subsidy payments—including support for nonrural ETCs, as calculated by the Cost Proxy Model, and the additional support for rural companies recommended by the RTF—have brought the Universal Service Fund to significant levels. USAC projected a need for $1.7 billion for the first quarter of 2005 alone; to generate that amount, telecommunications providers would need to contribute 10.7 percent of their interstate revenues to the fund.[56] Collection and disbursement of a fund this size has caused considerable controversy, both about who pays into the fund and who receives support from it. Only telecommunications providers contribute to the fund, which means the providers of new services such as cable modem service and Internet telephony are exempt. Because the contribution base for the USF is shrinking—largely because of declining long distance revenues caused by decreasing long distance rates and competitive inroads from new services like Internet telephony— the contribution factor continues to increase. As a result, providers of new technologies and services are at a competitive advantage because they are exempt from universal service obligations while the providers of traditional services are facing an increasing level of commitment. The ability of nontraditional providers, especially wireless carriers, to qualify as ETCs and therefore to be eligible for USF support has been another contentious issue.[57]

Hoping to limit the growth of high-cost support, the Joint Board recommended that only the provider of a subscriber's primary line be eligible for high-cost support and only for that primary line. This would be a change from the current system in which all providers of all lines, wireline and wireless, received high-cost support. The Joint Board further suggested that the basis for support be "rebased," and that CLECs who serve rural areas receive high-cost support on some basis other than the actual costs of the rural ILEC in that service area.[58]

Despite the controversies, the USF and the subsidy mechanisms which it represents do reflect the post-Telecommunications Act requirements for a universal service policy. All telecommunications providers help to support universal service; support payments are portable, rather than limited to incumbent telephone companies; support payments are explicit and identifiable. Subsidies have been wrung out of access charges, thus facilitating competition. The FCC has not been doctrinaire in its new approach to universal service, however. It has proceeded cautiously in providing support for rural telephone companies, retaining remnants of the old support systems in order to assure the continuation of affordable local rates in rural areas.

From Universal Service to Universal Access?

Some policymakers and consumer advocates have argued that the list of services defined as constituting universal service should be expanded to include a host of new and advanced services, including broadband; these requests have been in response to the Telecommunications Act's definition of universal service as "an evolving level of telecommunications services that the Commission shall establish periodically . . . taking into account advances in telecommunications and information services." Proponents of this approach call for a redefinition from universal *service* to universal *access*, calling for support of a wide range of services, including the ability to access and use online services, and warning against the possibility of a "digital divide" between information haves and have-nots.[59] Thus far the FCC has rejected these requests, limiting the list of supportable services in the interests of competition, hoping that by keeping the list short, a wider range of carriers will be able to qualify as ETCs.

The 1996 Act places some onus on the FCC to facilitate the widespread deployment of new technologies and services, stating that "it shall be the policy of the United States to encourage the provision of new technologies and services to the public."[60] To realize this policy objective, and at the request of Congress, the FCC conducts regular inquiries concerning the availability of advanced services and the need to intervene to facilitate their deployment. By 2004, the FCC had conducted four such inquiries, finding in each of these proceedings that the deployment of advanced telecommunications capability was "reasonable and timely on a general, nationwide basis"[61] so that no immediate regulatory action was needed.

Indeed, the prospects for the scope of universal service being expanded much beyond the basic telephone services that have been the traditional targets of universal service support are pretty slim. The more services that are included as supportable, of course, the more expensive universal service becomes. An expensive universal service policy could present a significant obstacle to the development of a competitive telecommunications marketplace.

9.5   THE FCC IMPLEMENTS NEW REGULATIONS

Just as universal service proved complex in its administration, so did the introduction of conditions for increasing local competition. Indeed on this issue swirled a host of lawsuits and finally a Supreme Court decision.

Interconnection

Perhaps the most important single FCC proceeding concerned specifying conditions of interconnection to ILECs by potential CLECs. It was part of what the FCC described as the "competition trilogy," the other two portions dealing with reform of universal service (*see section 9.4*) and access charges (*see section 7.3*).[62] As mandated by Sections 251 and 252 of the 1996 Act, the FCC was to establish national rules to guide carriers and state utility commissions in making local exchange access available to CLECs. How it should do this was argued in hundreds of comments (totaling some 17,000 pages) filed in the proceeding. Generally speaking, RBOCs and other ILECs and many state utility commissions argued the Commission should set loose general guidelines and leave details on specific cases up to state-level decision making. Interexchange and CLEC companies, on the other hand, felt the FCC would have to establish fairly specific national pricing guidelines for competitive access to work and to avoid a patchwork quilt resulting from different state regulatory regimes.

In August 1996 the FCC released its long decision on *Implementation of the Local Competition Provisions*, some 700 pages of discussion with thousands of footnotes.[63] The final rules came down on the side of nationwide specificity desired by CLECs seeking to compete with ILEC monopolies. The Commission made clear the concerns that drove it to such a lengthy and involved order:

> An incumbent LEC's existing infrastructure enables it to serve new customers at a much lower incremental cost than a facilities-based entrant that must install its own switches, trunking and loops to serve its customers.

Furthermore, absent interconnection between the incumbent LEC and the entrant, the customer of the entrant would be unable to complete calls to subscribers served by the incumbent LEC's network. Because an incumbent LEC currently serve virtually all subscribers in its local serving area, an incumbent LEC has little economic incentive to assist new entrants in their efforts to secure a greater share of that market.[64]

The Commission went on that its "obligation . . . is to establish rules which will ensure that all procompetitive entry strategies may be explored," thus allowing CLECs to choose among—or combine—construction of their own networks, use of some "unbundled" elements of the incumbent's network, or resale of bulk purchases of the incumbent's network.[65] The 1996 Act called for ILECs to "enter into such agreements on just, reasonable, and nondiscriminatory terms, and to transport and terminate traffic originating on another carrier's network under reciprocal compensation arrangements."[66] Further, CLECs would "likely need agreements that enable them to obtain wholesale prices for services they wish to sell at retail and to use at least some portions of the incumbent's facilities, such as local loops and end office switching facilities."[67] And lastly, it was clearly understood that such negotiations would be unequal, as the ILEC held all the cards and had little to gain by cooperating with a CLEC seeking to compete.

The Act envisioned three ways for CLECs to enter the local service market. The most desirable (but also the most expensive) was facilities-based competition in which the CLECs built their own lines and switches. CLECs could also enter the market through resale, purchasing local service from ILECs and then reselling these services to customers under their own brand name. Alternatively, CLECs could use unbundled network elements (UNEs), renting parts of the ILEC network and combining those parts with their own facilities to provide local service. Policymakers hoped that CLECs would use resale and UNEs as a transitional mechanism as they build their own facilities. As the years since passage of the Act show, resale and UNEs have not been a transition to facilities-based competition, but rather continuing bases for much of CLEC participation in the local service market; indeed, CLECs seem to have used resale as a transition to the use of UNEs. According to the FCC, in December 1999 CLECs provided 43 percent of their lines through resale, 24 percent through UNEs, and 24 percent through their own facilities; four years later resale had declined to 18 percent, facilities-based lines were at 23 percent, but UNE-provided lines had grown to 58 percent of the total.[68]

The FCC rules regarding UNEs have been among the most contentious in implementing the 1996 Act. This is not surprising, given the diametrically opposed interests of the ILECs and the CLECs. The more ILECs are required to break their networks down into UNEs, the more flexibility CLECs have in acquiring only the components they need to provide service. Conversely, the fewer UNEs, the fewer options for CLECs. In crafting the rules governing UNEs, the challenge for the FCC has been to provide CLECs with UNEs that allow them to enter the local service market, but that are not so attractive that the CLEC has no incentive to build its own facilities. In crafting the UNE rules, the Commission has had to interpret a key section of the Telecommunications Act which says that, when deciding what network elements should be made available as UNEs, the Commission "shall consider, at a minimum," if:

> (A) access to such network elements as are proprietary in nature is necessary; and
> (B) the failure to provide access to such network elements would impair the ability of the telecommunications carrier seeking access to provide the services that it seeks to offer.[69]

The Commission has been aggressive in its interpretation of the words "necessary" and "impair," as well as in other provisions regarding access to UNEs, all of which have led to a string of lawsuits in which the courts have sought to curb some of the FCC's aggressiveness.

In its massive 1996 *Local Competition Order*, the FCC established a relatively short list of UNEs which included local loops, network interface devices, local and tandem switching, interoffice transmission facilities, signaling and call-related databases, operations support systems, operator services, and directory assistance. In reviewing this order in 1997, the appeals court in *Iowa Utilities Board v. FCC* found that the FCC had been too aggressive in requiring ILECs to combine UNEs that had not previously been combined, arguing that it was up to the CLECs to combine these elements. The court also found that ILECs could not be asked to alter their networks in order to provide interconnection and access for CLECs that were superior in quality to what they provided for themselves. A year later, in *AT&T v. Iowa Utilities Board*, the Supreme Court found the FCC's interpretation of "necessary" and "impair" to be overly broad, finding that the FCC had not considered the availability of alternative sources of network elements and had assumed that any increase in cost caused by the denial of a UNE constituted impairment. According to the Supreme Court, "the Act requires the FCC to apply *some* limiting standard, rationally related to the goals of the Act, which it has simply failed to do."[70]

The FCC responded to the Supreme Court's decision with its September 1999 *UNE Remand Order* which added a few words to the interpretation of "necessary" and "impair," and then expanded the list of UNEs. Proprietary network elements were necessary if their lack would "preclude" a

CLEC from providing service; lack of access to nonproprietary network elements would present an impairment if it "materially diminished" a CLEC's ability to provide service. The list of UNEs was expanded to include access to dark fiber, subloops, inside wire, packet switching in limited circumstances, dark fiber transport, calling name database, E911 database, and loop qualification information. Two months later the FCC added yet another UNE in its *Line Sharing Order*. The high frequency portion of a loop that carried an ILEC's basic telephone service now became a UNE that could be ordered by a CLEC. In other words, in order to provide DSL service, a CLEC would not have to rent a whole loop from an ILEC, but could instead piggy-back its DLS service on the loop carrying the ILEC's local telephone service. In a January 2001 *Line Splitting Order*, the FCC expanded the list of UNE options even further, finding that one or more CLECs could provide voice and data over the same line.

Far from reducing its UNE requirements in response to court decisions, the FCC seemed increasingly demanding in its list of UNE requirements. The response from the ILEC industry was a request for further court review. In May 2002 the Court of Appeals, in *United States Telecom Association v. FCC*, vacated and sent back to the FCC for further consideration both the portions of the *UNE Remand Order* that dealt with the "necessary" and "impair" standards and the provisions of the *Line Sharing Order*.[71] The DC Circuit Court found that the FCC had not been "nuanced" in its approach to impairment, calling for unbundling in all markets, regardless of the state of competition in any specific market, and that, in requiring Line Sharing, had not taken into account the state of competition for broadband services from cable and satellite alternatives. The DC Circuit decision caused the FCC to embark on a strikingly different course in its approach to unbundling requirements.

Pricing Guidelines

Arguably the most controversial action taken by the FCC in implementing the 1996 amendments was its decision to establish specific pricing methodologies—and exact costs—for the states to use in arbitrating agreements between ILECs and their potential competitors. The Act required ILECs to provide their competitors with access to UNEs. A crucial issue was how much competitors would have to pay for the use of those UNEs.[72] The Act offered little guidance, other than to note that UNE prices should be "just and reasonable" and that UNE rates should be based on the cost of providing the UNEs; costs were to be "determined without reference to a rate-of-return or other rate-based proceeding" and "may include a reasonable profit."[73] The FCC determined that UNE pricing should be based on economic costs:

The Commission concludes that the prices that new entrants pay for inter-connection and unbundled elements should be based on the local telephone companies' Total Service Long Run Incremental Cost of a particular net-work element, which the Commission calls "Total Element and Long-Run Incremental Cost" (TELRIC), plus a reasonable share of forward-looking joint and common costs.[74]

In adopting TELRIC, the FCC was departing from decades of regulatory pricing tradition. Rates determined under rate-of-return methodology were targeted to cover a carrier's historical accounting costs. At least in theory, carriers would eventually recover all of the dollars they had actu-ally spent in providing service. But the Commission wished to adopt a cost methodology that would not require new entrants to pay UNE rates that reflected any ILEC inefficiencies—past or present. Instead, the Com-mission sought an approach that reflected the costs of providing ser-vice in the most efficient manner, using the most efficient equip-ment currently available.[75] The ILECs were understandably distraught by the TELRIC approach, claiming that such a method resulted in rates that didn't compensate their costs. By basing rates on the most efficient equipment, instead of the actual equipment in the ground and in the switching offices, the TELRIC methodology, argued the ILECs, would leave them with stranded investment, the cost of which would have to be absorbed by their stockholders.

In addition to adopting TELRIC for interconnection and UNE rates, the FCC also specified the manner in which resale discounts were to be deter-mined. The 1996 Act provided that wholesale prices be based on the ILECs' retail rates minus costs that "will be avoided" by the ILEC in selling the service at wholesale instead of at retail.[76] The FCC, however, adopted a more stringent standard, finding that wholesale discounts should be based on retail rates less any costs that hypothetically could be avoided. If states lacked the information to determine avoidable costs, the commis-sion provided the states with a default of a "discount range of 17–25 per-cent off retail prices, leaving the states to set the specific rate within that range, in the exercise of their discretion."[77]

FCC efforts to identify costing methodologies and discount strategies that appeared to tilt the balance in favor of new entrants, to the detri-ment of the ILECs, as well as the FCC's willingness to dictate how the states should approach the pricing of what were essentially local services resulted in a number of legal attacks. As explained below (see section 9.7), ILECs were initially successful in convincing the federal appeals court that the FCC had overstepped its authority in dictating how the states should proceed, and that there were some substantial questions about the use of the TELRIC methodology. The Supreme Court subsequently upheld the FCC's authority to both adopt the TELRIC methodology and to require that the states use it. Despite its vindication by the Supreme Court, however, the FCC began to show signs it was having second thoughts about its new pricing strategy. In late 2003 the FCC sought comments regarding changes in its application of TELRIC rules. In the first comprehensive review of its pricing rules since its adoption of TELRIC, the FCC noted,

> To the extent that the application of our TELRIC pricing rules distorts our intended pricing signals by understating forward-looking costs it can thwart one of the central purposes of the Act: the promotion of facilities-based competition. While our UNE pricing rules must produce rates that are just, reasonable and nondiscriminatory . . . they should not create incentives for carriers to avoid investment in facilities.[78]

Worried that its application of the TELRIC methodology might actually be encouraging competitors to provide service through low-priced UNEs, instead of building their own facilities, the FCC reiterated its commitment to the use of forward-looking costs, but asked if it should change the assumptions on which its TELRIC calculations were based. The Commission expressed concern that its network assumptions might indeed be resulting in rates below carrier costs found in an extremely competitive market, noting that, in the real world there is a mix of new and old technologies, and that it may not make sense to assume a market inhabited both by multiple competitors and by a ubiquitous carrier with a very large market share.[79] Asked if the actual ILEC network or a network that reflected ILEC plans could now be considered an appropriate basis for TELRIC calculations, the Commission concluded that after years of price cap regulation, ILECs could perhaps now be considered efficient, thus eliminating the need to worry about competitors paying for ILEC inefficiencies if rates are based on the actual ILEC network.[80] The questions asked by the Commission in this Notice of Proposed Rulemaking (NPRM) suggest that its earlier concern that UNE rates be as low as possible to encourage competition were being replaced by fears that low UNE rates may actually discourage competition.

Triennial Review

In December 2001 the FCC issued a *Triennial Review NPRM*, that sought information about all aspects of unbundling.[81] The order that resulted reflected the concerns of the appeals court and also reflected varied opinions among the commissioners. In its order released in August 2003, the FCC noted that it had been told twice (by the Supreme Court in 1998 and again by the appeals court in 2002), that it "had failed to implement unbundling in a reasonable manner because it did not adopt appropriate principles for limiting its application."[82] The FCC would now combine direction from court decisions, its own experience, and the experience of the telecommunications industry to reevaluate its approach, recognizing that unbundling "can serve to bring competition to markets faster than it might otherwise develop," but that "excessive unbundling requirements tend to undermine the incentives of both incumbent LECs and new entrants to invest in new facilities and deploy new technology."[83]

The major issues dealt with in this *Triennial Review Order* were the definition of "impairment," whether switching should continue to be a UNE, and whether UNE requirements should be applied to broadband services. The FCC chose not to change its definition of "necessary," keeping the definition established in its *UNE Remand Order*. However, the Commission did decide to be more specific in defining "impairment," finding that a CLEC is impaired by a lack of access to a UNE if such a lack "poses a barrier to entry, including operational and economic barriers, that are likely to make entry into a market uneconomic."[84]

The *Triennial Review Order* concluded that broadband services would not be subject to unbundling requirements. High capacity and fiber loops as well as packet switching would not be UNEs; only the voice grade channels on a hybrid loop would remain a UNE. By these actions, the FCC hoped to encourage ILEC investment in broadband facilities, since ILECs would not have to lease these facilities to CLECs at discounted UNE rates; and to encourage CLECs to build their own broadband facilities because of the lack of UNE alternatives.

In its treatment of switching, the FCC made perhaps its most controversial decision. In response to the DC Circuit Court of Appeals' finding that it had not been "nuanced" in its treatment of impairment by setting rules for unbundling in all markets, without considering the state of competition in specific markets, the Commission delegated to the states the task of deciding whether CLECs would face impairment without access to an ILEC switch.[85] The Commission determined that the states, because of their knowledge of specific local markets, could conduct more "granular" examinations of impairment issues. In its decision, the FCC also drew clear distinctions between business and mass market services (the latter being residential and small businesses).[86]

The *Triennial Review Order* was adopted by a severely split FCC. Only Commissioner Kevin Martin supported all of the provisions of the order; the other four dissented on several provisions, splitting along party lines. That Martin supported the whole order was not surprising since he had engineered the final outcome. Aligning himself with the two Democrats on the FCC (Michael Copps and Jonathan Adelstein), Martin, though a Republican, created a majority vote for maintaining switching as a UNE, for ending line sharing, and for delegating authority to the states. Unlike Martin, Chairman Michael Powell supported elimination of switching as a UNE, and had delayed the meeting on the *Triennial Review Order* a week while he tried to build a majority for his position.[87] The actual vote on the order was delayed 90 minutes as "Commissioners haggled over final details."[88] The final vote, which was seen by many as a "showdown" between Chairman Powell and Commissioner Martin, was "contentious."[89]

The lack of unanimity on the part of the FCC was troublesome to at least one of the Commissioners. In his comments, Commissioner Copps stated that he was "troubled by the less than satisfactory process that generated this decision" and expressed his disappointment that "we were not able to reach compromise on all of the questions and issue a unanimous decision as previous Commissions were often able to accomplish," citing the "different philosophical and regulatory approaches which exist among us" as the reason for this lack of unanimity.[90] Chairman Powell expressed his concern that the FCC was erroneously delegating primary decision-making authority to the states, and was creating a "litigation bonanza" by developing a system based on 51 state proceedings that would be litigated in 51 different federal district courts.[91] Indeed, Powell's concerns about the decision proved to be well founded; six months after the release of the order, the court of appeals struck down many of its provisions (*see section 9.7*).

Section 271 Proceedings

As discussed above, Section 271 of the 1996 Act specified, in a 14-point checklist, the requirements that the RBOCs had to fulfill in order to be allowed into the inter-LATA long distance market. The Section 271 process got off to a slow and rocky start. Ameritech was the first RBOC to formally seek inter-LATA entry in its own region, filing for FCC permission early in 1997. Under the rules established by the 1996 law, the 4,000-page (that's eight *reams* of paper!) application had to be acted upon within 90 days.[92] Almost immediately, MCI and AT&T both filed statements arguing against the Ameritech application, convinced that the RBOC did not have the requisite local competition in place. Ameritech counterargued that it had met the FCC's checklist requirements in the state of Michigan where CLECs were at work seeking heavy-spending big business customers.[93] Ameritech withdrew its request just weeks after the filing, promising to soon return with amended and improved applications. The carrier refiled for Michigan again in mid-1997, but the FCC denied the application; it was not until September 2003, that Ameritech, by then part of SBC, finally received approval to provide inter-LATA service in Michigan.

The process for such inter-LATA approval was lengthy and grueling and was to proceed state by state. An RBOC first had to file with the state utility commission and also had to present its case to the Department of Justice (DOJ). The FCC, however, had the ultimate authority in approving the Section 271 application, regarding the state commissions and the DOJ as serving only advisory roles. The FCC did not hesitate to deny an application that had already been approved by a state commission or to approve an application that had been found wanting by the DOJ.

Attempts by SBC in Oklahoma and by BellSouth (once in South Carolina and twice in Louisiana) were also denied early on. Faced with the frustration of being regularly turned down, RBOCs quickly turned to a broader attempt to overturn the whole approvals process. Immediately after the FCC turned down the SBC application to enter the long distance market in Oklahoma, SBC filed suit in Texas in July 1997 arguing that Sections 271–275 of the 1996 legislation were unconstitutional in the way they specifically focused on and limited RBOC actions.[94] Critics and members of Congress were divided in their reaction to the suit. Some argued the FCC had brought it on by misreading Congress and being overly harsh, while others concluded SBC was using yet another arena to at least delay local competition. One industry observer noted SBC had been among the most rigid RBOCs in resisting local competition, marshaling legal arguments against FCC and state decisions and charging excessive prices.[95] ·

Points of contention in these filings focused on RBOC pricing of UNEs and the status of local competition in the state in question. But the major disagreement centered on the RBOCs' Operation Support Systems (OSS). These included RBOC ordering, provisioning, and maintenance processes—they were the databases that were used to determine whether a local loop was available, and in what condition, or to report any need for repairs. To demonstrate that they had indeed opened their networks to competition, the RBOCs had to prove that their competitors had comparable access to these OSS databases. Indeed RBOCs had to show that there was no disadvantage to a subscriber ordering service from a CLEC who used RBOC facilities. If an RBOC could install service to a subscriber within 24 hours, a CLEC, providing service by using RBOC facilities, should also be able to promise 24-hour installation.

To show that they had opened their networks (including OSS access), RBOCs worked with committees of CLECs, long distance carriers, and other interested parties in each state, to determine what the CLECs expected and what metrics should be used to prove that the RBOC was meeting these needs. RBOCs also hired firms to audit their OSS procedures to determine that they were indeed accessible to competitors.

At the end of 1999, Verizon received authorization for New York and was the first RBOC allowed to provide inter-LATA service. Six months later, SBC received approval to provide inter–LATA service in Texas. After this slow start, the pace of successful applications began to accelerate. RBOCs in seven states received approval in 2001; 26 states were added in 2002; and finally, 13 states and the District of Columbia joined the list in 2003. While the 1996 legislation required RBOCs to proceed on a state-by-state basis, the RBOCs are regional companies. Arguing that they used the same OSS procedures and systems across states, RBOCs (after receiving approval from individual state commissions) began to make multiple filings before the FCC. SBC started this trend with a successful joint filing for Kansas and Oklahoma in January 2001. Qwest proved successful with a nine-state filing two years later. By December 2003 all four RBOCs had received Section 271 approval for all of their service territories. BellSouth was the first to complete, while Qwest was the last to begin and to complete, the Section 271 process (Arizona was the last state to be approved).

This process had taken almost eight years from the passage of the Telecommunications Act in February of 1996. At the completion of the Section 271 proceedings, the important question was whether RBOCs would prove more successful in adding long distance to their list of services than the long distance companies had been in adding local service to their long distance offerings. Many believed that the RBOCs would prove to be vastly more successful, and that instead of increased competition, the industry would see the development of four smaller parallels to the former Bell System in which substantial control of the local network could be used to gain dominance in the long distance market as well.

Subsequent industry developments suggest that the RBOCs did indeed win the battle with the interexchange carriers. By mid-2005, AT&T shareholders voted to merge with SBC. The former Bell System offspring would now incorporate what remained of the former parent. Meanwhile, Verizon emerged as the winner as Verizon and Qwest battled over MCI. Like its long-time rival, MCI would also become part of an RBOC (*see section 10.5*).

9.6 MERGER FEVER

Amidst all this development of detailed regulation, the weeks following passage of the 1996 amendments were punctuated by rumors and reports of mergers throughout the communications business. Many involved media outlets, especially radio stations, as the 1996 laws had eliminated most remaining restrictions on their ownership. But the underlying factors in mergers and takeovers were much the same no matter what sector was involved.

For one thing, there was a widespread belief that only larger firms would and could survive in a future of increased competition. Media and telecommunication companies were "bulking up" to better survive competitive pressures by bringing multiple businesses under one corporate roof. Ironically, only AT&T, once the largest player of all, seemed to be going the other way by breaking into three independent parts (*see section 7.2*). Another factor encouraging corporate mergers were the savings gained by the merged companies. The resulting integrated firms could seek the "streamlining efficiencies" (i.e., sometimes substantial personnel lay-offs) that would contribute to their bottom lines, but also to more general economic dislocations.

Further consolidation was forthcoming as this book went to press. While no mergers between telephone company and media companies were in the offing, the once unthinkable prospect of a merger between an interexchange and a local exchange carrier was fast becoming a reality. The viability of the interexchange business as a stand alone enterprise was definitely in question (*see section 10.5*).

All of this had been previewed long before the 1996 legislation appeared. In a glittering public announcement in October 1993, Bell Atlantic announced its intended acquisition of Tele-Communications Inc. (TCI), the country's largest multiple system operator (MSO) of cable television systems (*see section 8.4*). But the urge to merge did not dissipate because of one failure, though the demise of the huge Bell Atlantic/TCI deal several months after its announcement made everyone more cautious. Late in 1995, as the industry waited on Congress to complete action on the expected legislation, a lot of strategic thinking was evident off-stage: "Although telephone executives are still treading lightly around regulators for fear of derailing the bill, most are scrambling behind the scenes to line up partners and allies to help them compete."[96] The results of such activity became evident less than two months after the Telecommunications Act of 1996 was passed.

RBOC Mergers

First to enter the renewed merger arena was Pacific Telesis (Pac Tel), weakened by years of competition for its business services and close regulation by the California public utilities commission. In April 1996, SBC Inc. (formerly known as Southwestern Bell) announced its $17 billion acquisition of Pac Tel. The deal caught many industry observers off guard because it had come together quickly and in strict secrecy, and because SBC Inc. had been the most overlooked of the seven regional RBOCs. While CEO Edward Whitacre Jr. did not attract the media attention of some of his colleagues, he had led SBC to the strongest profit and growth position of the seven regional firms. His deal a decade before to purchase Metromedia's cellular franchises in several major cities had placed SBC firmly in a leadership position among RBOCs in that industry.[97] The takeover was unusual in that the two companies shared no contiguous borders, being divided by Arizona and New Mexico (then part of US West, now Qwest). Also unusual was that these partners announced plans to hire *additional* workers rather than pursue lay-offs. Further, they had pursued different nontelephone strategies, with SBC focusing on its core business plus cellular, and Pac Tel actively investing in potential home video entertainment enterprises. Combined, the partners would become the second-largest telephone company in the country (after AT&T) with some $21 billion in revenue.

A few weeks later, the long-expected "merger of equals" between Bell Atlantic and NYNEX was formally announced. The $22 billion deal set the stage for creation of the nation's second-largest telephone company. This marriage had been widely rumored in press reports even before the 1996 legislation passed.[98] Raymond Smith of Bell Atlantic was again at the center of this deal as he had been with the failed earlier merger with TCI (*see section 8.4*). But the synergies here seemed clearer and the soon-to-be-allowed entry of the RBOCs into long distance service fueled the merger in this region which provides 30 percent of the nation's long distance traffic. Further, the partners had worked together since mid-1995 on a cellular telephone operation. In part to deflect concerns among its own mid-Atlantic regulators, Bell Atlantic soon made clear it was buying NYNEX outright, rather than merging with it. But there was another reason, too: "The company did not need the approval of Congress, which has the right—through a 1913 law in the District of Columbia, where Bell Atlantic also operates—to approve any utility merger affecting the District."[99] Even had Congress not expressed concerns over the merger, the need for Congressional action would most certainly have delayed the action. Bell Atlantic made clear it would seek approval from all of the state commissions where it did business, not just those (apparently only Maine and Connecticut) with the apparent power to approve or deny the merger.

Coming on the heels of the earlier RBOC merger, however, and including such major markets as New York, Boston, Philadelphia, and Washington, this new announcement brought forth expressions of regulatory concern from several quarters. Both the New York Public Service Commission and the U.S. Justice Department made clear they would examine the merger's conditions and potential results carefully:

> In sharp contrast to the Justice Department's usual review procedures, the mergers now pending among four of the seven regional Bell companies will be scrutinized under both traditional antitrust law as well as under the terms of the sweeping new telecommunications law.[100]

All parties agreed that pricing conditions of competitive interconnection (*see section 9.5*) would be crucial in these regulatory findings. Though the DOJ cleared SBC's takeover of Pac Tel late in 1996, the Bell Atlantic–NYNEX merger came under more sustained attack. The Communication Workers of America came out against it but after the companies had announced a guarantee of the 70,000 jobs, represented by CWA, the union reversed its position.[101] Early in 1997, the New York attorney general and the Consumer Utility Board appeared to nix the deal. Both urged the state public utility commission (and other states in the region) to reject the deal as uncompetitive, especially for smaller users and consumers. In April an assistant U.S. attorney general met to persuade the New Yorker to soften his position, and in April the Department of Justice announced it would okay the deal without any special conditions.[102]

But the FCC was not as easy a regulatory barrier to overcome, as it made clear that it would require a variety of conditions before final approval. The conditions chiefly concerned local competitive entry conditions and various reporting requirements to make it easier for the federal agency to oversee the development of local service competition in the region.[103] In August, the FCC approved the merger after receiving a 10-page letter in which Bell Atlantic and NYNEX agreed to negotiate with competitors over technical standards on connecting to its network; develop a uniform operational support system allowing competitors to electronically order service for new customers; offer installment payments for smaller competitors; and not include historic network development costs when determining what to charge competitors for connections.[104] The FCC approval came some 16 months after the deal had been first announced, and the merger was finalized.[105]

Long distance carriers and other potential competitors understandably expressed misgivings about these RBOC mergers, arguing the growing concentration would extend and deepen the existing local exchange bottleneck. While the newly enlarged RBOCs could enter the long distance market after meeting the FCC's 14-point checklist (see Fig. 9.1) and could do so with limited expenditure, long distance companies seeking to break into the local exchange market would face a far more expensive and risky process and for some time were likely to serve but a small portion of the market.

Bellcore

Carved from parts of the old Bell Labs under the terms of the 1982 MFJ, Bell Communications Research (Bellcore) was cooperatively owned from 1984 into the 1990s by the seven RBOCs. It originally fulfilled three functions: emergency interconnection among the companies in case of national need (a role demanded by the defense establishment which had led to Bellcore's formation in the first place), cooperative research and development, and Washington representation of the seven regional companies. Bellcore had undergone substantial change in the decade after its creation. Its role as a policy representative in Washington was the first function to go as each RBOC set up its own lobbying office. With most of its facilities in northern New Jersey (a result of its Bell Labs origin), Bellcore increasingly concentrated on its shared research and development functions for which results and benefits were shared equally for its seven owners. But even this core technological role was threatened.

The seven RBOCs realized their increasingly competitive stance toward one another was poorly served by Bellcore's cooperative approach. Placed on the block early in 1995, it was sold late in 1996. With some 5,600 employees the Morristown (New Jersey) based operation became a subsidiary of Science Applications International for close to $700 million.[106] As a condition of the sale, Bellcore would change its name within a year after the deal was finalized. The RBOCs spun off and retained one part of Bellcore: its original emergency interconnection and reliability/security functions. Retitled as the National Telecommunications Alliance, it would continue to provide a single point of contact for national security/emergency preparedness agencies in the case of national or regional emergencies.

Rise of WorldCom

The RBOCs weren't the only players seeking to strengthen their market position through merger activity. Two large merger attempts in the long distance market, only one of which was successful, caught the attention of regulators both in the United States and in the European Union (EU). The successful takeover of MCI by its smaller competitor WorldCom resulted in the second-largest long distance company in the United States and the largest provider of Internet backbone capacity. The later unsuccessful bid by the merged MCI and WorldCom companies to take over Sprint showed that there were limits to the amount of industry concentration regulators were willing to allow.

WorldCom developed through a long series of mergers. Started by Bernard Ebbers, a former basketball coach and motel owner in Mississippi, WorldCom started as a small long distance reseller and eventually, through more than 70 mergers, became a $180 billion company,[107] which, because of shady bookkeeping and over-ambitious business planning, in July of 2002 filed the largest bankruptcy in U.S. history (*see section 10.2*). According to industry folklore, Ebbers, at a September 1983 meeting in the coffee shop of a Days Inn in Hattiesburg, Mississippi, with the help of businessmen Murray Waldron, Bill Fields, and David Singleton, sketched out on a napkin the plan for a discount long distance company, the name of which, as suggested by their waitress, would be Long-Distance Discount Service (LDDS). The firm's first customer was the University of Southern Mississippi, which it began to serve in January of 1984.

LDDS's strategy was to grow through mergers and acquisitions, as Ebbers acquired small long distance companies, spreading his reach across the United States. In 1994 LDDS significantly expanded its facilities by acquiring WilTel Network Services and then IDB WorldCom, an international company. The following spring, LDDS changed its name to WorldCom, Inc., with Ebbers as CEO. His acquisition strategy expanded WorldCom's holdings beyond long distance facilities to include the Internet backbone and alternative local service facilities. In August 1996 WorldCom merged with MFS Communications, one of the country's largest alternative local service providers for businesses in large metropolitan areas. In its merger with MFS, WorldCom also acquired UUNet, a major Internet backbone provider. WorldCom solidified its Internet position by acquiring Compuserve through a $1.2 billion stock deal in 1997. WorldCom kept Compuserve's data network, and traded its online services to AOL. Ebbers further strengthened WorldCom's presence in the local market in 1998 by acquiring Brooks Fiber properties for $3 billion.

All this paled in the face of its nearly $40 billion offer for MCI. Though MCI was the larger company, with $10 billion in revenues (WorldCom's were only $7 billion), it was the smaller firm that purchased its larger competitor. WorldCom was but one of three players seeking to acquire MCI. British Telecom (BT), which already owned 20 percent, had made an offer to purchase the remainder and GTE also put in a bid. MCI's board elected WorldCom's offer because, as MCI's CEO Bert Roberts noted, the two long distance companies had "'complementary strengths,' 'entrepreneurial roots,' and more than $5 billion in newly found 'synergies.' "[108]

Combining the second- and fourth-largest long distance carriers and the first- and second-largest Internet backbone providers, the merger provoked legal challenges. Both GTE and the Communication Workers of America sued to stop the merger, citing anticompetitive dangers. The merger had to gain approval from the Justice Department (DOJ) and the FCC in the United States and from the EU because of MCI's and WorldCom's provision of international Internet backbone facilities. In order to gain EU and DOJ approval, the two companies agreed to sell MCI's Internet assets to Cable & Wireless for nearly $2 billion.[109] DOJ approved the merger in mid-1998, finding that this "divestiture" of Internet assets preserved competition. The FCC followed suit a few months later, finding the merger to be consistent with the "pro-competitive" and "de-regulatory framework of the Telecom Act of 1996," and stating that the merger might produce tangible benefits for consumers.[110] The company resulting from this merger controlled nearly a quarter of the long distance market and more than half of Internet backbone traffic.

Perhaps emboldened by the success of the MCI merger, WorldCom next set its sights on another competitor. In October 1999, Ebbers and Sprint chairman Bill Esrey announced a proposed merger that was valued at up to $130 billion. This time, however, regulators found the prospect of a merger between the second- and third-largest long distance carriers to be anticompetitive. AT&T held a 42 percent share of the U.S. long distance market and with this merger, WorldCom would control another 37 percent, thus placing almost 80 percent of the market under the control of two companies.[111] When the DOJ decided in June 2000 to go to court to block the merger, WorldCom and Sprint withdrew their request for consideration by the EU, which had indicated that it would also block the merger, and then withdrew its request to the FCC to transfer licenses from Sprint to MCI WorldCom.[112]

WorldCom's merger fever was not totally extinguished by its failure to acquire Sprint. The company went on to establish a position in the wireless market by acquiring SkyTel and Wireless One. However, the unsuccessful effort to merge with Sprint seemed to have been a turning point. WorldCom's stock began to come under increasing scrutiny (*see section 10.2*).

9.7 APPEALS, DELAY, DECISIONS, AND IMPACT

Reaction to the FCC's complex and far-reaching interconnection rules was swift and predictable, following the positions of pre-decision commentary. ILECs and many state commissions were unhappy, arguing the FCC had exceeded its authority with such detailed rules which left far too little discretionary decision making to state agencies. Some ILECs argued further that the FCC's pricing policy was so draconian as to amount to "an uncompensated taking of property in violation of the Fifth Amendment."[113] Interexchange carriers and CLECs requiring access to the local exchange, on the other hand, applauded the Commission's findings. The losers (as seems all too common these days) appealed the new FCC rules, which were to go into effect on October 1, 1996.

Appeals

Numerous appeals filed by ILECs and many states piled up before various federal appeals courts in the month after the FCC rules were issued and requests to the agency for a stay of their implementation were denied.

Under federal rules calling for random selection when appeals are filed in multiple circuits, the appeals were consolidated early in September to be heard before the U.S. Court of Appeals for the Eighth Circuit, based in St. Louis.

A three-judge panel from that court placed a stay on significant portions of the FCC order, thus delaying their implementation while the court considered the issues raised.[114] This was a clear indication that the judges had concluded that the FCC's opponents were likely to prevail. Several members of Congress who had been deeply involved in passage of the 1996 Act also took the FCC to task for "exceeding its authority," thus adding weight to the appeals.[115] GTE and several other parties focused on the constitutional argument: that the FCC's pricing rules (with their sharp discounts for competitors) were akin to the government taking property without paying for it and were thus a violation of the Fifth Amendment. This "takings" argument had worked in a number of mid-1990s electric utilities cases. But the telephone appeals in St. Louis were different in that the ILECs were being paid (though, they argued, not enough) and the overall aim of the law and resulting regulation was *less* regulation, not more.[116] In the meantime, the stayed FCC rules could only be taken as "suggestions" to state commissions. Despite the legal shouting, by early 1997 more than 30 states were utilizing the FCC's TELRIC formula to at least guide their own determination of conditions for competition within their jurisdictions.

In July 1997 a three-judge panel released the appeals court decision—as expected, it was a unanimous finding that the FCC had exceeded Congressional intent by mandating a national guideline as to both interconnection procedures and pricing. The court found that the 1996 Telecommunication Act "plainly grants the state commission, not the FCC, the authority to determine the rates involved in the implementation of the local competition provisions of the act." FCC Chairman Reed Hundt responded to the decision by noting, "For the first time in the history of the Federal Communications Commission, a court has ruled that when Congress writes a specific statute that we do not necessarily have the authority to write a rule to implement it." While the FCC announced plans to appeal the decision to the Supreme Court, it seemed the Commission might have to rethink its reading of the 1996 Act, or that Congress would have to revisit the legislation.[117]

Observers concluded the appeals court decision would clearly further delay competition in the local arena, and might push off competition in both local and long distance service as investors held off on making decisions until the legal fight was resolved. Indeed, the RBOCs were taking a risk that by extending their local monopolies they were also increasing the time before they would be allowed into long distance services. They seemed to feel the risk was worth it, for in the interim, they moved ahead on plans to expand their local services to include such things as broadband Internet access (*see section 8.5*) designed to retain customers who might otherwise consider competitive offers. Observers argued the local exchange carriers would do far better agreeing on conditions for competitive local entry with state commissions which are often more sympathetic to their needs.

The Courts Rule

It was clear from the start of litigation that key issues in the 1996 law would eventually reach the U.S. Supreme Court. The appeals process had taken nearly three years to do so when in January 1999 the high court decided *AT&T Corp. v. Iowa Utilities Board*.[118] The 8–0 decision was a resounding victory for the FCC when the court, largely overturning lower court findings, held that the Commission *did* have the authority to set local telephone pricing standards, as per the 1996 act. While some parts of the lengthy FCC local competition rules were found to exceed the agency's authority, by and large the ruling vindicated the Commission as well as most provisions of the law itself.

In a very real sense, the landmark 1996 law only now could take effect, three years *after* it was signed, for most parties had suspended major business decisions while awaiting this final legal judgment. Observers argued that the Supreme Court decision might serve to reinforce ILEC decisions to not open their networks and simply take the resulting fines as a cost of doing business. Realizing they had nothing further to gain through court appeals, the incumbents could try to extend their monopoly control as long as they could. Or, as had so often been the case, there could be still more litigation—and that is exactly what happened.

The Supreme Court was far from through with this difficult legislation. During its 2001–2002 term, three more important issues (combined from nearly a dozen individual cases in lower courts) required decision. One dealt with the accounting method to be used by ILECs in charging CLECs for access, another with the rates charged by utility firms for use of their poles by cable companies providing Internet access service, and the third was more legalistic and concerned whether or not state actions under the 1996 Act could be appealed to federal courts.[119]

The first of these—on accounting rates—was brought amidst an atmosphere of failing CLECs as bankruptcies and service terminations afflicted the struggling competitive sector (*see section 10.3*). The FCC pricing regulations at issue had been written in six months, as Congress required, and they were then litigated for years thereafter. As the RBOCs continued legal delaying tactics even after the January 1999 Court decision, CLECs found it harder to raise the capital needed to develop their fledgling networks. The Court defined the issues of the relevant cases this way:

> The issues are whether the FCC is authorized (1) to require state utility commissions to set the rates charged by the incumbents for leased elements on a forward-looking basis untied to the incumbents' investment, and (2) to require incumbents to combine such elements at the entrants' request when they lease them to the entrants. We uphold the FCC's assumption and exercise of authority on both issues.[120]

The second set of cases re-raised the old issue of how to define and regulate the provision of Internet access—whether as a cable or telecommunications service or as an information service. The former is regulated, the latter is usually not. Further, cable (television) is far less regulated than are traditional telecommunications (telephone) carriers. If cable-delivered broadband access was defined as information, then cable providers could not be required to provide access to competitors. One of the cases concerned how much utility companies could charge cable firms for renting space on telephone poles as such rates are controlled for cable or telecommunication services only. Several lower court and one appeals court decision[121] appeared to define cable-provided Internet access as a telecommunication service, subject to equal access and regulation. The FCC appealed several of these decisions and in January 2002, the Court reversed the lower courts, holding that FCC decisions terming Internet as an information service—*not* subject to equal access requirements—were reasonable.[122] Later FCC decisions adhered to this finding.

The final set of cases raised an interesting constitutional issue not often heard. It concerned the Eleventh Amendment, which helps define what is state versus federal jurisdiction. Telecommunications policymakers generally favor federal jurisdiction to enable development of a consistent national policy. States-rights advocates are naturally more concerned about perceived federal encroachment on the ability of states to control business within their own borders. The 1996 Act set out certain actions the states may take. Proponents of the appeal argued that such actions should be appealable to federal courts as a federal law created the situation. In deciding the case, the Court summarized the question of law in this way: "whether federal district courts have jurisdiction over a telecommunication carrier's claim that the order of a state utility commission requiring reciprocal compensation for telephone calls to Internet Service Providers violates federal law."[123] The Court decided that the courts did have such jurisdiction given the federal law involved.

The Supreme Court's decisions upholding many of the FCC's actions in implementing the 1996 Act by no means marked the end of judicial activity. No sooner had the FCC issued its *Triennial Review Order* regarding network unbundling (*see section 9.5*) than virtually all parties filed appeals. Many of these were transferred to the District of Columbia Circuit of the U.S. Court of Appeals which consolidated them for review.[124] In March 2004, the Court found much of the FCC's Triennial Review Order unlawful.

In a decision that represented a significant ILEC victory, the appeals court required the FCC to once again rethink its approach to network unbundling. The court found that the FCC had erred in delegating to the state commissions the authority to determine whether switching should remain a UNE for the mass market (i.e., residential and small business users). According to the court, "case law strongly suggests that subdelegations to outside parties are assumed to be improper absent an affirmative showing of congressional authorization." When it comes to federal agencies, state commissions are outside parties, and Congress had made it clear that it was the FCC's responsibility to determine impairment. The court also vacated the FCC determination that mass market switching should remain a UNE, except in those specific markets in which a state commission found no impairment. The decision overturned the FCC's finding of nationwide impairment regarding transport facilities for the mass market and then vacated the FCC's finding that wireless carriers are impaired if they do not have unbundled access to ILECs' dedicated transport. In still another victory for the ILECs, the court upheld the FCC's decisions to exempt broadband loops from unbundling requirements and to phase out line sharing. The court found that the FCC had appropriately determined that fostering of broadband deployment outweighed any impairment experienced by the CLECs because of lack of access to unbundled broadband loops or to line sharing.

The court noted with some asperity "the Commission's failure, after eight years, to develop lawful unbundling rules, and its apparent unwillingness to adhere to prior judicial rulings." The court's decision left the industry in disarray, with no one certain about what the availability and pricing of UNEs should be. The Commission, seeking to buy time while it worked on a Triennial Review Remand order, asked the industry to negotiate UNE agreements, noting that "today we come together with one voice to send a clear and unequivocal signal that the best interests of America's telephone consumers are served by a concerted effort to reach negotiated arrangements."[125] State commissions interpreted the Telecommunications Act as requiring that negotiated arrangements be filed with state commissions to be effective.

It took the FCC eight months to issue its remand order. As with the Triennial Review Order, the Commission was unable to reach a unanimous decision. Voting along party lines, the three Republican commissioners, in December 2004, adopted a decision that eliminated circuit switching as a UNE; ended the availability of the UNE-P; established measures by which to determine markets in which CLECs would be impaired without access to DS1 and DS3 loops and transport facilities; and specified that impairment considerations would assume a reasonably efficient competitor.[126] Unlike the Triennial Review Order, which exempted broadband from unbundling, this order retained unbundled access to broadband loops and transport under specific conditions.[127] The earlier order had retained switching as a UNE in some markets; this order eliminated it in all markets. Commissioners were by no means certain that these new rules would satisfy the U.S. Court of Appeals.

Winners and Losers

Predictions about the impact of the 1996 amendments were widespread within hours of their passage and were initially favorable in the weeks following. As might be expected, politicians promised all things to all people. One senator predicted that the new law "will lower prices on local telephone calls through competition. It will lower prices on long-distance calls through competition. It will lower cable TV rates through competition. . . . This is the biggest jobs bill ever to pass this Congress."[128] Unfortunately, within a year it appeared that few of these promises would be forthcoming anytime soon—and the senator had lost his seat. Industry conferences and countless press articles reviewed the amendments on the sequential anniversaries of their passage and found the dragging process of legal wrangling preventing any substantial change. Still, a sense of winners and losers in the long drawn-out process became slowly evident. Who won and who lost depended, of course, on where you stood on the issues—and changed over time.

After years of lobbying, the RBOCs finally had succeeded in throwing off the shackles of the post-divestiture MFJ and were able to enter once forbidden businesses. And they generally excelled in this while accepting only a limited assault on their local turf from new competitors. That such local competition would *eventually* develop was widely assumed. But as long as the RBOCs continued to control an average of 80 percent of access lines in their individual regions, they enjoyed a massive subscriber revenue base on which to build assaults on the interexchange, manufacturing, and information sectors of the industry.

New competition *promised* more (and often less expensive) options, especially for large users—corporations and governments—on whom most CLECs would clearly focus. The new amendments provided considerable freedom for these large users to negotiate "one-stop shopping" and special price deals for all their telecommunications needs from an expanding number of potential providers. But most of this business was captured by the RBOCs in the first years after the Act's passage.

Though it was put through the policy wringer in having to complete so many complex dockets in a short period of time, the FCC appeared to be an institutional winner in that the basic scheme it created was largely upheld on appeal. Perhaps more centrally, the Commission reemerged as the primary regulator it had long been before the MFJ created a new locus of power in District Judge Harold Greene. However, even in its position as primary regulator, the FCC continued to face, and sometimes to lose, legal challenges.

While the CLECs looked to be huge winners at first—to some the Act appeared a declaration of their future role—they were, in fact, substantial losers. Despite a series of requirements in the Act and immensely detailed resultant FCC rules, the Act's regulatory promises merely lured in investors and ill-prepared and often undercapitalized companies that were ripe for failure when appeals delayed the full implementation of the law—and thus their expected revenues. When the telecommunication business began to collapse generally in 2001, the CLEC industry was an early victim (*see section 10.3*).

Generally rising prices pinpointed other losers—individual consumers. Despite predictions that consumer prices would decline as competition became more apparent, residential telephone (and cable) prices continued to rise into the early 2000s. Further, local telephone rates were predicted to rise even faster in smaller markets where cable/telephone mergers were now allowed. Local cable subscriber rates also rose in such markets immediately, and, under provisions of the law, they could rise in other cities after a three-year freeze. Many Democrats and consumer advocates argued that the 1996 amendments went too far in removing various "safety nets" that once protected consumers. Allowing RBOCs to provide competitive service as well as their largely still-dominant ILEC service, for example, was seen by some as reassembling the old Bell System model with its possibly illegal cross-subsidization—albeit a system now divided into smaller and more competitive pieces. With the supervisory roles of the Department of Justice and Judge Greene gone, only the FCC (often too weak and overworked to provide effective oversight) was left to determine whether local exchange competition did or could exist.

In the years after the amendments' passage, most telephone carriers pulled back from previously announced plans to enter the cable business, while cable system operators dropped most plans to enter local voice and data provision (though expanding VoIP options after 2000 turned cable around). Grand plans for wonderful new converged service offerings faded away as industries appeared to pull back to their established bases to fend off possible competitors in a "Maginot Line"[129] mentality rather than aggressively moving into a competitor's territory. The promised cornucopia of innovative new service grew slowly in the early years after the amendments' passage. And by the early 21st century, it was clear that the Act's provisions were ill-equipped to deal with cable-telephone competition over the Internet.

In part this delay was due to the industry's consumption in the late 1990s with an almost blind urge to merge (*see section 9.6*). Some critics argued the legislation was merely "ushering in the old world of telecommunications,"[130] while others cautioned their readers about the impact of tightening telephone industry concentration.[131] Some suggested that spin-offs had a better financial record than conglomerates or consolidations, and probably served the public better, too.[132]

Prospects

Not only did the 1996 amendments fail to bring "vibrant competition" to the telecommunications marketplace in the short term, the prospects for its development over the long term appear pretty dim as well. Implementation of provisions of the Act concerning RBOC entry into long distance, increasing concern about slow broadband deployment, and difficulties in formulating viable interconnection rules have all combined to strengthen the position of the ILECs, and especially the four RBOCs. With permission to provide interexchange long distance in all of their territories, they can rapidly take advantage of the synergies in both local and long distance service. Their long distance competitors have not proven able to build such synergies, finding it much more difficult to expand into local service. Weakened by unfortunate corporate strategies (in the case of AT&T) and by dishonest management (in the case of MCI WorldCom), the two largest U.S. interexchange carriers soon merged with their RBOC competitors (*see section 10.5*).

Hoping to encourage broadband deployment, the FCC weakened the unbundling rules for broadband, making it more likely that RBOCs will be able to invest in broadband facilities without having to share the benefits of that investment with their competitors. The Commission is likely to weaken those unbundling rules even further in the wake of a 2005 Supreme Court decision confirming that cable modems are an information service and thus are not subject to regulation or to unbundling requirements. In response to this decision in *National Cable & Telecommunications Association versus Brand X*, FCC Commissioner (soon to become chairman) Kevin Martin issued a statement noting his pleasure at the decision, saying that "This decision provides much-needed regulatory clarity and a framework for broadband that can be applied to all providers. We can now move forward quickly to finalize regulations that will spur the deployment of broadband services for all Americans."[133] It is far more likely that the RBOCs will have the resources to widely deploy broadband investment, suggesting that they have the potential to dominate the field. While cable companies have thus far dominated broadband Internet access, the RBOCs may yet prove formidable—if not dominant—competitors for that market.

Worried that its pricing rules are discouraging facilities-based competition, the FCC is drawing back from its policy of pricing unbundled network access at bargain basement rates. While such a move may encourage some CLECs to invest in their own facilities, it will also discourage other CLECs from entering or staying in the market. Indeed, the whole issue of unbundling, and of the FCC's inability to craft unbundling rules that meet with court approval, suggests that the regulatory balance will swing even more heavily in the RBOCs' favor, as they face the prospect of no longer having to provide switching to their competitors.

The failure of the 1996 amendments to deliver vibrant competition in all aspects of the industry has led many in Congress to call for a rewrite of the Act. Chair of the House Commerce Committee Joe Barton (R-Texas) told a meeting of the Federal Communication Bar Association that he believed "the best thing to do is just start from scratch."[134] Industry groups seem just as unhappy and ready for change. The U.S. Chamber of Commerce, the National Association of Manufacturers, the National Black Chamber of Commerce, and others formed a coalition to work toward passage of legislation that would be more deregulatory in nature.[135] Dissatisfaction with slow broadband deployment, escalating universal service fund contributions (*see section 9.4*), and the loss of jobs and capital suffered by the industry during the past 5 years (*see section 10.2*) suggest that the 1996 Act has been less than successful and that pressure is building on Congress to consider drafting new legislation.

NOTES

1. In substantially different form, the first portion of this chapter appeared as Christopher H. Sterling, 1996. "Changing American Telecommunications Law: Assessing the 1996 Amendments," in Lucien Rapp, ed. *Telecommunications and Space Law Journal, Vol. 3* (Paris: SERDI), pp. 141–165.

2. Telecommunications Act of 1996. Public Law 104-104. House of Representatives, Report 104-458, 104th Cong., 2nd Sess. (January 31), 214 pp. This is the essential official document which in its first 111 pages prints a full copy of the Act as passed, and in the remaining pages provides the conference committee rationale for various provisions. In an unusual situation, both the House of Representatives and the U.S. Senate acted on the same day—within two hours of each other. Both houses passed the bill by large majorities (414–16 in the House and 91–5 in the Senate).

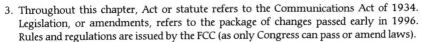

3. Throughout this chapter, Act or statute refers to the Communications Act of 1934. Legislation, or amendments, refers to the package of changes passed early in 1996. Rules and regulations are issued by the FCC (as only Congress can pass or amend laws).

4. *MCI Telecommunications Corporation v. American Telephone & Telegraph Company*, 512 U.S. 218 (1994).

5. See Public Broadcasting Act of 1967, Public Law 90–129.

6. See Cable Communications Policy Act of 1984, Public Law No. 98–549. See also, more generally, Charles D. Ferris, Frank W. Lloyd, and Thomas J. Casey, 1983. *Cable Television Law: A Video Communications Practice Guide;* later retitled *Telecommunications Regulation: Cable, Broadcasting, Satellite, and the Internet* (New York: Matthew Bender, four vols.).

7. See Cable Television Consumer Protection and Competition Act of 1992, Public Law No. 102-385, and Ferris et al.

8. The Communications Act of 1934 is divided into seven titles (six prior to 1984). Sections within each title begin with the same number; for example, Title III includes all sections 300–399. Thus knowing the overall content of titles allows one to quickly understand where a given section fits. The titles are:

   I   General Provisions (establishing the FCC and its operations)
   II   Common Carriers (regulations of telephone rates and services)
   III Special Provisions Relating to Radio (broadcasting and, after 1967, public broadcasting)
   IV   Procedural and Administrative Provisions (enforcement, FCC proceedings)
   V   Penal Provisions–Forfeitures (fines for infractions)
   VI   Cable Television (added in 1984, amended in 1992)
   VII Miscellaneous Provisions (emergency provisions)

9. Three congressional committee reports provide a summary of the issues and debates of the time. In the order of appearance, they are:

   House Commerce Committee, Subcommittee on Communications. *HR 3333, "The Communications Act of 1979" Section-by-Section Analysis.* Committee Print 96–IFC-11, 96th Cong., 1st Sess., April 1980.

   House Committee on Interstate and Foreign Commerce. *Telecommunications Act of 1980.* House Report 96-1252, 96th Cong., 2nd Sess., August 1980.

   Senate Commerce Committee. *Telecommunications Competition and Deregulation Act of 1981.* Senate Report No. 97-170, 97th Cong., 1st Sess., July 1981.

10. The Federal Telecommunications Policy Act of 1986, S. 2565, 99th Cong., 2nd Sess., June 18, 1986. See also Sterling and Kasle, 1988, pp. 2731–2768.

11. Despite the kind of publicity that suggested that a major legislative contender had stepped into the ring, the Dole bill not only failed to pass but, interestingly, was later repudiated by its sponsor, who claimed that he had not fully understood the impact that his bill would have had if it had passed. Members of Congress rarely engage in public disavowals of their handiwork; Dole's reversal was made more baffling by his failure to specify precisely why he changed his mind. Cynical observers suggested that he had been pressured by AT&T; others guessed that Dole, a lawyer, was reluctant to bring a regulatory agency into conflict with a federal court. Still others observed that the bill failed to generate the kind of support that Dole, as the chief sponsor, needed in order to carry on the fight for passage and, indeed, Dole let it be known that if he were going to lead the charge he expected to be able to turn around and see the congressional

troops dutifully following him. When the needed support failed to materialize, Dole moved on to other matters.

12. Every member of Congress has telephone company employees and customers as constituents; moreover, every member of Congress also has a potential election opponent who will claim that that member's telephone legislation led to rate increases.

13. For non-American readers a word of explanation: GOP stands for "Grand Old Party," by which term the Republican Party has been known for almost a century.

14. Both houses in Congress (House and Senate) work on legislation to the virtual exclusion of any consideration of what may be happening in "the other House." The result is that finished bills rarely coincide exactly and must then be committed to an appointed conference committee of key involved members from both houses. The conference process can and does tend to re-debate all the issues raised in original committee and then floor debate. And lobbies buy full-page newspaper ads in *The Washington Post* and *The New York Times* to try and influence the conferees. In a sense, the process means that any legislation must go through four levels of debate and approval before being sent to the President for his approval or veto: (1) committee hearings and bill mark-up in both houses, (2) floor action in both houses, (3) conference committee compromises to create a melded bill, and (4) final approval in both houses of those conference results.

15. To be historically correct, we are talking about *re*-introduction of competition. For much of the early 20th century, competing local exchange carriers operated side-by-side in many American cities. Most had been bought out by either AT&T or one of the larger independent (non-AT&T) holding companies by the early 1940s. The resulting "natural monopolies" were soon perceived as the norm.

16. Wholesaling by local exchange carriers of bulk facilities to other carriers for the latter to resell to end users.

17. The ability of customers to keep their assigned telephone numbers when switching carriers.

18. Essentially no dialing delays plus equal access to numbers, operator services, directory assistance, and directory listings.

19. Equal access by competing local carriers to telephone poles and other rights-of-way at rates and conditions subject to FCC review.

20. "Facilities-based" indicates a firm with its own switches and wire or cable connections to individual homes. Experts agree that initial competition to existing RBOC local exchange service will likely come from carriers buying RBOC capacity at wholesale prices and reselling to end users. The law requires such RBOC sales, though details on what will be "fair" wholesale prices are subject to court analyses.

21. FCC, 1986. *Report and Order in* Computer Inquiry III, 104 *FCC 2d* 958. The lengthy proceeding began in 1985 and stretched into the 1990s on various appeals and remands. It eventually developed an "open network architecture" scheme with an intricate set of requirements allowing RBOCs to enter competitive markets—but only under conditions of total and open equity and interconnection with other service providers.

22. And especially the big three—AT&T, MCI, and Sprint—who together control much of the national interexchange marketplace.

23. As defined in practice, a separate subsidiary means officers, employees, and financial records separate from the regulated parent RBOC.

24. The three-year period begins with the date an RBOC is first authorized to offer interexchange service. In any case, the separate subsidiary requirement sunseted four years after the enactment of the 1996 amendments—in early 2000.

25. These include *network nonduplication* (blanking out imported television signals that duplicate a local network affiliate), *syndicated exclusivity* (protecting a local broadcaster's

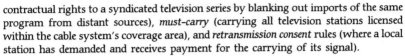

contractual rights to a syndicated television series by blanking out imports of the same program from distant sources), *must-carry* (carrying all television stations licensed within the cable system's coverage area), and *retransmission consent* rules (where a local station has demanded and receives payment for the carrying of its signal).

26. As wording revised Sec. 302 of the 1934 Act.

27. *Louisiana Public Service Commission v. FCC*, 476 U.S. 355 (1986) overturned an FCC attempt to preempt state public utility commission decision-making power concerning the setting of telecommunications carrier depreciation rates.

28. 47 U.S.C. 254 (c)(1)(A)–(D).

29. Ibid., 254 (b)(1)–(5).

30. See 47 U.S.C. 214 (e)(1)–(5).

31. Ibid., 214 (e)(2).

32. During the third quarter of 2002, 40 rural CLECs and 13 nonrural CLECs received such support, compared with 1,349 rural Incumbent Local Carriers and 87 nonrural Incumbent providers. See Universal Service Administrative Company, 2002. Quarterly Administrative Filings, Appendix HC01.

33. 47 U.S.C. 254 (j).

34. A Joint Board is made up of selected FCC commissioners and selected state utility commissioners. The Joint Board meets to discuss issues of joint concern between the FCC and the state commissions and then makes a recommendation to the FCC. The FCC can choose to accept the recommendation and issue an order enforcing it.

35. FCC, 1997. *Federal-State Joint Board on Universal Service*, CC Docket No. 96–45, Report and Order, 12 FCC Rcd at 8809, para. 61.

36. FCC, 2002. *In the Matter of Federal-State Joint Board on Universal Service*, CC Docket No. 96–45, Recommended Decision, FCC 02J-1 (Released July 10), para. 7. (Hereafter referred to as *Recommended Decision*)

37. Ibid., at para. 8.

38. Ibid., 254 (h)(1)(B).

39. FCC, 2004. Public Notice, "Proposed Second Quarter 2004 Universal Service Contribution Factor," CC Docket 96–45 (March 5), p. 2.

40. If the Commission has its way, even more local telephone companies will be recovering Lifeline revenues from the fund. In mid-2004, the FCC adopted a more inclusive and robust Lifeline/Link-Up program. This included income-based eligibility criteria for the program (a family of four with annual income of no more than $25,000 qualifies). The FCC estimated this would increase Lifeline/Link-Up subscribership by about 1.2 million. The Commission also adopted "outreach" guidelines to assure that those needing support received it.

41. FCC, 1997. "In the Matter of Access Charge Reform, Price Cap Performance Review for Local Exchange Carriers, Transport Rate Structure and Pricing, End User Common Line Charges," *First Report and Order*, CC Docket Nos. 96-262, 96-1, 91-213, 95-72, FCC 97-158 (Released May 16), at para. 92.

42. The Commission eventually allowed the SLC for second residential lines and for multi-line businesses to increase substantially, and allowed the local companies to recover some of their local loop costs through a new flat monthly charged levied directly on the long distance companies. Most long distance companies passed the charge directly along to subscribers, shifting even more of the local loop cost into subscribers' monthly payments.

43. Brian Hammond, 2000. "Consensus for Change: New Access Charges . . . and Debates That Remain," *Changing the Rules: The Future of Telephone Access Charges*, Telecommunications Reports International, p. 9.

44. FCC, 2000. *In the Matter of Access Charge Reform, Price Cap Performance Review for Local Exchange Carriers, Low Volume Long Distance Users, Federal-State Joint Board on Universal Service*, Sixth Report and Order in CC Docket Nos. 96-262 and 94-1, Report and Order in CC Docket No. 99-249, and Eleventh Report and Order in CC Docket No. 96-45, CC Docket No. 96-262, CC Docket No. 94-1, CC Docket No. 99-249, CC Docket No. 96-45. FCC 00-193 (released May 31).

45. According to the Act 47 U.S.C. 153 (37):

> The term "rural telephone company" means a local exchange carrier operating entity to the extent that such entity
>
> (A) provides common carrier service to any local exchange carrier study area that does not include either
>> (i) any incorporated place of 10,000 inhabitants or more, or any part thereof, based on the most recently available population statistics of the Bureau of the Census; or
>> (ii) any territory, incorporated or unincorporated, included in an urbanized area, as defined by the Bureau of the Census as of August 10, 1993;
> (B) provides telephone exchange service, including exchange access, to fewer than 50,000 access lines;
> (C) provides telephone exchange service to any local exchange carrier study area with fewer than 100,000 access lines; or
> (D) has less than 15 percent of its access lines in communities of more than 50,000 on the date of enactment of the Telecommunications Act of 1996.

46. The National Rural Telecom Association, The National Telephone Cooperative Association, the Organization for the Promotion and Advancement of Small Telecommunications Companies, and the United States Telecom Association.

47. See Petition for Rulemaking of the LEC Multi-Association Group, October 20, 2000.

48. Ibid., at para. 15.

49. FCC, 1999. *In the Matter of Federal-State Joint Board on Universal Service and Forward-Looking Mechanism for High Cost Support for Non-Rural LECs*, CC Docket No. 96-45 and CC Docket No. 97-160, Tenth Report and Order, FCC 99-304 (released November 2), at para. 22. (Hereafter referred to as the *Inputs Order*)

50. One scholar lists as many as seven different proposed models. See Steen G. Parsons, 1998. "Cross-Subsidization in Telecommunications," *Journal of Regulatory Economics*, 113:157–182, 174–175.

51. *Inputs Order*, at para. 8.

52. Rural areas were regarded as too difficult to model, and, as is discussed later in this section, the FCC decided to adopt a different approach to support for the small companies serving rural areas.

53. Payments for ETCs were not so simple to determine. The Telecommunications Act specified that there had to be "specific, predictable, and sufficient Federal and State mechanisms to support and advance universal service." In creating a Federal support mechanism, the FCC had to determine exactly what the role of Federal support should be in assuring that service rates were affordable across the nation, and also that rates were relatively comparable between urban and rural areas as required by Section 254(b)(3) of the Act. Jurisdiction complicated this question. The FCC has interstate jurisdiction; however, local service rates are intrastate, and so fall under the purview of the states. After a few false starts, in which the FCC proposed to tie Federal support to the amount of support each state could provide for universal service, the FCC finally determined the proper role for Federal support:

We find that, as a matter of policy, federal universal service high-cost support should be sufficient to enable reasonably comparable rates among states, while leaving states with sufficient resources to set rates for intrastate services that are reasonably comparable to rates charged for similar services within their borders. Because the Commission does not set rates for intrastate services, the decision we adopt today is intended to allow states to ensure that rates are reasonably comparable within their borders. We find that this is consistent with the goals that Congress established in section 254(b)(3).

See FCC, 1999. *In the Matter of Federal-State Joint Board on Universal Service*, CC Docket No. 96-45, Ninth Report & Order and Eighteenth Order on Reconsideration, FCC 99-306 (released November 2), at para. 7. (Hereafter referred to as the *Methodology Order*)

54. FCC, 2000. Public Notice, "Common Carrier Bureau Announces Procedures for Releasing High-Cost Universal Service Support Amounts for Non-Rural Carriers and Revised Model Results, DA 00-110 (January 20). To avoid disruption while the new procedure was implemented, the FCC created an interim hold-harmless provision for each carrier on a line-by-line basis. Support would be targeted to lines served in the highest cost wire centers and would be portable. Interim hold-harmless support would be phased out by reducing the support by $1.00 per line per year for each non-rural carrier until the hold-harmless support was eliminated.

55. FCC, 2001. *In the Matter of Federal-State Joint Board on Universal Service and Multi-Association Group (MAG) Plan for Regulation of Interstate Services of Non-Price Cap Incumbent Local Exchange Carriers and Interexchange Carriers*, CC Docket Nos. 96-45 and 00-256, Fourteenth Report and Order, Twenty-Second Order on Reconsideration and Further Notice of Proposed Rulemaking in CC Docket No. 96-45 and Report and Order in CC Docket No. 00-256. FCC 01-157 (released May 23). (Hereafter referred to as the *RTF Order*).

56. FCC, 2004. Public Notice, "Proposed First Quarter 2005 Universal Service Contribution Factor" (December 13).

57. For example, a group of rural telephone companies in Nebraska appealed the Public Service Commission's decision to designate Western Wireless as an ETC. The Nebraska Supreme Court upheld the Nebraska PSC's decision. See 264 Neb. 167.

58. Ibid., at paras. 3 and 4.

59. The National Telecommunications and Information Administration, during the Clinton Administration, issued a series of reports warning about the dangers of such a digital divide. The Bush Administration was less convinced that a digital divide existed.

60. 47 U.S.C. 157 (a).

61. FCC, 2004. *Availability of Advanced Telecommunications Capability in the United States*, GN Docket No. 04-54, Fourth Report, FCC 04-208 (released September 9), p. 8.

62. FCC, 1996. Report and Order in CC Docket 96-98 as reprinted in *Federal Register* (August 29, Part II), paras. 6–8.

63. Ibid. The interconnection decision and regulations were said to be the longest federal rules issued by any regulatory agency in American history. More than a few commenters have suggested that the widespread use of computerized word processing was making for more complex government decisions.

64. Ibid., para. 10.

65. Ibid., para. 12.

66. Ibid., para. 13.

67. Ibid., para. 14. In addition, the FCC had earlier required LECs to ensure that customers who changed local service providers would be able to retain their existing telephone numbers. FCC *Report and Order* in CC Docket 95-116, "Telephone Number Portability."

68. FCC Press Release, 2003. "Federal Communications Commission Releases Data on Local Telephone Competition" (December 22).

69. 47 U.S.C. 251, (d)(2)(A) and (B).

70. *AT&T v. Iowa Utilities Board*, 525 U.S. 366-390 (1999).

71. *United States Telecom Association v. FCC*, 351 U.S. App. D.C. 329 (2002).

72. UNEs priced too low could encourage competitors to use the ILEC network, instead of building their own facilities. UNEs priced too high could discourage competitors who were not quite ready to build their own facilities from even entering the market. Theoretically, UNE prices should reflect the ILECs' costs so that the ILECs are adequately compensated for the use of their networks and the competitors are sent the appropriate economic signals about market entry. The issue of what constitutes costs, as both economists and accountants will attest, is not a simple one.

73. 47 U.S.C., Section 252(d)(1).

74. FCC, 1996. "In the Matter of the Implementation of the Local Competition Provisions in the Telecommunications Act of 1996, and Interconnection Between Local Exchange Carriers and Commercial Mobile Radio Service Providers," CC Docket Nos. 96-98 and 95-185, FCC 96-235 (released August 8), at para. 29. (Hereafter referred to as the *Local Competition Order*)

75. The TELRIC costing method built rates based on an efficient network at optimal capacity, instead of the actual costs incurred by the ILECs in building and maintaining their networks and the actual level of usage of those networks. New, more efficient technologies tended to be less expensive than the older technologies still in use. Basing rates on forward-looking costs that assumed full network capacity, instead of actual usage, resulted in lower rates as well. Instead of the fully allocated cost approach of rate-of-return regulation, in which the carrier recovered all of its overhead costs, the TELRIC methodology allowed ILECs to recover only that portion of overhead costs allocatable to a specific service.

76. 47 U.S.C., Section 252 (d)(3).

77. *Local Competition Order*, at para. 32.

78. FCC, 2003. *In the Matter of the Commission's Rules Regarding the Pricing of Unbundled Network Elements and the Resale of Service by Incumbent Local Exchange Carriers*, 2003. Notice of Proposed Rulemaking, WC Docket No. 03-173, FCC 03-224, para. 2. (Released September 15). Hereafter referred to as TELRIC NPRM.

79. TELRIC NPRM, at paras. 50–51.

80. Ibid., at para. 58.

81. FCC, 2001. *Review of the Section 251 Unbundling Obligations of Incumbent Local Exchange Carriers, Implementation of the Local Competition Provisions of the Telecommunications Act of 1996, Deployment of Wireline Services Offering Advanced Telecommunications Capability*, CC Docket Nos. 01-338, 96-98, 98-147, Notice of Proposed Rulemaking, 16 FCC Rcd 22781.

82. FCC, 2003. *In the Matter of Review of the Section 251 Unbundling Obligations of Incumbent Local Exchange Carriers, Implementation of the Local Competition Provisions of the Telecommunications Act of 1996, Deployment of Wireline Services Offering Advanced Telecommunications Capability*. CC Docket Nos. 01-338, 96-98, 98-147, Report and Order on Remand and Further Notice of Proposed Rulemaking, FCC 03-36 (released August 21), at para. 2. (Hereafter referred to as *Triennial Review Order*)

83. Ibid., at para. 3.

84. In making this determination, such issues as an ILEC's scale economies, sunk costs, first-mover advantage, absolute cost advantages, and other barriers within the control of the ILEC should all be considered.

85. Whether CLECs would be impaired from competing in the local market if switching were no longer a UNE had become a strongly debated question. CLECs had been providing service through what was called the unbundled network element platform, or UNE-P. The UNE-P was a combination of UNEs that included switching; in effect, by combining switching with other UNEs, CLECs were leasing the ILEC network at steeply discounted prices. In the ILEC view, switching could be removed as a UNE without causing CLECs impairment; a significant number of CLECs were providing their own switches, especially in metropolitan areas. Moreover, the steeply discounted cost of providing service through use of the UNE-P provided an incentive for CLECs to use the UNE-P, instead of providing their own switching facilities. As FCC Chairman Powell noted in his comments on the *Triennial Review Order*, between the beginning of 2000 and June of 2002, the number of lines CLECs were connecting to their own switches had declined by over half a million lines per year, suggesting that the UNE-P, instead of encouraging facilities-based competition, was actually encouraging a transition away from facilities-based competition. FCC, "Separate Statement of Chairman Michael K. Powell Approving in Part and Dissenting in Part," 2003. *Re: Review of the Section 251 Unbundling Obligations of Incumbent Local Exchange Carriers, Implementation of the Local Competition Provisions of the Telecommunications Act of 1996, Deployment of Wireline Services Offering Advanced Telecommunications Capability*, CC Docket Nos. 01–338, 96–98, 98–147, Report and Order on Remand and Further Notice of Proposed Rulemaking, FCC 03–36 (released August 21), p. 6.
86. For CLECs seeking to serve the mass market, there would be a presumption of impairment; however, each state commission would have nine months to conduct proceedings, on a "granular" basis, to decide otherwise. If a state commission found no impairment in a specific market, there would be a three-year phase out of the UNE-P in that market. For CLECs serving business customers through the use of higher capacity loops, there would be a presumption of no impairment, and thus no UNE-P; state commissions would have 90 days to find otherwise, again on a market-by-market basis. State commissions would be free to define "market" for these analyses. In addition to switching, the state commissions would also make determinations about whether certain types of loops and transmission facilities would continue to be UNEs. The lack of access to UNEs for DS1, DS3, and dark fiber loops and transport facilities would be regarded as an impairment, unless state commissions found otherwise.
87. "Commissioners Adopt 'Triennial Review' Order, with Martin Lone Backer of Entire Decision," 2003. *TR Daily* (February 20).
88. Ibid.
89. Ibid. While Commissioner Martin supported all provisions of the order, the two other Republicans on the FCC, Chairman Powell and Commissioner Kathleen Abernathy, dissented on ending line sharing, continuing switching as a UNE, and delegating decision making to the states. Both supported exempting broadband from unbundling, and continuing to require unbundling for narrowband loop and transport facilities. The two Democrats, Commissioners Copps and Adelstein, supported an end to line sharing, continuation of switching as a UNE, and delegation to the states. Both dissented on exempting broadband from unbundling requirements.
90. "Separate Statement of Commissioner Michael J. Copps Approving in Part, Concurring in Part, Dissenting in Part" (August 21, 2003). *Re Review of Section 251 Unbundling Obligations of Incumbent Local Exchange Carriers*, CC Docket No. 01–338, FCC 03–36, p. 3.
91. "Separate Statement of Chairman Michael K. Powell Approving in Part and Dissenting in Part" (August 21, 2003), *Re Review of Section 251 Unbundling Obligations of Incumbent Local Exchange Carriers*, CC Docket No. 01–338, FCC 03–36, pp. 7–8, 16.
92. "First Baby Bell Seeks Entry Into Long Distance," 1997. *New York Times* (January 3), p. D1.

93. "Ameritech Seeks Access to Long-Distance Market," 1997. *Washington Post* (January 3), p. B11.

94. "SBC Files Suit Challenging Telecom Act's Local Entry Provisions," 1997. *Communications Daily* (July 3), p. 1.

95. Peter Elstrom. 1997. "Why SBC Shouldn't Be the First Bell in Long Distance," *Business Week* (July 21), p. 33.

96. "Communication Legislation Sirs Talk of Mergers by Baby Bells," 1993. *New York Times* (December 19), p. D4.

97. "2 Bell Companies Agree to Merger Worth $17 Billion," 1996. *New York Times* (April 2), p. A1.

98. "Communication Legislation Stirs Talk of Mergers by Baby Bells," 1995. *New York Times* (December 19), pp. D1, D4.

99. "Now, Bell Atlantic Plans to Buy Nynex, Not Merge With It," 1996. *New York Times* (June 17).

100. "Justice Dept. Vows Scrutiny of Bell Deals," 1996. *New York Times* (April 29), p. D1.

101. "CWA Joins NYNEX in Rebutting N.Y. Attorney Gen.'s Opposition to Merger," 1997. *Communications Daily* (February 7), p. 2.

102. *Wall Street Journal*, 1997. (July 8), p. A3. Klein's own approval to become the head of the Antitrust Division (he was an acting Assistant Attorney General and had been since 1996) was held up for weeks because of deals like this. Senate Democrats were concerned that the Department of Justice had become a mere pushover for industry mergers and was not considering the deals carefully enough.

103. Ibid.

104. "Bell Atlantic–NYNEX Merger," 1997. *Washington Post* (July 20), p. A17 (boxed feature).

105. "Bell Atlantic, Nynex Reach Merger Accord with FCC," 1997. *Washington Post* (July 20), p. A17.

106. "Sale of Bellcore Reported to Big Research Company," 1996. *New York Times* (November 22), p. D2.

107. "The Company Sketched Out on the Back of a Napkin that Grew to $180bn," 2002. *The Guardian* (June 27), accessed on April 19, 2004, at http://www.guardian.co.uk/interntional/story/0,3604,7444723,00.html

108. "MCI's Board Accepts New Merger Bid from WorldCom," 1997. *TR Daily* (November 10), accessed on April 19, 2004, at http://www.tr.com/tronline/trd/1997/td7217/td721701.htm

109. "With Divestiture Promised, DOJ Oks MCI–WorldCom Merger," 1998. *TR Daily* (July 15), accessed on April 19, 2004, at http://www.tr.com/tronline/trd/1998/td8135/td813501.htm The sale included MCI's nationwide Internet backbone, including 22 domestic nodes, 15,000 interconnection ports, 40-plus peering agreements, various pieces of equipment, 3,300 Internet access corporate accounts, about 1,300 Internet service provider customers, MCI's dial-up Internet access business, and MCI's Web hosting and managed firewall services.

110. "WorldCom Completes Merger with MCI After FCC Approves," 1998. *TR Daily* (September 14) accessed on April 19, 2004, at http://www.tr.com/tronline/trd/1998/trd8177/trd817701.htm

111. "Telecom Mergers Proceed—Cautiously: FCC Takes an Open Stance on Competition in Big Deals," 1999. *Telephony Online* (October 18), accessed April 18, 2004, at http://www.internettelephony.com

112. FCC, 2000. "In Re: Applications of Sprint Corporation, Transferor, and MCI WorldCom, Inc. Transferee for Consent to Transfer Control of Corporations Holding

Commission Licenses and Authorizations Pursuant to Section 214 and 30(d) of the Communications Act and Parts 1, 21, 24, 63, 73, 78, 90, and 101." Order, CC Docket No, 99-333, DA 00-1771 (Released August 4).

113. FCC, 1996. "Joint Motion of GTE Corporation and the Southern New England Telephone Company for Stay Pending Judicial Review," CC Docket 96-98, (August 28), p. 6.

114. Petition for Review Pending and Partial Stay Granted, sub nom. *Iowa Utilities Board et al. v. Federal Communications Commission.* Case No. 96-3321 and consolidated cases, U.S. Court of Appeals, Eighth Circuit, filed July 18, 2000.

115. "House Members Blast FCC for Exceeding its Authority in Interconnection," 1996. *Communications Daily* (November 20), p. 2.

116. "Is Telecom Deregulation Unconstitutional?" 1996. *Business Week* (November 4), p. 98.

117. "Court Sets Back F.C.C. Efforts to Open Local Phone Markets," 1997. *New York Times* (July 19), p. 1.

118. 525 U.S. 366 (1999). The Supreme Court consolidated seven related cases in this decision. The Court also noted the ambiguity in much of the law and other critics argued it was one of the most poorly written pieces of major legislation, this naturally making the appeals process more complicated and prolonged.

119. "Slew of Supreme Court Cases to Focus on '96 Telecom Law," 2001. *New York Times* (October 1), p. C8.

120. *Verizon Communications Inc. et al. v. Federal Communications Commission et al.,* 2002. 535 US 467.

121. See for example, *Gulf Power Co. v. FCC,* 208 F. 3d 1263, April 11, 2000.

122. *National Cable & Telecommunications Association v. Gulf Power Co. et al.,* 2002. 534 US 327.

123. *Verizon Maryland Inc. v. Public Service Commission of Maryland et al.,* 535 U.S. 1076 (May 2002).

124. *U.S. Telecom Association v. FCC.* 2004. United States Court of Appeals for the District of Columbia Circuit, 359 F.3d 554.

125. "Five FCC Commissioners Endorse Call for ILEC/CLEC UNE Rate Negotiations," 2004. *TR Daily* (March 31). Accessed on April 27, 2004, at http://www.tr.com/online/trd/2004/td033104/td033104.htm

126. "TRO Remand Sets Impairment Thresholds Higher Than Expected," 2004. *TR Daily* (December 15). Accessed on December 28, 2004, at http://www.tr.com/online/trd/2004/td121504/td121504.htm

127. Chairman Powell included provisions for broadband UNEs in hopes of brokering a compromise with the two Democratic commissioners (Copps and Adelstein); this effort failed. "ILECs, CLECs Both Voice Discontent with TRO Remand," 2004. *TR Daily* (December 15). Accessed on December 28, 2004, at http://www.tr.com/online/trd/2004/td121504/td121504.htm

128. Senator Larry Pressler (R-SD), on February 7, 1996, as quoted in the *Washington Post* (January 19, 1997), p. H5. Clearly the bill was not useful for *his* job—Senator Pressler lost his bid for re-election eight months later.

129. The Maginot Line was built at huge expense by the French in the 1930s to defend their borders against Fascist Italy and Germany. When invasion finally came in 1940, the lines proved useless as enemy troops went around them. The French army, weakened by their trust in invincible fixed defenses, were no match for the fast-moving enemy forces, and France surrendered in six weeks.

130. See Allan Sloan, 1997. "Landmark Legislation Ushering in the Old World of Telecommunications," *Washington Post* (June 3), p. C3.

131. See Daniel Kadlec, 1997. "Phone Pranks: Why We Should Hang Up on a Plan to Rebuild Ma Bell," *Time* (June 9), p. 46.

132. Ibid.

133. Chairman Kevin J. Martin's Announcement Regarding the Supreme Court's Decision in *Brand X*," *FCC News*, June 27, 2005.

134. Todd Shields, 2005. "Pol Calls for Telecom Act Rewrite," *MediaWeek.com* (April 12). Accessed on July 11, 2005, at http://www.mediaweek.com/mw/search/article_display.jsp?schema=&vnu_content-id-1000876061

135. "TeleCONSENSUS Group Formed to Lobby for Telecom Act 'Update,' " *TR Daily*, April 12, 2005. (online version)

build a fiber network covering much of the world.[10] A new trans-Atlantic cable that entered service in the winter of 2002–2003 (developed by Britain's Cable & Wireless and France's Alcatel) carries 30 percent more information than *all* existing trans-Atlantic capacity, satellite or cable, combined.[11] The consulting firm TeleGeography estimated that fiber cable capacity across the Atlantic by early 2003 totaled 3.5 gigabits per second (gbps) versus predicted demand of only .5 gbps.[12]

These high-capacity international fiber links will *eventually* be taken up and used. But while predictions differ, it will take years before further expansion is called for. In the meantime, in place but underused fiber facilities will act as a substantial drag on industry recovery.

Wireless Networks

Wireless technology was also improving and greatly increasing system capacity as well as options. In the five-year period from 1997 to 2002, no fewer than six national mobile communications networks (in order of size by 2002 they were Verizon, Cingular, AT&T, Sprint, Nextel, and VoiceStream) began operation, dramatically increasing competition and forcing prices and thus revenue down. Profits were scarce which limited attempts to improve networks to eliminate dropped calls and poor quality service[13] and make various networks sufficiently compatible to allow text and other advanced message sharing. Limited profits have also slowed development of the digital third generation (3G) services needed to attract new customers and more revenue. Given the limited development funds available to any wireless carrier, controversy has developed as to whether they are best spent on patching up existing networks for basic services, or offering new service which will place even more pressure on network infrastructure.[14]

Amidst such competition technological progress both helps and hurts. Recent developments may further diffuse customers. For example, Wireless Fidelity (WiFi) transmission technology (also called 802.11b [or, more recently, g] for the technical standard used) provides fast Internet links and is extending its presently limited range—but may add to woes of wireless data providers as its capacities force prices down.[15] The number of available networks expanded from just over 1,000 in 2001 to more than 15,000 by the end of 2003.[16] About the same time, more than two-thirds of all laptop computers were WiFi compatible. A veritable free network underground has developed that encourages use of the urban-based WiFi systems by as many people as possible. A variety of companies have begun to offer paid access. Starbucks provides such a service in more than 4,000 of its ubiquitous coffee shops, illustrating how varied businesses can meld offerings. Many Internet cafes are working on the same princi-

Chapter 10

Meltdown . . . and the Future (Since 2000)

"The fall of telecom ranks as the biggest wreck in corporate history"
Fortune[1]

The overheated telecommunications and information technology boom of the 1990s began to unravel rapidly beginning in March 2000. In the space of a few months, and deepening over the next several years, stock prices dropped precipitously, employees were laid off, and by 2001 some fairly large firms were nearing bankruptcy. 2002 revealed several scandals and the resulting collapse of huge carriers including WorldCom and Global Crossing. The amounts of money and jobs lost were astonishing. Investors all but disappeared.

By mid-2002 what was widely being termed a telecommunications "meltdown" already ranked as the biggest business debacle in U.S. history. Controlling for the effects of inflation over the years, it surpassed the financial collapse of the railroads in the 1890s, the savings and loan crisis of the 1980s and grew to be 10 times larger than the "dot-com" crash of 2000–2001 which itself had begun the meltdown that soon expanded to telecommunications. Predictions from most experts were that things would continue to worsen for at least two more years before they turned around. Even the huge semi-annual Comdex trade show in Las Vegas—the largest in the computer and communications business—had shrunk by half in just two years following 2000.[2]

What could have gone wrong so quickly? How serious was the slump, and how long would it last? This chapter, written near the bottom (or seeming bottom) of the telecommunications meltdown, provides an ini-

tial survey of what happened to the domestic telecommunications industry and why. There was no single villain, nor was any one decision or
company responsible. The precipitous decline of a once luminous industry was due to a variety of interlocking factors that implied considerable
caution for the future.

10.1 OVERCAPACITY

In a segment of Walt Disney's classic animated movie *Fantasia* (1940), a
sorcerer's apprentice (played by the youthful Mickey Mouse) seeks to au-
tomate his tedious task of fetching water from a well. The "broom crea-
tures" he organizes to undertake the work do so well, however, that they
soon exceed Mickey's demand for water. His several attempts to slow
them down and regain control of the situation only lead to still more carriers and thus more water—and rising floods (vast overcapacity) soon result. This tale can be seen as a metaphor helping to illustrate how technological breakthroughs and investment in the 1990s helped to create the
telecommunications floods of the early 2000s.

At its most basic, the information/telecommunication industry meltdown that began in 2000 resulted from dramatically increased capacity—far more than demand growth warranted. Substantial overpre-
dictions of likely demand (*see below*) contributed to the overcapacity
problem. So did the growing number of providers. One or two companies
building capacity was one thing, but dozens pursuing similar strategies
of rapid expansion spelled trouble. Some critics have argued that the famous "Moore's law" concerning integrated circuits (that capacity dou-
bles every 18 months while prices decline) is now just as applicable to
telecommunication services. In turn, the increase in capacity has driven
prices down and eliminated profitability (the margin of error) for many
carriers and manufacturers.

Fiber Links

Chief among telecommunication developments has been a dramatic increase in the capacity of what an individual fiber optic link can carry.
This is due largely to development of dense wavelength division multiplexing (DWDM), which increases the capacity of a given fiber line a
thousand fold without need for amplification, thus dramatically cutting
construction and operating costs. Normally to be welcomed as a valuable
cost-cutting breakthrough, DWDM unfortunately arrived just in time to
make a bad overcapacity problem hugely worse. "Prior to 1995, telecom
carriers could send the equivalent of 25,000 one-page e-mails per second

over one fiber-optic line. Today they can send 25 million such e-m
over the same fiber strand, a 1,000-fold increase."[3] Development and
stallation of faster switches made the DWDM fibers even more efficien

Even without DWDM, however, construction of far too many hi
capacity links quickly outstripped existing as well as projected deman
dozen new national data fiber backbone networks were built within
a handful of years. Streets were torn up in cities across the countr
companies competed to place their networks into service first. Jus
tween 1998 to 2001, terrestrial optic fiber lines in the United States
five-fold while (thanks to constantly improving technology) actua
pacity growth was perhaps 100-fold.[4] All told, the 1997–2001 p
saw the installation of (by some estimates) 100 million miles of fibe
enough to reach the Sun. Unfortunately for carriers, however, der
merely quadrupled and thus by 2002 the vast majority of that fiber
"unlit" or "dark" (not in use). Indeed, *less than three percent* of all th
ber capacity was reported to be in service.[5] Despite continued der
growth, this capacity may not be effectively taken up for a decade or
ger. One industry consultant went so far as to claim that more tha
years of capital investment had been expended in just the late 19
Naturally, depressed by such excess capacity, prices dropped sharp
major intercity routes.[7]

Yet despite substantial evidence of excess capacity, the "herd" or
tion" mentality persisted among at least some carriers. In mid-2002
were new calls for development of still more metropolitan fiber sys
a seeming example of the "build it and they will come" idea flying
face of reality. So was another proposal to lay a 300-mile fiber link
the floor of Chesapeake Bay.[8] A central aspect of the problem was
too many different firms were being driven by competition to d
same thing, thus resulting in the overbuilding of parallel transm
links, often along the same or similar routes.

Overcapacity soon defined many major international trunk rou
well. Trans-Atlantic fiber optic cable and satellite capacity grew in
short years by a factor of 19. While satellite links had outpaced t
pacity of traditional undersea cables into the 1980s, the roles of th
transmission modes were reversed in the late 1990s by new
capacity fiber cables. Here, too, the jump in available capacity gr
much faster than demand led to price cuts. For example, the cos
leased trans-Atlantic line dropped after 1997 from about $125 tho
to just $10 thousand a year. Just in the nine months it took one fi
design a trans-Atlantic cable, technology allowed a doubling of its
ity at little increase in cost.[9] In the face of a dearth of customers t
up the glut, however, one of the largest cable firms, Global Crossi
nally declared bankruptcy in January 2002 after spending $15 bill

ple.[17] While WiFi networks are very limited in range and suffer potential security drawbacks (though more attention is being paid to encryption add-ons), they are already making digital subscriber line (DSL) and cable modem providers nervous given WiFi's ability to accommodate numerous users on a single "line." Further, WiFi is becoming increasingly available for home use in the form of internal wireless links among computers in different rooms or on different floors, thus avoiding the need for hard wiring or separate accounts. By late 2002 some 15 million users were making regular WiFi connections.[18] Likewise, mobile phone providers are worried that WiFi expansion could effectively foreclose their hoped-for expansion into provision of Internet services. As WiFi spectrum is unregulated and thus virtually free, its expansion calls into question the huge auction prices paid for 3G wireless spectrum by incumbent operators around the world (*see section 10.4*). The one thing going for mobile service operators is that WiFi is not a "mobile" service per se, though it is becoming so widely available in urban areas that it may provide a real competitive option for mobile users. Indeed, several mobile providers (Voice Stream and AT&T Wireless among them) have become WiFi investors in a protect-your-flanks move.[19] On the other hand, some local carriers are threatening users that make their connections available to others by means of WiFi links.[20] AT&T, IBM, and Intel combined efforts to establish a nationwide WiFi network known as Cometa Networks, offering 20,000 access points by the end of 2004.[21] Intel announced plans to include wireless link capability in its future microprocessors. Cometa executives made clear theirs was a wholesale approach, selling access to stores and shops, not to end users. They foresee WiFi working hand-in-hand with 2.5G or 3G wireless networks, not directly competing. Industry observers were less sure of that, especially given the excess capacity problem evident everywhere.

Even more promising, though just as short-ranged for now, is ultrawideband (UWB) service which was experimentally approved for commercial operation by the FCC early in 2002. Based on technologies pioneered by the American military back in the 1950s, this new version is presently limited to low power, limited range, and very high frequencies (ranging from 3.1 to 10.6 GHz). It may become a serious competitor to traditional wireless links.[22] Rather than the traditional mode of sending one signal on one frequency, UWB breaks a signal into pulses and sends them over a wide range of frequencies at the same time using very low power. The FCC rules limit range to about 40 feet, but if interference concerns of traditional spectrum users can be assuaged, greater distances may be allowed.[23]

Other technologies are in the wings or being tested, including "smart" antennas that would allow a mobile transmitter or receiver to work different frequencies at the same time, so-called mesh networks, and what some have labeled "ad hoc networks."[24] Any or all may provide a very flexible means of gaining broadband access to homes.

Another wireless service, domestic satellite carriers (domsats), were also in growing trouble. Faced with strong price competition from the collapsing fiber cable carriers and from terrestrial mobile service providers, the domsat industry began to postpone launches of new satellites while slowing research and orders for later generations. Further, several planned global mobile satellite services (such as Iridium and Teledesic) collapsed, some before launching any capacity into orbit. According to one consulting firm, "Of the 675 launches expected between 1998 and 2002, only 275 satellites reached space."[25] The shrinking number of satellite transponders was having serious impact on the U.S. military's information and operational capacities, an unintended national security impact of changes in the commercial industry.[26]

Broadband Services

In the early 2000s, customers for one heavily promoted telecommunication revolution developed more *slowly* than initial projections suggested. Despite the many benefits of broadband Internet connection (e.g., far faster connections both up and downstream, elimination of the need for second telephone lines, and the quality of "always on") only about 10 percent of U.S. households utilized such connections by 2001, though they had grown to between 22 and 32 percent (depending on region of the country) by the beginning of 2003.[27]

Regional Bell Operating Companies (RBOCs) and their supporters argued that regulatory barriers were behind broadband's slow roll-out. They called for the elimination of 1996 telecommunications act provisions which forced Bell companies to provide access to their DSL facilities by competing carriers (*see Tauzin-Dingell discussion at section 10.4 below*). They argued that such provisions suppressed RBOC interest in investing in DSL in the first place. Of the relatively small proportion of households that *do* subscribe, two-thirds utilize cable modems. Thus broadband has benefited local cable monopolies rather than telephone carriers—another argument used by the latter in their quest for policy relief. As noted above, WiFi and related technologies may provide a lower cost breakthrough in provision of broadband services, thereby heating up competition for and thus growth of that market.

Critics suggest that the real broadband problem is its pricing—that broadband's slow growth resulted from consumers rejecting the high prices charged for the service. Despite the fact that a large majority of U.S. households have DSL (or cable modem) service options available, most have chosen not to subscribe, concerned about growing monthly payments. No wonder—combined local and long-distance telephone, cable subscription, cellphone fees and Internet connection costs often total well over $1,500 a year per household. And that's not counting additional expense to subscribe to the promised cornucopia of new services waiting in the wings.

Poor Predictions

If huge increases in trunk telecommunications capacity provided the basis for the ongoing meltdown, massive overestimation of telecommunications demand created the immediate cause. Projections of huge Internet-driven growth were widespread but, in retrospect, fantastic. Reputable consultants with solid track records in the past were reporting that Internet demand would continue to expand at 50 to 100 percent per *month*—when cool-headed analysis would suggest that 100 percent a *year* would be more accurate. In its initial commercial years (1995–96), use of the World Wide Web *did*, in fact, grow at something close to 100 percent per month. But predictions based on that period were carried over to later years when the actual pace of expansion had already declined. The 1997 projections of one company—WorldCom subsidiary UUNet (a primary Internet backbone carrier)—were repeated until they became widely accepted mantra.[28] Naturally they became a somewhat self-fulfilling prophecy, with more investment sought to keep ahead of the presumed demand growth. While there was general agreement that demand *was* growing and would continue to do so, what too few observers noted was that system capacity was growing even faster. The result was that telecommunication carrier profits began to decline just as their construction and other debt was ballooning.

With the 2000–2001 collapse of much of the dot.com industry (which had created many high-traffic Web sites—*see section 10.3*), the growth pace of Internet usage slowed well under what earlier projections suggested. At the same time, overestimation of how quickly the public would adopt convenient and widely touted (but expensive) broadband connections—and thus make even more use of interactive and video services—contributed to continuing glowing projections of demand growth that simply did not materialize.

Slower growth than predicted, ferocious competition, and resultant price cutting led to cost cutting including trimmed spending on research and development just when the industry needed new services and products to attract users. Product development takes years, suggesting that those new products may be some time in coming, thus contributing to the dim economic outlook. By the turn of the century it was also becoming increasingly evident that most computer and telecommunications manufacturing output was serving replacement or upgrade purchases— a far smaller market than the go-go growth years of the 1970s through 1990s.

10.2 ECONOMIC PRESSURE

There are a host of ways to measure the economic debacle that befell telecommunications and the economy after 2000. The end of readily available investment cash was an early indicator, caused by growing concern about profitability among banks and other funding sources. Another was the precipitous drop in stock values affecting virtually every player in the industry. Certainly of most concern to many was the growing wave of personnel layoffs that soon totaled a half million lost jobs. The incidence of accounting and other fraud uncovered by 2002 suggested pressure was rising to desperate levels at some companies. Finally, of course, spreading bankruptcy made vividly clear the deep crisis the industry was experiencing.

Investment: Frenzy to Freeze

For a time the marketplace seemed to offer almost a license to print money. Tales of early investments in seemingly promising start-ups resulting in huge returns helped to fuel an investment frenzy. Everyone seemed to say "you couldn't go wrong." Small companies with breakthrough hardware or software ideas could count on being bought out at a huge premium. The result would be more investment funding seeking a good home.

The combination and interaction of two vital factors helped to inflate the investment bubble: the availability of more investment money and the Internet as a breakthrough technology. More homes were investing in stocks—and had the income to do so. Money also flooded in from abroad. Stock mutual funds, for example, expanded from $125 billion in new money in 1995 to some $300 billion at the peak just five years later.[29] At the same time the Internet integrated the computer, software, and telecommunications industries into new combinations that rapidly built up a huge public appeal and seemingly had endless growth.

Potential investors were egged on by market analysts' reports promising golden returns over short periods of time. Star among them was Jack B. Grubman, a one-time AT&T employee who arrived on Wall Street in 1985 and moved to Salomon Smith Barney in 1994. He was soon heading their telecommunications team. In the late 1990s, Grubman's associates raised $190 billion for 81 companies, earning $1.2 billion in fees. But by then Grubman was far more than a mere analyst. He earned his $20 million dollar annual income by becoming a kingmaker in the industry, advising companies on who their CEO should be, sitting in on board meetings, and often tendering advice. He became deeply involved in such companies as WorldCom, Qwest, and Global Crossing, moving far beyond any definition of the "distinterested advisor" that many investors counted upon. And Grubman continued to issue glowing "buy" reports well after other analysts turned tail—in considerable part, argued his critics, because of his conflicted involvement with many companies. He encouraged buys on stocks such as WorldCom only weeks before the firm collapsed into bankruptcy.[30] In mid-2002 he was called to testify on his activities before Congress, and early in the fall he left Salomon Smith Barney. News of what appeared to be inflated Grubman ratings of AT&T stock (in return for a complex deal getting his child into a competitive preschool in Manhattan) broke late in 2002, casting still more doubt on his activities.[31] Some sort of legal action for securities violations seemed increasingly likely by mid-decade.

Simple greed was clearly one factor behind the stock frenzy. Investment sources, whether banks, investing institutions, or venture capitalists, often pushed companies into expansion more rapidly than conditions warranted. Why? Because floating stock deals, mergers and acquisitions, or other arrangements generated substantial fees for all concerned. Along with the investment, of course, came the constant pressure for revenue growth. "This fixation on revenue growth, demanded by the investors, is what has ultimately caused the downfall of so many companies."[32] Many new paper millionaires or billionaires were widely touted. "And it was during this period that stock analysts began to think and act like investment bankers, investment bankers like venture capitalists and venture capitalists like masters of the financial universe."[33]

Of course the rush was too good to last. Too much money, given too quickly and easily, and without sufficient safeguards in place—all combined to end the venture capital boom in the Internet and technology sector. Noted one observer, "Every one of the new companies got to borrow money with ease. That's an invitation for disaster, and that's what we have."[34] In retrospect either of two things would surely have helped: fewer companies pursuing the same customers, or far more viable business plans that took into account just how many competing and overlapping players there were in most sectors of the business. And another element—effective action by government regulators—might have forestalled the worst. But too many possible critics were looking the other way, and too many might-be logical gatekeepers were themselves being paid well to enjoy the ride and suspend their judgment.

Banks had provided most of telecommunication's expansion loans. More than $650 billion was loaned between 1996 and 2001 alone, many loans based on company stock as collateral as banks joined the investment frenzy. They often banded together into syndicates to provide the huge sums sought by telecommunication companies. J. P. Morgan Chase, for example, held $66 billion of the telecommunication debt while Bank of America Securities held nearly $39 billion and Goldman Sachs (*see above*) held another $20 billion.[35] Insurance companies often bought up telecommunication bonds as investments and were thus left holding the bag when the bonds declined along with their issuing companies. Prudential had once owned nearly $5 billion and Metropolitan Life just under $4 billion in telecommunication bonds that had dissipated most of their value.[36] When the bubble began to lose air and company stock prices fell, all of these investors found themselves with huge loan liabilities. Naturally they declined to invest further, drying up an important source of industry funding.

Stock Price Declines

Investment dried up because company stock values were plummeting. But the early 2000s' Internet and telecommunications stock price debacle is also better understood when compared to what came before. To take one stellar example: WorldCom stock rose some 7,000 percent in the course of the 1990s as Bernie Ebbers built the conglomerate through acquisition of dozens of smaller firms to become an important national player.[37] Other companies' growth was also often spectacular (Metropolitan Fiber Systems stock grew 32-fold over just two years before the company was gobbled up by WorldCom). Virtually everyone investing did well in an expanding market. The options seemed endless; the chances of winning big were especially shiny for early venture capitalists who would encourage others to buy in. Speculative investment drove the sector in the late 1990s as strong and rising stock prices attracted further investors. Companies were encouraged to think big, often floating huge stock issues.

One unfortunate by-product of this excitement was that far too many senior managers became more concerned about their stock prices than the day-to-day operation of their companies. Too often decisions were made that were designed to raise the stock price, even if in the long run such decisions were not in the best interest of the corporation. They focused on stocks as a key part of their own income package. From 1995 to 2001, the average annual compensation for CEOs at large firms more than doubled—from nearly $6 million to almost $12 million—a good deal of that in stock options.[38]

But then, beginning in March 2000, and soon accelerating rapidly, came the realization that any reasonable return (profit) for most Internet-based companies was increasingly unlikely. Their capitalization often vastly exceeded actual results, especially slow rates of customer growth. The revelation was a bit like another old fairy tale: that of the emperor's new clothes. As soon as a few people questioned the high stock prices for firms that did not appear very promising, then more questions followed and progressive stock declines were soon experienced by most dot.com companies. The Internet bubble had burst and the reckoning was swift and severe.

Beginning on the technology-laden NASDAQ index, but soon spreading to the New York Stock Exchange, Internet and then high-technology stock prices fell throughout 2000 and into 2001. At first analysts (more than a few themselves heavily invested in the sector) claimed this was a "healthy readjustment" or "rationalization" of market values. But as fears about one company or aspect of the business were realized, the push to sell rapidly spread to others. Such a cascading effect is often observed by those who follow the markets. The investment community suffers the same "herd" mentality often seen in auction situations. A steady drumbeat of lowered earnings expectations can hasten any decline which then becomes something of a self-fulfilling prophesy. The decline is often very difficult to stop.

The stock price drop often seemed almost vertical, with many firms experiencing declines of 60 to 90 percent in the year following the March 2000 peak. For example, Net2000 dropped from $40 to less than $3; PSINet, declined from nearly $61 to just over a dollar in the same period; and Teligent fell from $100 to $1.62.[39] By 2002 more than $2 *trillion* in telecommunication stock value had been wiped out (in 2002 dollars, the savings and loan bailout of a decade earlier totaled "only" $250 billion).

In seeking explanations, analysts found that too many modern telecommunications companies had taken on conflicting characteristics. They shouldered huge indebtedness to expand (both by acquisition and construction), seeking rapid growth in their size and income based on sky-high predictions of growing demand. Yet at the same time they faced a marketplace of steadily declining prices due to increasing competition.[40] The problem infected both new and traditional companies, though to varying degrees. And the daunting financial collapse soon had a more immediate impact on peoples' lives.

Layoffs

To many in the general public, the first sense of the depth of trouble in Internet and telecommunications companies came from news reports of soon massive layoffs of workers. Individual companies began to thin out at an often drastic rate, and often quite quickly, reducing themselves to a half or third of their former size. Thousands of employees were shown the door thanks to mergers. When Qwest won the expensive takeover fight for US West in 1999, for example, some 11,000 employees (many of them middle managers) declared "redundant" were dropped in a move to cut costs and help meet the costs of the takeover. At the same time, hundreds of new technicians were hired.[41]

The outstanding example of employment trimming has been Lucent, the former AT&T manufacturing and research (most of the former Bell Labs) arm, spun off in 1996 (*see section 7.2*). Initially a glowing success, Lucent went into a nosedive as customer companies began to cut back on planned purchases of telecommunication equipment and services. From employment levels near 153,000 workers at its peak in 1999, the company began in 2000 what became a series of layoffs to cut costs. These continued into 2002 (totaling 62,000 by April)—including yet another cut of 10,000 announced in October. By 2004 Lucent declined to less than a quarter of its size just five years earlier.[42] Three-quarters of its work force had been terminated.

Another indicator of both company financial trouble and layoffs was what happened to the pensions and health coverage of retired workers. All too often heavily (or totally) invested in the company's own stock— and thus imperiled when the company fell on hard times—pensions were cut or in the worst case disappeared entirely. With far more retirees than current workers Lucent, for example, had to cover pensions of 275,000 workers though only 35,000 active workers were paying into the pension trust fund by early 2003.[43] Low rates of interest merely worsened the problem, as the trust funds barely grew at all, if indeed, they did not decline. Thus the growing labor problem in telecommunications included those no longer working.

The overall *proportion* of total labor layoffs attributable to the telecommunications sector also kept increasing: from just 6 percent in 2000, to 16 percent a year later, rising to more than 21 percent by mid-2002.[44] By the fall of 2002, the total number of telecommunications job losses in all companies totaled more than a half million.[45] Given such a number (and proportion), of course, the telecommunication sector turndown was having a dragging effect on the entire economy.

Frustration over the employment issue soon spread to the media. Companies and unions both took out advertisements to put forth their views. The Communication Workers of America, for example, excoriated Verizon in ads that ran in the *Washington Post* and other papers. The ad featured a pile of cash bundles and complained the carrier was giving one executive $78 million and "more than $365 million in salary and bonuses to just nine executives over the last five years," just as pink slips terminating another 4,000 workers were going in the mail only weeks before the holiday season.[46] Another RBOC placed ads in magazines claiming jobs were at risk because FCC rules were requiring them to sell service below the cost of providing (and maintaining) it. While unlikely to change any company's action, such ads were a clear attempt to reach policymakers. So were television spots featuring a wolf in sheep's clothing (SBC) as AT&T argued against the rates charged by the RBOC to competitors. Late in 2004, another rather sad ad attacked Lucent for "turning its back" on retirees' health and pension benefits as a part of its cost cutting.

Fraud

Scandal will always grab media headlines and thus public (and sometimes regulator) attention. After nearly two years of steadily worsening telecommunication news, what had been rumors became confirmed stories of impending scandal, reaching the headlines in 2002. Of course, the bad news was not restricted to telecommunications—the huge fraud-driven Tyco and Enron collapses of 2001 riveted many to their television news services and made many question American business ethics.

The first major telecommunications company to fall due to scandalous behavior was Global Crossing Ltd. which declared bankruptcy in late January 2002. With more than 100,000 miles of undersea cables and other facilities, the conglomerate headed by Gary Winnick had become a major domestic and international voice and data carrier. The company showed liabilities of $12.4 billion in what was the fourth-largest bankruptcy in American history. Many of its balancing assets, however (including Rochester, NY-based phone company Frontier, purchased for $11 billion in 1999) struck observers as overpriced. Demand for voice and data transmissions on Global Crossing's network linking 200 cities in 27 countries had been waning steadily. To top off the stories of scandal came reports (some from internal whistle blowers) that Global Crossing had inflated revenue and cash flow figures to convince investors the company was healthier than it really was. Yet Congress moved only slowly to investigate, for Global Crossing, as well as other telecommunication companies, had been a major political donor. In August 2002, Global Crossing announced sale of a 60 percent share to companies based in Hong Kong and Singapore for about $750 million.[47] Investors lost billions and thousands of company employees lost their jobs and pensions.

What finally pushed Global Crossing off business front pages was the developing saga of WorldCom which had first become an important player with its purchase of Metropolitan Fiber Systems in August 1996 for $12 billion. Having purchased control of the much larger MCI in 1998 and then peaked at a market capitalization of some $175 billion in

1999 (when it made an offer to buy Sprint which the government eventually blocked on antitrust grounds), the company had become increasingly troubled as market growth ceased and revenues flattened in 2000. CEO Bernie Ebbers appeared to lose his way and lacked any back-up plan other than cutting costs, often in markedly petty ways. After company stock had declined by 80 percent in the first four months of 2002 alone, Ebbers resigned under pressure, by which time his once high-flying company was worth (in stock valuation) a mere *4 percent* of its peak level.[48] His successor John Sidgmore was expected to slash capital spending and personnel, and sell one or more divisions of the firm to keep WorldCom afloat.[49]

Then came the truly shocking late June 2002 announcement that the company had "misaccounted" for some $4 billion in expenditures—errors blamed on the Arthur Anderson accounting firm (which soon collapsed). Still more errors of the same kind were announced just a week later. Presumed profits for the previous several years evaporated under examination. Heads rolled and the search for scapegoats began (Anderson, already under investigation for its role with Enron where it was fined for destroying relevant documents, soon closed down entirely). WorldCom's eventual collapse into bankruptcy in July 2002 became—at $107 billion—by far the largest in American history.[50] Sidgmore's focus switched to trying to retain Worldcom's major customers to keep the company's core viable. Meanwhile, the WorldCom accounting fraud only grew worse as auditors kept probing, the total of earnings inflation exceeding $9 billion by late in 2002.[51] Seeking to clean house of all those leaders even around at the time of the accounting fraud, WorldCom named a new outside chairman—one with no telecommunications experience whatever—former Compaq (and briefly Hewlett-Packard) president Michael D. Capellas, in November 2002.[52] At the same time, WorldCom took out full-page ads in leading newspapers trumpeting the beginning of their financial turn-around—cash on hand, new accountant, available financing, and the like.[53]

Of the four remaining RBOCs, only Qwest had attracted significant regulatory attention. The former US West had been taken over by Qwest in mid-1999 in a complex $55 billion dollar deal backed by billionaire Philip Anschutz who placed former AT&T official Joseph Nacchio in charge of the new company.[54] The company expanded by both construction and sometimes expensive acquisitions. Then as telecommunication financial indicators turned negative, Qwest's troubles increased. The stock price dropped while debt rose (to more than $26 billion) and questions were raised on some accounting practices, leading to Securities and Exchange Commission and Justice Department investigations. Finally, in June 2002, Nacchio was forced out, to be replaced by Richard Notebaert, a former CEO of Ameritech.[55] Critics focused on the largest individual investor (18 percent of company shares), Philip Anschutz, and what he knew about the company's financial dealings, as they continued to assess Qwest's past and more importantly, its future.[56] In a bid to survive by growth, Qwest fought hard but lost a 2005 bidding war to take control of MCI.

Amidst these often complex events, public attention was focused upon stories of senior company executives selling out their shares before or as their companies turned sour. For example, Global Crossing CEO Gary Winnick made well over $700 million, while Qwest's Philip Anschutz and Joseph Nacchio reaped $1.9 billion and $248 million respectively. World-Com's Bernie Ebbers was "loaned" more than $400 million by the company's compliant board of directors before he was eased out—still owing the bill.[57] All of this was occurring while the companies were laying off thousands of workers, sometimes raiding pension funds, and sliding toward bankruptcy. (Later in 2002, Winnick made headlines by a commitment to pledge $25 million toward those who had lost their retirement funds. Critics noted the difference between the pledge and what Winnick had made in his stock sales. But he was also the only such executive to "step up to the plate" with any offer whatever.) Many of these stock deals were under continuing government investigation by late 2002.

Bankruptcy

While the scandal-driven bankruptcies were most newsworthy, many other companies were being forced to the financial wall by the parlous state of telecommunication. In addition to operational and construction costs, merger and acquisition activity contributed to massive debt loads and many companies could not service their debt. Lowered stock prices and financial rankings made it even harder and far more expensive to seek loans or capital investment. As one early 2001 report noted, new companies were especially at risk:

> The closing of the capital markets has left dozens of half-built upstarts gasping for cash. They have huge debt loads, they can't get new financing, there's nobody to buy them out, and their businesses are generating less money than expected because of brutal price wars.[58]

The list of telecommunications and related companies seeking protection under the bankruptcy laws rapidly grew longer. Among the first to go were many of the once-promising start-ups such as Teligent, headed by former AT&T official (and once seeming heir-apparent) Alex Mandl. Despite often charismatic leadership, sometimes viable business plans, and often great press coverage, the ill-financed companies could not survive the investment drought that first appeared in 2000 and deepened in the years to follow.

Once under the protection of bankruptcy laws, companies seeking to reorganize made life that much harder for other carriers struggling in the rising waters. In an attempt to meet creditor demands, assets were often sold off at near fire sale prices:

> A $4.5 billion investment in equipment by Winstar Communications, for example, brought just $15 million in bankruptcy a few years after its network was built. Global Crossing's $20 billion network sold for $750 million; Viatel's $2 billion in equipment went for $15 million.[59]

Buyers of such fully depreciated assets could offer service or equipment to end users at very low prices, upping the pressure on companies still laden with debt. When AT&T took over Velocita in late 2002, for example, it gained a $350 million fiber network for but $37 million in cash and (primarily) stock. The 20,000-mile network was 90 percent complete at the time of the sale.[60] Or companies could be broken up and sold piecemeal as happened with once-golden Comsat, taken over by Lockheed Martin.[61] The result was an often short-term windfall for customers, but only at serious risk of a far less competitive market down the road as more firms failed.

10.3 WAVES OF DISASTER

Taking into account what has been discussed thus far—rapid and sometimes destabilizing technical development, poor market predictions, and a growing financial meltdown, it is important and useful to analyze just how the disaster has spread across various elements of telecommunications. Students of industry have long noted that the high technology and telecommunications industries are closely interconnected—with each other and with the economy in general—in a variety of ways. Research and development of equipment is essential to utilize the various transmission links. Growth or decline in one business or subsector is soon felt in others.

> The reality is that in an investment bubble as big as this one, what begins in one or two sectors eventually spreads through the rest of the economy. Flush with cash pushed at them by investors, all those dot.commers and telecom executives not only bought computers and build office buildings— the things picked up as "fixed investment" in the national income accounts—they also used the invested cash the fly on airplanes, stay in ho-

tels, retain investment bankers and consultants, take out ads in newspapers and magazines, and buy themselves fancy cars and vacation homes. . . . Before long it was impossible to tell where the Internet money stopped and the rest of the economy began.[62]

This symbiotic relationship has served to increase the pace of development and investment in good times, but has often had a negative snowballing impact when times are bad—the proverbial domino effect.

After the beginning of the meltdown in early 2000, the close interrelationship became positively poisonous, with problems in one part of the sector rapidly spreading to harm another. A few really dour observers argue that this negative spiral will have a lasting impact as industries try and pull out of the tailspin, poorer sectors holding the others back. Because of this uniquely intertwined situation, something of a domino or "wave" sequence has developed where concerns that crop up in one sector soon impact another. And these waves can develop in a relatively short time.

There is one other vital way the various businesses are interconnected—their key executives and board members. Many of the same people show up at different companies. "Telecom is the most incestuous industry anywhere" complained one consultant.[63] Though some critics argued that inexperienced managers and new companies brought about the industry's decline by trying to grow things too fast, careful analysis suggests the opposite—experienced telecommunications managers let things get out of hand.

First: Dot.coms

The shift from information sector growth to decline began with the initial turndown in overblown "dot.com" Internet industry stocks in March 2000. That downturn was initiated by several things, primarily the realization that far too many of the new companies lacked any viable business plan—or even a likelihood of earning a profit in the near-term future. Too many ill-considered companies were seeking (or being pushed by investors into trying to seek) golden profits too quickly. A few commentators have argued that the U.S. District Court findings of law against Microsoft on a huge pending federal/state antitrust case (issued on April 3, 2000)—increasing the threat that the company would be broken up—were enough to "spook" investors that the easy money era was over (see section 10.4).[64]

In the new and closer examination of the Internet-based firms, even such well-known and growing companies as Amazon.com came under the microscope as investors asked whether a profit was in even the mid-term future. While its sales were expanding—even spectacular for an online retailer—the costs of developing supporting warehouses and ordering systems dragged profits into losses. Online auction service ebay.com was one of the few dot.coms said to be making money and thus escaping most concern. Many others were overpriced but undercapitalized or suffered poor management or internal financial controls because they had expanded too quickly.

Because dot.com investors were also often deeply involved in other technology or telecommunication firms (or their investors often were active in all of these areas), the Internet decline began to spread.

Second: Manufacturers

Soon companies that purchased communications technology (today, of course, this means virtually every telecommunication carrier as well as most businesses) began to grow cautious as they viewed the spreading dot.com market carnage. Plans to expand or replace equipment were delayed or canceled while managers awaited developments. This triggered a second wave of declines and sell-offs as the equipment suppliers (the largest being Lucent and Nortel) began to feel the squeeze as their hardware orders dropped off. Soon news of layoffs at the manufacturer added to the gloomy financial news.

The core of the problem was at Lucent, the former AT&T manufacturing arm which had been spun off in 1996 (*see section 7.2*). Once the country's premier telecommunication manufacturer, Lucent enjoyed growth and expansion for the first several years after it became independent and its stock price reached above $70. Lucent was riding the wave of both its independence from AT&T and the thriving telecommunications sector. It provided more than half of the industry's switching equipment, always a big-ticket purchase, and ran the renowned Bell Labs organization. But as the dot.com turndown and growing business caution began to cut into equipment orders, Lucent's outlook turned darker. Telephone company equipment purchases declined and were projected to continue doing so. Lucent was seen as too dependent on its traditional telephone equipment business and too slow to pursue new lines of business.

Growing pressure to perform led to some questionable financial practices (excessive customer discounts and some accounting decisions), which forced a change in Lucent's executive office in the fall of 2000 and delayed the search for a permanent replacement leader.[65] Yet even a weakened Lucent still held appeal for a potential partner. A projected $23 billion merger with the French Alcatel concern, briefly touted as the solution to both company's problems (each would gain new markets), was called off in May 2001 shortly after negotiations had become public, over disagreement concerning control of the resulting behemoth. Lucent would have to face its deteriorating situation alone. For not only were Lucent's overall sales dropping, pushed by competition and the weak economy, prices of telecommunications equipment were also declining by as much as 30 percent a year.[66]

In a desperate search for operating funds, Lucent announced in 2000 that it would spin off its business systems division (primarily PBX manufacture), which became Avaya Communication in 2000. Two years later the microelectronics components business (and some 10,000 employees) was also spun off to become Agere Systems. Analysts argue that Lucent and the Canadian-based Nortel Networks (formerly Northern Telecom) surely *will* survive—if only because telephone carriers value "the people who understand their legacy networks."[67] Indeed, by late 2004, Lucent was chalking up profits over the year before—the first time this had happened. But the still-large companies may offer a narrower offering of product lines for perhaps about a third of the revenues achieved before the meltdown devastated customers and suppliers alike.

Faced with all this bad news, investors turned away from the technology sector and increasingly sought options elsewhere—or holed up to await an upturn. A few bright spots remained—for example, Cisco and Agere as two of the prime manufacturers of the hot WiFi technology devices (*see section 10.1*). But the general drying up of available capital had a devastating impact on yet another sector of telecommunications.

Third: CLECs

The decline in available investment capital soon triggered a crisis among the struggling competitive local exchange companies (CLECs), many of which had counted on continued borrowing to develop their networks and services to (and thus penetration of) business and residential users. In the immediate aftermath of the 1996 law's passage, of course, things looked rosier. The law made specific provision for the new competing local carriers and appeared to require the incumbent monopolies to interconnect them. And competition did, indeed, appear. By the turn of the century, most Americans had at least potential access to one CLEC option. The number of CLECs rose from only about 30 in 1996 to well over 700 by 2000, though relatively few were facilities based. Of the latter, perhaps 300 had been created and were operating by 2000. While many were relatively tiny start-ups focused on resale as a basis for provision of discounted service, both AT&T and WorldCom acted as CLECs in many markets—indeed they accounted for three-quarters of the revenue of the 20 largest CLECs in 2001.[68]

Unfortunately for investors, however, only about 70 of the facilities-based start-ups were still providing service early in 2002.[69] CLEC stock values had fallen through the floor, with assets often selling for pennies per dollar of investment. Sometimes too thinly funded, often facing substantial debt repayments, and suffering delays in shared revenues due to them from RBOCs (among other problems), countless CLECs fell by the wayside. Their customers often had to scramble to find replacement carriers and that confusion added to the dark cloud gathering above the entire CLEC business. When several economist consultants assessed the financial performance of 24 publicly traded CLECs from 1996 to 2001, they found the firms had collectively raised (through stock sales or loans) $43 billion for construction or operations. However, only one of the companies had turned a profit, and all but seven had declared bankruptcy or been bought out by other firms.[70]

Especially in times of economic stress, communications facility managers will often turn to the tried and true—in this case their familiar RBOC service provider—and newer companies often have a harder time breaking into a marketplace. CLEC bankruptcies helped to reinforce the already strong position of the incumbent RBOC monopolies. By 2002, the RBOCs still retained nearly 92 percent of local customers (though CLECs had managed to achieve a 20 percent share of the market in New York). The CLECs' lack of customer access contributed to the poor investment climate even though about 20 percent of businesses now (still) have competitive service options.

So why did most of the smaller CLECs fail? Their trade association argues regulatory problems were the chief cause—policy delay, not holding RBOCs to the letter of the law, or (rather ironically) that policymakers moved *too fast* and encouraged resale as a means of market entry rather than building networks.[71] But a careful analysis of the remaining wreckage of the industry argues otherwise. Investors pushed new companies to develop far-reaching business plans. The plans were driven by the need to raise capital (first) and only then developing a facility or providing a service. Far too many companies entered the market at one time—sometimes the very same market—and yet marketing and business plans were never changed to meet such competitive conditions. Finally, some CLEC management just did not understand the complexity of the business they had entered.[72]

McLeodUSA offers one example.[73] The Cedar Rapids, Iowa-based CLEC had been founded in 1991 (with money Clark McLeod had made from the sale of an earlier company to MCI) to provide local voice service. After 1996 when its shares were first sold to the public, the company embarked on an expansion spree, taking up smaller firms and expanding into data transmission. Investors flocked to his vision of steady growth (both reselling and building its own network, the company provided service to two dozen states), not noticing that McLeodUSA was not earning a profit. Instead, McLeodUSA along with countless other telecommunication and dot.com companies touted its "EBITDA" (earnings before interest, taxes, depreciation and amortization) figure which actually hid more than it disclosed.

> Winning market share and acquiring networks became essential for sustaining analyst support and investor confidence. A company's stock could not rise, otherwise. Debt was viewed less as a liability in the late '90s than as growth's instrument, a necessary precondition for building the networks and subscriber bases that, theoretically, would lead to extraordinary profitability later. The outlook fueled short-term stock gains but courted long-term disaster.[74]

As long as the "free flow" of investment recommendations and subsequent dollars continued, all was well. But with the dot.com turndown expanding into technology stocks generally, investment dried up and CLECs like McLeodUSA had passed their peak. By late 2000, McLeodUSA was already in trouble with high debt, comparatively little revenue, and no further expansion possible—and the stock price continued to slide down into 2001. McLeod was forced out in August and the firm entered bankruptcy early in 2002—with its stock worth about 18 cents compared to the more than $100 high of early March 2000.

One glimmer of hope remained as a number of substantial resale companies were not only surviving as CLECs but actually making a profit. The number of such resold lines rose from about a half million to a projected five million from late 2001 to the end of 2002.[75] The fly in this ointment is the resellers' dependence on very low wholesale rates which are regulated by the FCC and the state public utility commissions. Regulators were under fierce RBOC pressure to raise those rates, their attorneys and economists arguing that the wholesale rates do not cover the cost of maintaining or operating the basic network. SBC, for example, was reported to be losing 12,000 customers to resellers every *day*, a number expected to climb to close to 30,000 as more resellers swarm to the market.[76] Why have resale-based CLECs survived and in some places thrived? Because they have taken expansion one step at a time, not pursuing their own facilities development until they have developed a business or residential market sufficient to warrant the expenditure. Experience suggests this takes an average of at least five years. That such firms as Reston, Virginia-based Talk America served a quarter of a million customers spread across half the states—and made a profit for several quarters in a row—suggests the approach can work.

Fourth: Interexchange Carriers

As the economic downturn spread, interexchange carriers both here and abroad were soon dragged into serious trouble. Here the problem was declining prices amidst fierce and growing competition. Some analysts have argued there will soon be no such thing as separate "long distance" service (and mergers announced in 2004 and 2005 support this view) and that flat rate calling (as is already the case for many mobile users) will dominate the business. While at one time a relatively low-cost service to provide, and one promising substantial returns, interexchange links are saturated and hotly competitive, forcing prices down and making profits thinner or scarce.

Outright fraud (*see section 10.2*) didn't help things any. The financial crises that brought down Worldcom and Global Crossing in 2002, on top of many smaller company failures, threatened to flood the market with low-priced assets for sale, increasing pressure on the "survivors." Bankruptcies forced assets on the market at a fraction of their cost to build. By 2002 the telecommunications industry guideline for such purchases was as low as 3–10 cents paid per dollar of actual construction cost.

One indication of how serious things had become appeared when the FCC announced in the summer of 2002 that it would entertain the idea of an RBOC taking over an interexchange carrier. The previous fall, when an AT&T–BellSouth merger was rumored, the commission had remained silent. AT&T had touted its consumer and business telephone businesses to three of the RBOCs—BellSouth, Verizon, and SBC, the first two of which achieved annual revenues larger than their former parent. Back in 1997 then-FCC chairman Reed Hunt made quite clear his feeling that a combining of AT&T with one of its former regional carriers was "unthinkable,"[77] thus shutting down further discussion of a rumored merger that might have combined AT&T and SBC.[78] In the face of continuing decline, however, monopoly fears gave way to concerns about saving viable infrastructure and by 2005 the prospect of an SBC-AT&T merger became all but a certainty—as well as the takeover by Verizon of MCI.

And where was AT&T, the one-time industry leader and driver, amidst this chaos? Under chairman Michael Armstrong (1997–2002), AT&T had pursued a strategy of buying cable systems to obtain both broadband and voice access to end users (*see section 8.4*). This cost billions and, to company critics, diverted its attention from the changing voice and data market. When investors finally made clear they would not wait for the cable strategy to pay off—which might well take years—Armstrong was forced to retreat, and in July 2001 announced AT&T would sell the cable holdings (by then the majority of AT&T) to Comcast for $47 billion in stock plus the assumption of some $20 billion in AT&T debt. As part of the deal, Armstrong went to Comcast with the sale. Four years later, AT&T itself was sold to SBC, thus ending the independent status of the once-dominant AT&T.

AT&T Wireless had earlier been spun off as an independent company (emphasized by relocation of its headquarters to Seattle, far from the familiar northern New Jersey AT&T "campus" so familiar to generations of AT&T leaders). What remained on consummation of the cable sale in early 2003 was a much smaller interexchange carrier (that also served as a CLEC in some local markets) about 10 percent the size of the pre-divestiture AT&T of two decades before. And its core business would continue to decline as it had for years.

And what role did the traditional number two interexchange carrier, MCI play in all this? It had become part of Bernard Ebbers' growing WorldCom empire in an $87 billion deal announced in 1997. Becoming part of the larger and not very well managed firm caused MCI to lose a good deal of its once-vaunted competitive edge. And WorldCom's role—prior to its fraud-induced collapse discussed earlier in this chapter (*see section 10.2*)—was harmed by its too-rapid expansion through the takeover of dozens of companies in the late 1990s, chief among them MCI and Internet backbone provider UUNet. But WorldCom never really integrated those many subsidiaries into a cohesive whole. The growth process came to a halt early in 2002 when federal antitrust officials rejected WorldCom's attempt to take over number three long distance carrier, Sprint, arguing excessive market concentration.

Generally alarmed by the general business downturn, let alone the sharp decline in telecommunications, by 2001–2002 businesses large and small sought to cut costs in any ways possible. These included layoffs, less spending on research and development, and slower replacement or upgrading of telecommunication links.

Fifth: Wireless Providers

Even the seemingly glamorous wireless business was dragged down. The wireless industry's problems by 2002 were at least four-fold: slowing mobile phone and service contract sales (as the most likely users were already customers), the disappointingly slow rollout of the highly touted 3G services, fierce price-cutting competition in a finite marketplace, and a growing variety of technical challenges.

Industry growth after 2000 appeared to be slowing substantially from the heady levels of the 1990s. While about 60 percent of American households owned and used a cell phone by 2004, penetration levels in many European nations varied from 60 to 75 percent.[79] Contributing to the difference is the lack of an agreed technical standard here (five are in use, each incompatible with the others), in the face of the fact that two-thirds of the world uses Global System for Mobile Communication (GSM) technology allowing considerable transnational phoning. That means that future sales will be largely limited to more marginal users, or replacements or upgrades, the latter dependent on the offering of attractive new features or services. Newer users will not be especially profitable as they will likely make less use of the network.[80] Still overall usage keeps rising—from 12 percent of all telephone minutes in 2000 to a doubling of that proportion by 2003.[81] At the same time, however, service quality complaints are up, and as the rate of growth slows down, preventing customer shifts to a competitor is becoming more important.

So-called third-generation (3G) or graphic cellphone services, theoretically the next major stage in wireless development, have been delayed worldwide. While Asia leads (DoCoMo, for example, offers a hugely popular 3G service in Japan), the European 3G rollout has been delayed by some problems with the agreed-upon technical standard and the lack of funds for construction. Many European carriers, including those in France and Germany, for example, expended a total of more than €100 billion to buy spectrum licenses at auction and now lack the wherewithal to actively develop their networks. So do most potential customers who thus far in the United States have resisted the price levels proposed. And in the wings are a variety of developing technologies suggesting far lower priced means of accomplishing much of what the 3G networks may offer (see section 10.1).

Sometimes excessive competition has led to sharp drops in prices—by 25 percent for most service in early 2002 compared to a year earlier. Typical cellphone costs averaged 14 cents a minute early in 2002, compared with 56 cents in 1995.[82] While the drop was clearly good for customers, at the same time average wireless carrier stock prices dropped 45 percent just in the first three months of 2002.[83] Whether customers would pony up to pay the increased fees for more capable services to come was anyone's guess.

The price and stock declines combined with slower rates of growth, low levels of investment, and heavy debt loads spell a tightening financial squeeze. Of the six mobile carriers offering national service, only the two largest firms—Verizon and Cingular—are making a profit. Most agreed four companies would be the ideal number but debt loads precluded further consolidation in the short term (but see section 10.5).

Finally, just as technology is changing the face of fiber optic links, so it is already modifying the outlook for wireless links as well (see section 10.1). Improved antennas, so-called mesh networks, ad hoc system architectures, and ultra-wideband transmissions, or even any one of them, may dramatically change the capabilities and capacities of wireless systems in the near-term future.[84] While economics pushes the industry toward consolidation, technology may prompt a more decentralized industry structure somewhat paralleling that of the Internet.[85] It is simply too early to tell.

Smaller Waves?

Here discussion of "waves" moves from assessing past events to assessing likely future developments. The fundamental question was whether further damage would spread. There were clear indications that damage *could* spread, further deepening the crisis and delaying a return to stability, let alone growth. Widely discussed was the impact of all this on the most substantial sector of the field: the incumbent local exchange carriers (ILECs).

Several Regional Bell Operating Company ILECs appeared to be winners amidst the carnage. They still control the vital "last mile" connection to end users and enjoy more stable funding (and revenues) than any of their competitors. At the same time, solid firms such as Verizon and BellSouth have seen their cost of borrowing rise just as their revenue base flattens due to the economic pressure forcing rates down across the board. Mobile telephone expansion has led to the first declines (since the Depression) in the use of landline telephones, and more particularly of traditional pay phones which were being removed across the country. Relatively minor competition from cable companies and growing VoIP (Voice over Internet Protocol) from Internet service providers contributed to the continuing decline in landline use. Local exchange carriers face being gradually left with an expensive and aging "legacy" technology base that serves a slowly declining customer base. Faced with this, many RBOCs by 2004 were actively marketing their own VoIP services or those of others.

Among the regional firms, by far the weakest was Qwest Communications, the former US West. A takeover battle had added greatly to the company's $25 billion debt load. By mid-2002, Qwest (the primary carrier in 14 western states) was under criminal investigation for financial dealings, and had replaced its aggressive CEO for an experienced and lower key telephone manager. Richard Notebaert (former chairman of Ameritech) cut debt to $16 billion by late 2003, in part by selling Qwest's Yellow Pages subsidiary. He also joined the wireless world in a partnership with Sprint, and began to focus more on mid-size business users. But the firm's core telephone business continued to suffer from competitors.[86] And Notebaert's attempt to take over MCI in 2005 failed while further exposing Qwest's precarious finances.

Some observers predicted during the worst of the crisis that perhaps as many as 24 of the top 30 telecommunications firms would have to seek bankruptcy, though that seemed less likely by 2005. Critics worried that most of the debt the industry has accumulated (upwards of $900 billion just since the beginning of 1998) might never be repaid. Consolidations and merger and acquisition activity added a huge debt load to many companies (the inability to meet such interest payments helped to send several firms over the edge). Companies operating under bankruptcy, and therefore free of normal debt repayments, can cut prices or sell assets, both of which increase pressure on firms operating under heavy debt loads. WorldCom alone was laboring under a $32 billion debt when it declared bankruptcy in mid-2002.

The meltdown effect also hit cable carriers—perhaps a negative effect of the convergence so widely discussed. Cable stocks had dropped nearly 60 percent in just a year by late 2002 (far more than the Standard & Poor 500 stock index which was down about 20 percent over the same period), pushed down by fears about the lack of subscriber growth and competition from satellite services.[87] Many cable companies had shouldered substantial debt in 1990s' consolidations and system rebuilds to increase capacity, and many still faced immense construction costs to upgrade their systems to offer broadband service.

Only at the end of 2002 did the first hints of a possible upturn in the telecommunication business appear. Prices were stabilizing, a few firms reported higher profits (or lower losses) than expected, and new products were appearing after months or years of promises. Observers noted that the now data-dependent industry was increasingly tracking the overall economy, which was also showing signs of looking up.[88] By early 2004, widespread reports of increased capital spending in telecommunications suggested the debacle might have bottomed out. Several RBOCs and cable carriers began to aggressively offer VoIP service by 2004 and others were soon to follow.[89]

10.4 POLICY CONFUSION

Industry critics—and their number naturally grew as the telecommunications crisis deepened—argued that the post-2000 meltdown was made both worse and longer by the widespread deregulation of prior years which weakened government's ability to monitor or respond to, let alone attempt to alleviate, the crisis. What was needed instead, some argued, was *more* and more effective government action, not less. What was too often missing was a better understanding of how telecommunication markets worked amidst rapid technical change. While by no means the

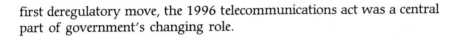

first deregulatory move, the 1996 telecommunications act was a central part of government's changing role.

"Promise" of the 1996 Act

Overreliance on the "new world" promised by the 1996 Telecommunications Act (*see section 9.2*) appears to have been a substantial contributing factor to the late 1990s' telecommunications investment boom and subsequent crash. The legislation appeared to promise a new world of local competition along with a host of new competitors—hundreds of them, all seeking investors and a market niche. Industry pundits spoke of the "promise" of a "war of all against all" with no safe havens. Companies would survive only on the quality of what they offered and the prices they charged—and the astuteness of their managers and investors. After what was presumed to be a fairly short period of transition from the existing local telephone and cable monopolies to widespread competition, there was a feeling that regulators could stand back and let the marketplace do its thing. (Some critics argue that is exactly what the FCC is already doing.)

For a while, it certainly looked as though these predictions were correct. New companies encouraged both by the legislation and the seemingly bottomless demand for Internet access and capacity appeared—and were funded—almost overnight. Competitors appeared in virtually every sector of the business, soon to be joined by many more. It appeared that the customer-friendly "war of all against all" in telecommunications *was* driving down prices while encouraging investment and new entry as well as extensive research activity—a seemingly ideal world.

As usual in this complex industry, however, the truth was more convoluted. First and foremost, intensive litigation delayed full implementation of the new law for three years (*see section 9.7*) and then some as RBOCs took a firm stand to defend their home turf no matter what. The new competitors found gaining access to the RBOC-controlled customer connections was far harder than had been expected.

Tauzin–Dingell Debate

Persuaded by the sector's sharp downturn that the RBOC restrictions embodied in the 1996 Act were holding up recovery, two key congressmen cobbled together legislation to undo some of the earlier law's provisions. Convinced that broadband service would roll out faster if RBOCs were not required to provide access to their digital networks by would-be competitors (such as the growing cable role of AT&T—*see section 8.4*), Congressmen W. J. "Billy" Tauzin (R–LA), and John Dingell (D–MI) introduced their "Internet Freedom and Broadband Deployment Act" in 1999 that would have allowed just that—free RBOC expansion into the digital subscriber line (DSL) marketplace without the 1996 law's requirement of letting competitors have access to the same facilities.

The bill's sponsors argued that the Bell companies needed support in their fight against AT&T and other national giants. They claimed that the bill's passage would unleash new expansion in the telecommunication industry, thus putting companies and their employees back to constructive work. As the RBOCs were more likely to quickly invest in broadband facilities, the bill was aimed to ease their means of (and benefits from) doing so. It was also argued that consumers would gain faster access to the benefits of broadband interconnection from the RBOCs.

Critics, of course, saw things quite differently. Passage of the Tauzin–Dingell bill, they claimed, would merely increase the already huge advantage held by the RBOCs, further limiting the tenuous toe-hold held by the few surviving CLEC firms. Competition would totally disappear, reverting carriers and their customers to the high-price oligopolistic structure that had once characterized telecommunication. CLECs (including AT&T and WorldCom) and others lobbied hard against the bill, arguing that the 1996 Act had to be given a full chance to work.

Lobbying efforts by both sides heated up, as did newspaper advertisements, as final debate in the House neared. Finally, in late February 2002, the House passed the bill by a vote of 273 to 157 with votes cutting across party lines. The bill was sent to the Senate where its chances were nil given Democratic control of the upper house and especially the negative views of Senator Fritz Hollings (D–SC), the influential chair of the commerce committee. Republicans regained control of the Senate in the November 2002 elections, however, and talk of reviving legislation was heard in various corners. By 2005, the talk had become action and several Congressional committees held hearings to reconsider aspects of the 1996 law.

FCC Deregulatory Mantra

As the telecommunication market began to slide in the early 2000s, the FCC seemed almost at a loss for what to do, usually concluding that "nothing" might well be the best option. There were virtually no FCC decisions and precious few words of any kind despite all the industry failures and layoffs even though Chairman (as of early 2001) Michael Powell admitted (somewhat reluctantly, it seemed) that there was a crisis afoot. Yet for several more years, the FCC did nothing concrete (and said remarkably little) about the worst crisis to hit the telecommunications industry since the agency's founding.

As discussed in earlier chapters, deregulation beginning in the 1970s had by the 1990s transformed the FCC's role. The Commission had become more focused on oversight (some argued short-sightedness was closer to the mark) rather than regulation. Regularly issuing reports, filling its massive Web site with data and trends, and on occasion talking with the requisite Congressional committees, the FCC seemed more and more irrelevant to the debacle happening in the marketplace.

Aided by a Bush Administration that let the Commission crawl along with but four members (in 2001–2002, and again in 2005), commissioners were regularly quoted about the positive impact of deregulation including reference to the old saw that government was best that governed least. In a sense, one cannot blame them. Having all but exhausted its staff with a six-month tornado of regulation writing mandated by the 1996 Act (*see section 9.5*), the FCC then had to defend most of its actions in court. This it did successfully (in large part) such that by 2000 or so, the FCC *thought* it was largely free of post-1996 Act work, save only for dealing with a flow of applications from RBOCs (under Section 271 of the Act) to offer long distance service. And even those were largely granted by early in 2003 (Bell South becoming the first to provide long distance in all of the states where it also provided local service).

Of course the Commission was wrong about being out of the post-1996 regulatory woods. The fight between incumbent and competitive telephone carriers merely spread to new venues—full-page newspaper advertisements pressing Congress or the FCC to "do something" to help get the industry out of crisis. Of course opinions on what should be done varied hugely depending on who was buying the advertising space. Incumbents led by the four surviving Bell companies sought to have the post-1996 pricing rules rescinded, thus removing discounted access rates for competing carriers seeking to use the Bells' last-mile connections. Competing providers of local service (some of them small, but with ad campaigns largely bankrolled by AT&T and the bankrupt WorldCom operating as "Voices for Choices") sought retention of the discounts, arguing that to drop them would kill competition just as it was beginning to take hold. (Local competitors had seen their national share of telephone lines rise from 4.3 percent in late 1999 to 11.4 percent by mid-2002.[90]) Early in 2003, the badly divided FCC effectively split the difference with two decisions. The first retained the discounted access prices for competing carriers, but shifted the battle to state public utility commissions, theoretically to allow local conditions to rule policymaking, but in effect merely shifting the battlefield. The second decision held that broadband service competitors were *no longer* (after a three-year phase-out period) to receive discounted access to digital lines of incumbent carriers.[91] The effect of all this, of course, was not only to "split the difference," but to create more confusion and fertile ground for yet more litigation.

Auction Debacle

Perhaps ironically, implementation of marketplace thinking into the spectrum assignment and licensing process hastened the financial crisis for telecommunication carriers both here and abroad. After years of arguing that traditional government block allocation of frequency spectrum was inefficient (nor did it retain any of the value for government), regulatory economists finally won the day with 1990's authority for the FCC to auction spectrum licenses for some services (*see section 8.4*). While most aspects of traditional services (such as radio and television broadcasting) would remain under the old regime, new services seeking spectrum would now be subject to auctions among competitors rather than the deadening hand of administrative comparative hearings.

Auctions were put into place by the FCC for the distribution of spectrum for digital personal communication services (PCS) in the early 1990s. Initial PCS spectrum auctions raised more than $7 *billion* for the U.S. Treasury, tantalizing Congressional and budget officials with the chance to generate more revenue "out of the air" and encouraging further auction experimentation. But the authorizing legislation for auctions mixed economic and social priorities. The FCC was instructed that auctions had to ensure licenses to "a wide variety of applicants, including small businesses, rural telephone companies, and businesses owned by members of minority groups and women."[92]

To comply, the Commission reserved the so-called C block of frequencies specifically for auction limited to such groups, requiring merely a 10 percent downpayment and installment payments for the rest of what was bid. While further revenues were seemingly generated for the treasury, one major bidder (NextWave) found it could not pay the exorbitant amount bid and entered bankruptcy, initiating years of legal appeals. In the meantime the FCC re-auctioned the spectrum for billions more to new bidders (chiefly the RBOC wireless arms). When NextWave appealed that under the bankruptcy laws the FCC could not take away its licenses, the frequencies were thrown into legal limbo, unavailable to anyone.[93] After the Supreme Court agreed to hear the case early in 2002, the FCC finally refunded the substantial downpayments the "secondary" bidders had put down. In the meantime, however, those tied-up funds had provided no service and merely added to growing RBOC financial woes. The high court finally decided against the FCC (in other words, that the Commission could not rescind NextWave's licenses under federal bankruptcy laws), returning the fought-over authorizations to NextWave.[94] The company pursued development of its frequencies in 2004, though many still expected it to sell them for a huge profit.[95]

Further U.S. auctions for 3G PCS spectrum were originally scheduled for the early 2000s, but kept being pushed back by the FCC given the parlous state of the industry (and at the industry's request). The money for bidding simply does not exist, and they had still not taken place by early 2005. Overseas, massive auction investment ($100 billion) by incumbent European carriers for projected mobile 3G service spectrum created a similarly dangerous debt load—with little balancing revenue likely for years to come.

Antitrust as Policy

A totally different regulatory approach—antitrust—has also had a substantial impact on this field. Two huge federal antitrust cases provide evidence of both the strengths and weaknesses of that field of law in creating what effectively become sectorwide policies. First of course is the 1974–84 AT&T case (*see chapter 6*) that led to the breakup of the giant telephone monopoly and laid the ground for all that is discussed in the last half of this volume. The different impacts of that decision have already been discussed. The other, only noted briefly above, was the case brought in 1998 against Microsoft by the Justice Department and 20 state attorneys general. Not the first attempt to rein in the Washington state-based software giant, as with the AT&T case, this one also was caught up in the changing of the White House guard from Democratic to Republican control.

This is not the place to relitigate all the many arguments in the Microsoft case. But several issues arise from it which are directly relevant to the discussion of what has happened in recent years to the information and telecommunications sectors. The first concerns the validity of such proceedings as de facto creators of policy (an example of the "law" of unintended consequences), and the second (noted briefly above), focuses upon the impact of such cases and their many appeals and decision points on financial markets. A third point (evident in Microsoft but not AT&T) is whether the huge legal cost and effort are worth it when, in the end, the status quo is largely continued.

Federal and private sector antitrust cases are brought under the provisions of the brief (it is but a page long) Sherman Act which was passed in the very different world of 1890. Modified by the Clayton Act of 1914, these two statutes remain the foundation of the many and often far-reaching antitrust cases brought in the years since.[96] Many (lawyers and otherwise) have observed, however, that applying antitrust law is not generally a good way to make basic national policy. To cite but a single example, the settlement of the AT&T case in 1982 did not take into account its far-reaching impact on the nation's balance of trade. Focused entirely on the reordering of the domestic industry, the decision not to allow RBOCs to manufacture equipment forced them to seek it elsewhere, and overseas providers were the least expensive option. While the migration of at least consumer telephone equipment would eventually have followed the earlier trend of consumer electronics to overseas (largely Asian) manufacture, a look at trade statistics shows that the U.S. balance of telecommunications trade swung into the red in 1984, just as the antitrust settlement was going into effect. Telecommunications manufacturing in the United States has never recovered as is evidenced by the state of Lucent and other equipment firms today.

More fundamental, of course, is the bifurcated industry created by the 1982 Modified Final Judgment agreement. Local firms would focus on "natural" monopoly exchange service while other players would take part in the less (or non) regulated portions of the industry. The deal looked neat and logical on paper (and seemed to address the reasons the case had first been brought), but it turned out to work far less well in the real and fast-changing marketplace. And its limitations and barriers were a fundamental reason for passage of the 1996 Telecommunications Act—policy legislation seeking to overcome the limits of a narrowly defined antitrust resolution.

The point here is that antitrust *is* a relatively narrow branch of law. It seeks to halt a particular kind of market-controlling corporate behavior without taking into account what other impacts such legal action may create. Antitrust actions can be far-reaching, as with the breakup of Standard Oil in the early 20th century, or the breakup of AT&T. They can also come up relatively empty-handed either because the industry involved has changed or the court loses control of what often becomes a massive flow of paper (as in the failed 1969–84 antitrust case against IBM), or because the welter of facts and arguments undermine a clear sense of direction (as may have been a factor in the Microsoft case). In any case, such actions are rarely brought with an overall sense of sector policy in mind. The Microsoft case was fought over differing definitions of software and Internet sectors and their control—not on any broad conception of national policy for information industries. Private antitrust cases are usually even narrower in their conception—to punish a competitor for seeming malfeasance that has hurt the company bringing the action. The FCC, created as the federal expert agency in this field, was merely an observer in the AT&T case brought by Justice in 1974 (as well as in the welter of private antitrust cases brought by MCI and others against the Bell System).[97]

Antitrust actions can be seen as a final threat by a federal government (or individual states as in the Microsoft case) that has been frustrated by its inability to effect change in other ways. To that extent, antitrust indicates a failure of policy to constructively shape a sector of the economy, and thus a focus on the punishment of one key player in an attempt to force change. When such drastic action fails, as in the Microsoft case, the sector of the economy involved may be in worse shape than had no case been brought in the first place.

For one thing, financial markets are easily confused or stampeded by antitrust actions. A review of Microsoft and other stocks at various points along the way of the antitrust proceedings will demonstrate that the market usually reacts quickly to perceived threats to specific firms. Less obvious is a more serious problem—that the whole tenor of investment in a sector (in this case, the dot.com world) can be thrown off by the uncertainty that pending antitrust actions create. The beginning of the dot.com decline in the spring of 2000 had many causes, one of which may well have been the antitrust uncertainty which then hung over the largest player in the field.

10.5   LOOKING TO THE FUTURE

Effects of the information/telecommunication sector meltdown described in this chapter will continue to dominate the field for the remainder of the 2000s decade. Further bankruptcies may make things worse and drag out the recovery, pushing more companies to the financial brink. Only slowly will things get better—and the shape of the emerging survivors will soon look quite different.

Clearing the Glut

Unfortunately the fundamental problem of oversupply will not disappear quickly. "A survey of 20 major long-distance and local trunks shows that networks are running at about half of ideal capacity."[98] Most analysts expected the slump to continue before the industry stabilized and then very slowly resumed growth. They may have been a mite too pessimistic.

There were small signs beginning as early as 2002 and slowly accelerating in the years that followed that telecommunication prices might be steadying or even inching up (though weakness in European and Asian markets could push the spiral down again).[99] Future rates of expansion will probably hover around 2 to 3 percent, far from the peaks achieved in the heady past.[100] A major general economic upturn, with resulting higher levels of business communication, would shorten telecommunication's hard times. But whatever the overall picture, until the accumulated capacity glut is effectively used, the industry cannot begin the fundamental changes that are needed for the future.

Users were warned that they all need to develop back-up disaster plans—not only for natural (or terrorist-caused) disasters, but for service cut-offs due to continued industry unrest—especially the bankruptcy or shutdown of providers.

Consolidating Further

Most observers agree that as (and certainly after) the existing glut is overcome, the telecommunication industry will undergo substantial consolidation. This may take time to come to full fruition due to the high debt levels carried by nearly all players in the field, but it clearly will occur. The result will be a likely return to the telecommunication industry's former oligopoly structure. The move to consolidate, of course, began some time earlier.

Within months of the passage of the 1996 Act, for example, the number of RBOCs had begun to decline (*see section 9.6*), leaving four players where seven had stood before. On the national level, Bernie Ebbers' WorldCom seemed well into its voracious takeover of multiple telecommunication companies. It purchased Metropolitan Fiber Systems (MFS) for $14 billion to gain control of its developing web of fiber cable in major cities that would allow WorldCom to bypass monopoly local telephone companies. A year later it purchased another fiber optic company for $7 billion—several times what the physical facilities were worth.[101] Prompted by the predictions of huge Internet demand growth, it seemed that "if you build (or buy) it, they will come" was indeed the case. WorldCom continued gathering companies large and small under its wing.

By early 2000, however, neither WorldCom, the newly merged AOL Time Warner (*see section 8.4*), nor AT&T could predict the sharp decline in both technology and telecommunication company stock values that would begin that spring, placing growing pressure on all three firms. At AT&T, for example, management came under increasing pressure to make its huge cable investment (*see section 8.4*) pay off—especially as the company's core long distance telephone business continued to drop, and at an increasing rate. As early as May 2000, investors were made uneasy by an AT&T earnings report that was 5 percent lower, despite the fact that the company's cable systems rebuilding project was meeting its targets, albeit at considerable cost. But company campaigns to sign up cable-based telephone and Internet users were running well behind intentions, making the short-term outlook decidedly cloudy.[102] Financial pressure continued to grow over the next several months, reducing the company's room to maneuver as the stock declined by two-thirds.

Finally Armstrong had to concede his strategic plan for a unified AT&T, as it would take too long to come to fruition amidst a worsening financial outlook. After considering a variety of options, in October 2000 AT&T announced that it planned over the next two years to break the company into four separate units, one each focused upon business, consumer, broadband, and wireless services.[103] The first two accounted for 75 percent of the company's 1999 revenues, despite their declining base, while the broadband business that had cost so much to develop was bringing in less than 8 percent.[104] Ironically, most analysts credited Armstrong with having the right idea in 1998–99 as he tried to build his company into what was intended to become an integrated "one-stop shopping" source providing services across the spectrum of telecommunications. But Wall Street got cold feet, made chillier by the overall declining economic scene in 2000–2001.

Thanks to its large cable system holdings and forthcoming spin-off status, AT&T Broadband became the target of a bidding war by firms seeking to add systems to their existing lineup. In July 2001, Comcast bid nearly $45 billion for the company but the AT&T board rejected the offer and opened up to competing bids. AOL Time Warner and Cox Cable were among the numerous suitors. Comcast finally won the bidding war for $47 billion in December 2001 creating a combined cable ownership unit almost twice the size of AOL Time Warner, and thus able to directly reach a fifth of the nation's homes.[105] The combined number one and three cable companies would operate as AT&T Comcast and own cable systems in more than 40 states. AT&T's Armstrong left to become a "non-executive chairman" of the new larger broadband firm when the deal was completed in November 2002.[106] There he quickly disappeared as others actually ran the merged operation. In the meantime, revenues at the still-integrated AT&T continued to decline and the company experienced its fifth consecutive quarter of losses.[107]

The next move came with the sale of AT&T Wireless to Cingular in 2003, a deal only finally approved in late 2004.[108] By 2005, AT&T had shrunk to only 6 percent of its pre-1984 size and, though still the largest interexchange carrier, as its prospects to survive as a free-standing company continued to dim, its shareholders on June 30, 2005, approved its acquisition by SBC Communications for $16 billion. SBC shareholder approval was not required because the number of SBC shares issued to acquire AT&T was less than 20 percent of the outstanding SBC shares.[109]

The irony of the merger is underlined by the relative size of the two companies. The once-mighty AT&T will be swallowed up by one of its former offspring. And regulators are lined up to approve this merger as a good thing; by the time the AT&T shareholders approved SBC's bid, regulators in 26 states had given it the green light, with approval still needed from 10 more states and the federal government.[110]

AT&T's bitter rival, and the architect of long distance competition, the second-largest interexchange carrier, MCI, faced a similar prospect of becoming part of an RBOC. A bidding war between Verizon and Qwest, in which Qwest bid as much as $9.75 billion or $30 per share, ended when Qwest dropped out of the bidding after MCI's board accepted an $8.44 billion, $20 per share, offer from Verizon.[111] MCI cited several reasons for its acceptance of Verizon's lower offer, including Qwest's lack of a wireless network and the stockholder lawsuits that resulted from Qwest's accounting scandal.[112]

These mergers reflect the RBOCs' efforts to seek new lines of revenues as they watch their local service market being eroded by cable companies offering VoIP telephone service and by wireless service providers. The result will be further lay-offs and the re-emergence of an industry structure that harkens back to the days preceding the break up of the Bell System.

Reforming Management

Especially after the battering the industry received due to corporate malfeasance, at least some reform in corporate governance appears necessary. The revelations of misdeeds at WorldCom, Global Crossing (both forced into bankruptcy) and Qwest (which got close), woke up consumers and regulators alike. Clearly changes are needed. Corporate boards of directors need to be made up of more outside (and independent of management) members and clearly need to ask far more questions. Boards may well meet more often and question executives more closely. They may call more often for their own outside accounting of what is going on below. They will also be far more active in both the selection and supervision of senior executives.

Some critics argue that future telecommunications executives may lack the go-go charisma of a Michael Armstrong (former AT&T chairman), for example, but demonstrate stronger managerial talents and aptitude. WorldCom's Bernie Ebbers in his cowboy boots reminded stockholders where he came from—and perhaps suggested his approach to the takeovers he loved to develop—but said little for his limited ability to manage a conglomerate, especially one in a flat or declining industry. He eventually gave way to the decidedly uncharismatic John Sedgecoe. Likewise the often arrogant Joseph Nacchio of Qwest, who was finally forced out (and replaced with a veteran telephone manager) when the company's financial performance continued to decline. Ego, not to say greed, drove too many of the telecommunications firms in the 1990s' climb, and over the financial edge after 2000.

At the same time, there is almost surely going to be closer government surveillance of accounting by the Security and Exchange Commission and other agencies. A first step was taken in 2002 when chief executives of the largest publicly held companies were first required to personally attest to the validity and reliability of their financial reports to the federal government.

Continuing Research

In the long run, of course, any technology-based industry survives only by continuing research and development. Companies under stress all too often trim research and development expenses just when they need them most. This is exemplified by the sad decline of Bell Labs.

Spun off with Lucent in 1996, the Labs originally continued their combined work in both basic and applied research. But Lucent's financial troubles (*see section 10.2*) led to a shrinking of the labs from about 25,000 employees in 1997 to around 16,000 by 2001,[113] and not even half that three years later. While this was in part due to Lucent divestitures, and part to attrition, there was clear downsizing going on. More fundamentally the basic role of the Labs was changing. Research was increasingly focused on applied product or service development rather than the basic research for which the Labs were known. As Noll and other researchers point out, of course, this trend was larger than any one company.[114] Many other firms have also trimmed or reshaped their research efforts, and some have been closed entirely.

An indicator of Bell Labs' sad decline came in the late summer of 2002 when the Labs announced that a promising young researcher had been fired and most of his published research papers (some of them co-authored by respected colleagues) had to be withdrawn after it turned out the findings could not be replicated and had apparently been fabricated. Observers shook their heads at how far the mighty had fallen, perhaps in the quest for rapid development of practical innovations, as opposed to the fundamental research on which the Labs' reputation had been originally built up over decades of effort.

Another factor can slow innovation. In the computer software field, there is growing evidence that the Microsoft monopoly has generated huge returns but slow innovation (why improve a product or service when you already largely control the market?). While operating systems and word processing, for example, have changed little in the past five years (the same is true of Internet browser programs)—all fields now dominated by Microsoft—changes in other areas where the company is not so pervasive have been greater and come faster.

Reviving Policymaking

A growing number of industry observers are concluding that what happened to American telecommunications in the early 2000s presents a strong argument for *more* rather than less public policy. Excessive deregulation left industry executives to focus upon short-term profits rather than long-term growth and stability. Some argue that the telecommunications and information sector is both too large and too central to the nation to be left only to profit-driven managers. Just one example suffices to underline how some management needs to be saved from itself.

The late U.S. District Judge Harold Greene, when overseeing the aftermath of the AT&T divestiture, refused for years to let the RBOCs expand into other than regulated local exchange service. He feared that freed of any restrictions they would quickly dissipate shareholder equity with ill-conceived ventures into seeming high-profit ventures *(see section 7.1)*. How right he often was! As the restrictions were gradually lifted in the late 1980s and early 1990s, RBOC ventures into computer stores and some other ill-considered options quickly turned sour. (Admittedly many others worked spectacularly well, too.) Greene's years as the overseer of a large segment of American telecommunications partially overlap the British creation of an Office of Telecommunications (Oftel) to supervise the liberalization of British Telecommunications and its growing gaggle of competitors. There, too, one man (Oftel's director) held extraordinary government-backed power to control or limit change in the telecommunications business.

While virtually nobody argues anymore in favor of actual government operation of all telecommunication services (the old PTT model has all but disappeared from the rest of the world), something other than almost complete deregulation does seem in order. Just as financial markets operate within rules and regulations of the Securities and Exchange Commission (among other agencies), many industry observers (and not all are theory-driven academics) feel that a more proactive FCC would be useful for consumers and industry alike.

Incumbent regulators in the early 2000s surely did not seem to share such views. FCC Chairman Michael Powell seemed generally unperturbed by the upheaval in the industries the Commission regulates. He was quoted in a variety of news stories as maintaining oversight, but not planning any specific response to the meltdown. What a change from the commissioners of three decades earlier, who had to be urged by the Department of Justice to introduce competition and at the same time allow some competitors to fail (*see section 5.4*).

A decade later, experience suggests that the 1996 telecommunications legislation (*see chapter 9*), which was intended to remake American telecommunications, took (or allowed) some seriously wrong turns. The many general guidelines left for the FCC to define opened up huge areas for industry lobbying and appeals—and delay. Some have argued that passage of the law actually increased confusion and legal activity on the part of communication carriers, exactly the opposite of what Congress intended.[115] That certainly seemed to be the case in a pair of FCC decisions made in February 2003 when, with sharply split votes, the Commission decided to turn over to state regulators the question of how much and under what conditions competitors should pay incumbents for access to their telephone lines. Observers argued this would continue hugely expensive lobbying campaigns and legal costs—and might well end up with a variety of different decisions. At the same time, the FCC agreed with the incumbents that they did not have to provide low-cost access to their competitors for broadband service.[116] Such decisions send very mixed messages to the marketplace—and to the courts. Indeed, in March 2004, the federal appeals court unanimously overturned the first decision, saying the FCC lacked authority to turn such responsibility over to the states.[117]

The 1996 law's attempt to introduce and encourage local competition to monopoly telephone and cable carriers was doomed from the start as no matter what safeguards were installed (such as the vaunted 14-point list—*see Fig. 9.1*), the law could not overcome the economic reality of incumbent carriers unwilling to compete. Far more willing to hire expensive legal talent and litigate (or even merely threaten to litigate), the RBOCs successfully staved off most would-be CLEC competitors for years—long enough to force most of them out of business.

Summing Up

This book spans the history of the U.S. telecommunications industry from the introduction of the telegraph to the current arguments regarding Internet access. It analyzes an industry that has vacillated between competition and monopoly; examines the technical breakthroughs that fueled the development of new services; and reviews the efforts of regulators, policymakers, and courts to shape industry structure. It is only fitting to end this survey by taking stock of where the industry stands as this book goes to press.

FCC statistics suggest that the level and type of competition envisioned by Congress when it drafted the Telecommunication Act of 1996 is far from being realized. Competitive local exchange carriers (CLECs) are not building their own facilities to deliver local service in competition with the incumbent local exchange carriers (ILECs), as Congress had hoped. At the end of 2004, 84.5 percent (or 145.1 million) of the total 177.9 million wireline local service access lines were provided by ILECs. After 9 years of regulatory activity and a multitude of court cases, CLECs provided only 32.9 million, or 18.5 percent of the lines. Perhaps even more discouraging is that CLECs used their own facilities to deliver only 26 percent of these 32.9 million lines; rented or resold ILEC facilities accounted for 74 percent of their business.[118]

Wireline local service did lose ground to wireless service in 2004, with more wireless subscriptions (181.1 million) than wireline connections (177.9 million).[119] And ILECs continued to lose the battle for Internet access services. At the end of 2004, of the total 37.9 million high-speed connections, 21.4 million (or 56 percent) were provided by cable modems, while only 13.8 million (36 percent) were over DSL lines and the remaining 7 percent were provided over satellite or other connections. The rate of growth of DSL lines for 2004 was 45 percent, compared with the 30 percent growth rate for cable modems, suggesting that DSL deployment may begin to gain on cable modems.[120]

What do these numbers suggest about the future direction of the telecommunications industry? They may suggest quite a bit, given the service profiles of the major industry players and the impact of recent merger activities. The pending (as this is written) mergers of AT&T with SBC and Verizon with MCI will put back together what the Modification of Final Judgment (MFJ) tried so hard to keep apart: local and long distance services. Two underlying assumptions of the MFJ were that the two services are separable and that long distance is a viable stand-alone service. Those assumptions have, for all practical purposes, been proven false. The difference between local and long distance services matters if distance is a concern; however, fiber optics, digital switching. VoIP, and wireless technologies have made distance virtually irrelevant. The difference in the cost of hauling a call 500 miles or 5 miles is negligible, whether that call is hauled over the Public Switched Network, a wireless network, or the Internet. Wireless providers and cable companies providing long distance through VoIP understand this and so they bundle all minutes together—local and long distance—for a flat fee. All of which is appealing to consumers who don't care about regulatory boundaries between local and long distance or between state and interstate. In order to compete with wireless and cable VoIP services, telephone companies need to bundle local and long distance; that's hard to do with two separate companies. Hence the logic of the AT&T-SBC and Verizon-MCI mergers.

Not only will the mergers of the two largest interexchange carriers (IXCs) with two of the RBOCs deal a death blow to the independent interexchange industry, they will have dire implications for the future of the CLEC industry as well. As a former FCC chief economist pointed out, "AT&T and MCI are the industry's largest local competitive providers to business customers. . . . If these mergers are approved, we'll have effectively turned the telecom clock back more than 20 years."[121] In other words, these CLECs will become ILECs. And what of the remaining CLECs? Their prospects may not look too rosy, if they have to depend on their local service competitors for long distance services. All of this seems to harken back to lessons about the interdependence of local and long distance services that should have been learned from the very earliest years of the telephone business. The telephone was originally envisioned as providing local service that would cover a range of 20 miles or so, while the telegraph would take care of long distance communication (*see section 2.4*). It didn't take long, however, for the builders of the early Bell system to recognize that there were major benefits in expanding their range of services beyond those 20 miles to include long distance—a lesson whose potency the MFJ seems to have underestimated.

Given the implications of the pending RBOC-IXC mergers for the CLEC industry, the FCC's 2004 statistics for CLEC market penetration may well mark the high point for that industry. The FCC statistics also underline an interesting point about the effectiveness of the 1996 Telecommunications Act and the future focus of telecommunications policymakers. The FCC's 2004 statistics indicate that there is indeed competition in telecommunications; it just hasn't quite taken the form envisioned by Congress in 1996. Congress assumed that competition would come from the cable companies re-engineering their networks to provide telephone service and from CLECs who would first rent parts of the ILEC's network and then build their own telephone facilities. Competition would be wireline and would be voice-based. Instead, real competition seems to have developed because of Internet data and wireless services.

Voice services did not entice the cable companies into competing with the ILECs. It was data, in the form of cable modem access to the Internet, that brought the cable companies to the table. And with the development of VoIP, cable companies are now expanding on that cable modem service to compete for local and long distance. Meanwhile, wireless is emerging as a strong competitor for wired local service, with wireless subscriptions now (2005) exceeding wireline connections for the first time; and, because wireless providers bundle local with long distance with flat-rate pricing, they are also effective competitors for long distance service. With the development of broadband wireless, wireless companies will be able to compete with wireline DSL and with cable modems in the Internet access market as well.

Competition is occurring across platforms, with wireline telephone companies, cable companies, and wireless companies all seeking to provide similar services. The current regulatory regime is platform-specific, however, rather than service-specific. Wireline, cable, and wireless providers may be striving to deliver the same services, but they are all regulated differently. For example, wireline telephony (local and long distance) is considered a common carrier service and so subject to universal service charges, oversight by state utility commissions, and other regulatory provisions. On the other hand, the VoIP-based local and long distance services provided by the cable companies are considered information services and so not regulated at all. The challenge for policymakers will be to develop a service-specific regulatory framework that allows cross-platform competition to continue to evolve. Hopefully the results of these efforts will be more successful and enduring than the 1982 MFJ and the 1996 Act proved to be. But legislating policy in a technology-based industry has never been easy.

NOTES

1. Stephanie N. Mehta, 2002. "Birds of a Feather: Who Wrecked Telecom?" *Fortune* (October 14), p. 197.
2. "This Year, It's Comdex Lite," 2002. *Business Week* (November 25), p. 8.
3. Ibid.
4. Yochi J. Dreazen, 2002. "Behind the Fiber Glut," *Wall Street Journal* (September 26), p. B1.
5. Ibid. The source cited for the figure (actually 2.7 percent) is Telegeography, a Washington, DC, research firm.
6. Steven Pearlstein, 2002. "Fiber-Optic Overdose Racks Up Casualties," *Washington Post* (May 2), p. A10.
7. Ibid., see especially "Too Much, Too Soon" table showing prices between eight selected city pairs from the first quarter of 2000 to the first quarter of 2002. The most substantial drops came in the longest routes.
8. Matthew Mosk, 2002. "Underwater Cable Plan Called Risky," *Washington Post* (August 9), p. B3. Concerns were also raised about the possible environmental impact of the project which would connect Baltimore, Washington, and the Tidewater cities.
9. Dennis K. Berman, 2002. "Innovation Outpaced the Market," *Wall Street Journal* (September 26), p. B8.
10. Simon Romero, 2002. "5 Years and $15 Billion Later, A Fiber Optic Venture Fails," *New York Times* (January 29), p. A1.
11. "The Telecom Depression: When Will It End?" 2001. *Business Week* (October 7), p. 68.

12. TeleGeography estimate as reproduced in Kas Kalba, 2002. *Telecom in the Time of Crash*. (Cambridge, MA: Harvard Program in Information Resources Policy, November), p. 22. Accessed February 15, 2003, from http://pirp.harvard.edu/pubs_pdf/kalba/kalba-i02-2.pdf

13. Simon Romero, 2002. "Cellphone Service Hurt by Success," *New York Times* (November 18), p. A1.

14. Ibid.

15. Roger O. Crockett et al., 2002. "All Net, All the Time," *Business Week* (April 29), pp. 100–104.

16. Ibid., p. 100.

17. Ibid., p. 102.

18. John Markoff, 2002. "Businesses, Big and Small, Bet on Wireless Internet Access," *New York Times* (November 18), p. C1.

19. Ibid., p. 103.

20. Heather Green, 2002. "You Say You Want a Wireless Revolution?" *Business Week* (April 29), p. 108.

21. John Markoff, 2002. "High Speed Wireless Internet Network Is Planned," *New York Times* (December 6), p. C1.

22. Catherine Yang, 2002. "The Pulse of the Future," *Business Week* (April 22), pp. 92–93.

23. "Watch This Airspace," 2002. *The Economist* (June 22), p. 21.

24. Ibid., pp. 15–16.

25. Greg Jaffe, 2002. "Military Feels Bandwidth Squeeze as the Satellite Industry Sputters," *Wall Street Journal* (April 10), p. 1.

26. Ibid.

27. "Joining the Fast Lane," 2003. *Business Week* (January 20), p. 12.

28. Dreazen, p. B8.

29. Steven Pearlstein, 2002. "In Euphoria, Key Players Looked Away," *Washington Post* (November 10), p. A19.

30. Steven Rosenbush et al., 2002. "Inside the Telecom Game," *Business Week* (August 5), pp. 34–40.

31. Patrick McGeehan, 2002. "More Details on Message By Ex-Analyst for Citigroup," *New York Times* (November 15), pp. C1, C4.

32. Darby et al., p. 7, quoting Martin F. McDermott, 2002. *CLEC: Telecom Act 1996, An Insider's Look at the Rise and Fall of Local Exchange Competition* (Penobscot Press), p. 96.

33. Pearlstein, p. A19.

34. Mehta, p. 202, citing Tom Evslin, a former AT&T employee and later CEO of ITXC.

35. Heather Timmons, 2001. "Feeling the Telecoms' Pain: Insurers and Banks," *Business Week* (April 23), p. 110.

36. Ibid.

37. Mehta, p. 202.

38. Steven Pearlstein, 2002. "From Heroes to Villains?" *Washington Post* (November 10), p. A19.

39. Yuki Noguchi, 2001. "Falling Stock Values Befall Telecom Firms," *Washington Post* (February 28), p. G8.

40. Mehta, p. 202.

41. Frank Rose, 2001. "Telechasm," *Wired* (May), p. 133.

42. Simon Romero, 2002. "Lucent Narrows Quarter Loss; Revenue Decreases Further," *Washington Post* (October 24), p. C8.

43. Mary Williams Walsh, 2002. "Another Cloud on the Horizon for Lucent Retirees," *New York Times* (November 20), p. C10.

44. U.S. Bureau of Labor statistics.

45. Mehta, p. 197.

46. "Verizon Gave One Executive $78,000,000 . . . Now They Want to Give 4,000 Workers Pink Slips. . . ." 2002. *Washington Post* (December 6).

47. Accessed October 28, 2002, from http://www.globalcrossing.com/xml/news/2002/august/09.xml

48. "Yesterday's Man," 2002. *The Economist* (May 4), p. 64. Ebbers was eventually found guilty of securities fraud, conspiracy, and filing false documents with regulators. Dan Ackman, "Bernie Ebbers Guilty," *Forbes.com*, March 15, 2005. Accessed July 12, 2005, from http://www.forbes.com/home/management/2005/03/15/cx_da_0315 ebbersguilty.html. In mid-2005, he was sentenced to 25 years in jail, essentially the rest of his life.

49. Charles Haddad, 2002. "Saving WorldCom: An Impossible Dream?" *Business Week* (May 13), p. 49.

50. Simon Romero and Riva D. Atlas, 2002. "Worldcom Files for Bankruptcy; Largest U.S. Case," *New York Times* (July 22), p. A1.

51. Kurt Seichenwald and Seth Schiesel, 2002. "S.E.C. Files New Charges on WorldCom," *New York Times* (November 6), p. C1.

52. Seth Schiesel and Steve Lohr, 2002. "Ex-President of Hewlett to be WorldCom Chief," *New York Times* (November 16), p. B1.

53. See, for example, "WorldCom Wants You to Know," 2002. *Washington Post* (November 13); the same ad also appeared in the *New York Times* and *Wall Street Journal*.

54. Laura M. Holson, 1999. "Qwest's Game of Aggressive Phone Tag," *New York Times* (July 25), pp. 3:1, 3:11.

55. Kris Hudson, 2002. "Nacchio Exit in Works for Weeks," *Denver Post* (June 18), p. C1.

56. Christopher Palmeri et al., 2002. "What Did Phil Know?" *Business Week* (November 4), pp. 114, 116.

57. Steven Rosenbush et al., p. 37.

58. Elstrom, p. 108.

59. Gretchen Morgenson, 2002. "Deals Within Telecom Deals," *New York Times* (August 25), p. 3:10.

60. Yuki Nogouchi, 2002. "AT&T Gets Network, Cheap," *Washington Post* (November 8), p. E5. Velocita had been formed in 1998 and was funded with $1 billion, and was chaired by Robert Annunziata, a former senior AT&T official. The company's position had been progressively weakened as other fiber optic providers went bankrupt, thus lowering the value of its own network.

61. Yuki Noguchi, 2002. "Telecoms Feast on Leftover Firms," *Washington Post* (October 18), p. E5.

62. Pearlstein, "In Euphoria, Key Players," p. A19.

63. Mehta, p. 198, citing Rich Nespola, CEO of The Management Network Group, and himself, a veteran of both MCI and Sprint.

64. As with all such cases, this one spread over several years, having originally been filed in 1998 by 20 states and the federal government, and finally ending in early November 2002 in what many termed a soft settlement with the company, more of a slap on the wrist than the possible breakup of the firm that loomed in 1999–2000 and which may have helped push the dot.com industry into recession and then worse. The Microsoft antitrust case and the issues surrounding it—merely the latest in a string of legal actions against the software giant—provide considerable grist for another book.

65. Seth Schiesel, 2001. "The Genesis of a Giant's Stumble," *New York Times* (January 21), pp. 1, 13.

66. "Lucent: One Step Forward, Two Steps Back," 2002. *Business Week* (April 8), p. 74.

67. James Slaby, cited as an analyst with Giga Information Group, as quoted in Barnaby J. Feder, 2002. "F.C.C. Chief Says Telecom Isn't Doomed by Cutbacks," *New York Times* (October 21), p. C1.

68. Darby, p. 4, citing "Comments of BellSouth et al." 2002 in FCC Docket 01-338 (April), p. V-7.

69. Larry F. Darby et al., 2002. "The CLEC Experiment: Anatomy of a Meltdown," *Progress on Point Paper 9.23*. (Washington, DC: Progress and Freedom Foundation, September), p. 1.

70. Ibid., p. 2.

71. Ibid., p. 6, summarizing the arguments of the Association of Local Telecommunication Services (ALTS).

72. Ibid., pp. 16–17.

73. This paragraph is based on a long investigative story, Michael Leahy, 2002. "A CEO's Lesson: What Goes Up. . . ." *Washington Post* (November 10), pp. A1, A20–A21.

74. Ibid., p. A20.

75. Yuki Nogouchi, 2002. "Revival of a Local Phone 'Reseller'," *Washington Post* (November 1) p. E5.

76. Ibid.

77. "In Unusual Move, F.C.C. Chief Criticizes a Possible Deal," 1997. *New York Times* (June 19), p. D1.

78. "AT&T Is Said to Break Off Merger Talks with SBC," 1997. *New York Times* (June 28), p. 25.

79. "Mobile Telecoms: The Tortoise and the Hare," 2002. *The Economist* (March 16), p. 63.

80. Peter Elstrom et al. 2002. "What Ails Wireless?" *Business Week* (April 1), p. 82.

81. Ibid., p. 60.

82. "Cutting the Cord in a Wireless World," 2002. *Business Week* (March 18), p. 40.

83. Elstrom, p. 50.

84. "Watch This Airspace," 2002. *The Economist* (June 22), pp. 14–21.

85. Ibid., p. 21.

86. "Quest: Off the Critical List, But . . ." 2003. *Business Week* (November 3), p. 73.

87. Tom Lowry, 2002. "A New Cable Giant," *Business Week* (November 18), p. 110.

88. Steve Rosenbush and Peter Coy, 2002. "Light at the End of the Fiber? *Business Week* (December 2), pp. 86, 88.

89. "Telecommunications: Strong Signals the Bad Times Are Over," 2004. *Business Week* (January 12), pp. 100–101.

90. "Outpacing Inflation," 2003. Chart in *New York Times* (February 21), p. C5.

91. Stephen Labaton, 2003. "Local Phone Rules to Stay in Place," *New York Times* (February 21), p. 1.

92. Brock, 2003, p. 455, citing 37 USC 309(j)(3)(b), the revised Section 309 of the Communications Act of 1934.

93. See *NextWave Personal Communications v. FCC*, 254 F.3d 130 (2001).

94. *FCC v. NextWave Personal Communications*, 537 U.S. 293 (2003).

95. Jennifer Lee, "Next Wave: Airwave Rich but Cash Poor," 2003. *New York Times* (August 4), p. C3.

96. *Congress and the Monopoly Problem: History of Congressional Action in the Antitrust Field, 1890–1966: Seventy-Five years,*" 1966. U.S. Congress, House of Representatives, Select Committee on Small Business, 89th Cong., 2nd Sess., Committee Print.

97. One of the authors of this book (Sterling) was a senior staffer at the FCC when the AT&T case settled—and can attest that the Commission was informed of the settlement only on the day it was to be announced, and by an AT&T official, not the Justice Department. To a considerable degree, Justice ignored the FCC, which was treated as merely another commentator in the legal proceedings.

98. Steve Rosenbush et al., 2002. "The Telecom Depression: When Will It End?" *Business Week* (October 7), p. 70.

99. Ibid., p. 72.

100. Ibid., p. 66.

101. Elstrom, "Telecom Meltdown," p. 108.

102. "Ma Bell Gets Mauled," 2000. *Business Week* (May 15), pp. 52–54.

103. Seth Schiesel, 2000. "AT&T, in Pullback, Will Break Itself Into 4 Businesses," *New York Times* (October 26), p. A1.

104. Floyd Norris, "AT&T Realigns Its Planets," 2000. *New York Times* (October 26), p. C1.

105. Seth Scheisel and Andrew Ross Sorkin, 2001. "Comsact Wins Bid for AT&T's Cable," *New York Times* (December 20), p. A1.

106. Lowry et al., "A New Cable Giant," pp. 108–118. See also Barnaby J. Feder, 2002, "U.S. Clears Cable Merger of AT&T Unit with Comcast," *New York Times* (November 14), p. C7.

107. Seth Schiesel, 2002. "AT&T Lost Almost $1 Billion but There Were Bright Spots," *New York Times* (April 25), p. C2.

108. Laura Rohde. 2004. "Cingular Buys AT&T Wireless," *PC World* (February 17). Accessed July 21, 2005, from http://www.pcworld.com/news

109. Associated Press, 2005. "AT&T Shareholders Approve Acquisition by SBC," *Advanced IP Pipeline* (June 30). Accessed July 12, 2005, from http://www.advancedippipeline.com/164904352

110. Ibid.

111. Jeff Smith, 2005. "Qwest Calls It Quits," *RockyMountainNews.com* (May 3). Accessed July 4, 2005, from http://www.rockymountainnews.com/drmn/business/article/0,1299,DRMN_4_3747252,00.html

112. Ibid.

113. A. Michael Noll, 2002."Telecommunication Basic Research: An Uncertain Future for the Bell Legacy," pp. 8–9. Prepared for Delivery to the 30th Telecommunications Policy Research Conference. Accessed October 28, 2002, from http://www.intel.si.umich.edu/tprc/papers/2002/39/NollRD.pdf

114. *Insert reference to Mike Noll TP papers on Bell Labs.*

115. Martha A. Garcia-Murillo and Ian MacInnes, 2002. "The Impact of Incentives in the Telecommunications Act of 1996 on Corporate Strategies," School of Information Studies, Syracuse University, no date. Accessed October 29, 2002, from http://arxiv.org/ftp/cs/papers/0109/0109071.pdf

116. Stephen Labaton, 2003. "Local Phone Rules to Stay in Place," *New York Times* (February 21), p. 1.

117. See, for example, Griff Witte, "FCC Rule on Local Phone Service Rejected," 2004. *Washington Post* (March 3), p. E1.

118. Federal Communications Commission, Wireline Competition Bureau Industry Analysis and Technology Division, 2005. "Local Telephone Competition: Status as of December 31, 2004," July, pp. 1–2.

119. Ibid.

120. Federal Communications Commission, Wireline Competition Bureau Industry Analysis and Technology Division, 2005. "High-Speed Services for Internet Access: Status as of December 31, 2004," July, pp. 2–3.

121. Simon Wilkie is quoted in "Business to Face Higher Prices, Fewer Choices if SBC/AT&T and Verizon/MCI Merger Requests Approved; Former FCC Chief Economist Argues Mergers are 'Bad for Business,' " 2005. *Business Wire* (June 14). Accessed July 4, 2004, from http://home.businesswire.com/portal/site/google/index.jsp?ndmViewId =news_view&newsId=20050614005672

Regulatory Concepts in Telecommunications Economics, Finance, and Accounting

A.1 REGULATORY REGIMES

Rate-of-Return Regulation Rate-of-return regulation (RoR) was, until fairly recently, the dominant form of regulation for utilities like gas and electric as well as telecommunications services (*see section 4.2*). RoR is a cost-plus method of regulation in which the service provider is allowed to set its service prices at a level high enough to recover its expenses of providing service and to generate a profit adequate to pay its debts and to attract investors. The company's expenses plus profit are known as its revenue requirement. Service prices are targeted to recover this revenue requirement. The RoR formula in which the revenue requirement is calculated is as follows:

Revenue Requirement = (Rate Base – Accumulated Depreciation) × Allowed Return + Allowed Expenses

The RoR formula may appear to be simple and straightforward; its implementation, however, has tended to be contentious, with regulators, telephone companies, and various interest groups arguing over each element in the formula. In the RoR process, the burden is on the regulator to balance the interests of the subscriber—adequate service at just and reasonable prices—with the interests of the shareholder—adequate return to cover expenses, pay debt, and attract investors.

In order to facilitate comparability between telephone companies, and also to establish the investment and expenses over which the regulator

360

has jurisdiction—that is, those investments and expenses that go into the revenue requirement calculation—regulators require companies to maintain their books according to a specified chart of accounts. The Federal Communications Commission (FCC) has designated the chart of accounts for telephone companies in Part 32, "Uniform Systems of Accounts for Telephone Companies," of its Rules and Regulations, and most state commissions have recognized this accounting plan. Part 32 outlines the account structure that a telephone company is to use, designating the numbering scheme for the accounts and specifying into which accounts assets, expenses, and revenues are booked. For example, Part 32 requires companies to book investment in cable and wire facilities in account 32.2410, and to book the associated maintenance expenses for cable and wire facilities in account 32.6410. Even though many telephone companies are no longer regulated by RoR, they are still required to maintain the Part 32 uniform system of accounts in order to facilitate regulatory oversight of their operations.

Perhaps the most important element in RoR is the rate base, because it is the return on the rate base that constitutes the profit earned by the company. The rate base consists of all of the assets (buildings, trucks, etc.) and equipment (switches, copper wire, fiber optic cable, central office equipment) used in the provision of regulated services. The rate base is valued at historical cost. In other words, instead of listing an asset on the books at replacement cost (i.e., the cost of the item if it were purchased today), assets are listed at the cost of the item when purchased. In order to qualify for inclusion in the rate base, an asset must be deemed to be "used and useful." In other words, the asset had to actually have been used in the provision of service; retired switches, for example, cannot be included in the rate base. Because there are constant upgrades and additions to the telephone network, the issue of plant-under-construction has been an issue in RoR, with regulators and telephone companies arguing over when plant-under-construction should be included in the rate base.

Another issue related to the idea of "used and useful" is the matter of "above and below the line" investments and expenses. Investments and expenses actually used in the provision of regulated services are considered "above the line." In other words, they are included in the RoR formula that calculates the revenue requirement, and therefore, are paid for by the subscriber through regulated service prices. Investment and expenses that regulators regard as not being useful in the provision of service are considered "below the line," and excluded from the RoR formula. As a result, these expenditures are absorbed by the company's stockholders. An example of a below the line expenditure has been advertising expenses. During the era of monopoly, many regulators did not see a need for advertising by the monopoly telephone company.

Depreciation has been another expense fraught with controversy. Depreciation is an accounting procedure that seeks to match the recognition of the cost of an asset with its useful life. In order to understand depreciation, it is helpful to keep in mind the distinction between investments and expenses. Expenses are expenditures whose benefits are enjoyed in the current year; payroll is such an expense and is therefore recognized—that is, deducted from revenues—in the current year. Investments, on the other hand, are expenditures whose benefits are enjoyed over several years. For example, a company that buys a $2 million piece of equipment that it will use for 20 years does not recognize the full $2 million as expense in the year in which the equipment is purchased. Such recognition would distort the company's earnings calculation during the whole 20-year period by grossly overstating expenses in the first year and understating expenses during the next 19 years. Instead, the company recognizes the $2 million over the full 20 years of the asset life by recognizing $100,000 in depreciation expense each year.

In terms of the RoR formula, depreciation plays two roles. Accumulated depreciation reduces the rate base. In the above example of the $2 million piece of equipment, after five years, the asset would be valued at the historical cost of $2 million less five years of depreciation, or $500,000. In terms of the rate base, depreciation tends to reduce the revenue requirement. Depreciation also represents an expense, however. In the revenue requirement calculation, expenses are covered dollar for dollar. This means that each dollar of depreciation expense is included in revenue requirement. In the RoR calculation, depreciation expense has a greater impact than accumulated depreciation on the size of the revenue requirement that is calculated. Because regulators have sought to keep service prices as low as possible, they have sought to keep depreciation expenses low by requiring telephone companies to use long depreciation lives for equipment. If instead of using a 20-year useful life in the above example, a company used a 40-year life, depreciation expense would be $50,000 instead of $100,000. Telephone companies were traditionally required to maintain long depreciation lives for equipment, thus keeping their revenue requirements lower. As new technologies developed, telephone companies argued that long depreciation lives did not provide revenue requirements sufficient to fund innovation and that long depreciation lives discouraged them from upgrading facilities.

Another contentious element in the RoR calculation is the allowed return. The allowed return represents the company's cost of capital. The cost of capital is the return on investment that the company needs to pay its debts and to attract investors. It is the weighted average cost of capital for the company. A cost of capital calculation is as follows:

Weighted	Percentage of Total Capital	Annual Cost	Cost of Capital
Long-term debt	40%	8%	3.2%
Preferred stock	10	9	.9
Common stock	50	12	6.0
			10.1%

In this example, the cost of capital is calculated as 10.1%, the return on rate base that the company needs to satisfy its debtors and also to attract shareholders. In the cost of capital calculation, the annual cost of debt is easy to determine; the company has borrowed money at 8%. The annual cost assigned to the preferred and common stock is not so simple to determine, however, because it is subject to debate.

A low allowed return results in a lower revenue requirement and therefore lower service prices. Those wanting lower service prices—consumer interests, for example—tend to argue for a lower allowed return. Telephone companies argue for a higher return, reminding regulators that, under the Fifth Amendment of the U.S. Constitution, a company cannot be deprived of its property without adequate compensation. Regulators in the RoR process seek to set an allowed return that is both fair to the subscribers, by generating just and reasonable service prices, and fair to the company's owners, by generating sufficient revenue requirement so that the company can meet its obligations and remain viable. In determining an appropriate return on preferred and common stock, regulators have tended to use either a comparable earnings approach, which looks at how companies have performed, or a market approach, which looks at investor expectations. In the comparable earnings approach, the regulator tries to estimate the average level of returns on earnings by firms throughout the economy, adjusted for the level of risk in the telephone industry. In the market approach, the regulator refers to the securities market in trying to calculate the rate of return that investors would require in order to invest.

In most instances, telephone companies under RoR file rate cases with state commissions to ask for a rate increase when they do not earn their allowed return. The formula for calculating a company's actual earnings is as follows:

Earnings = (Service Revenues Billed – Expenses)/(Rate Base – Accumulated Depreciation)

For example, if a company has billed $100 million in regulated services revenues, has expended $80 million in allowable expenses, and has a rate base, less accumulated depreciation, of $200 million, that company's earnings would be:

$$(\$100 \text{ million} - \$80 \text{ million})/\$200 \text{ million} = 10 \text{ percent}$$

If the company's allowed return is more than 10%, the company asks to raise its service prices in order to generate sufficient revenues to attain the allowed return. In the above example, if the allowed return were 11%, the company would need $102 million in revenues to reach its allowed return:

$$(\$102 \text{ million} - \$80 \text{ million})/\$200 \text{ million} = 11\%)$$

The company would raise its service prices to a level that would generate $102 million. Even if companies do attain their allowed return, they can file rate cases to ask for increases in that allowed return. Many telephone companies filed for such increases during the inflationary period of the late 1970s.

Incentive Regulation   A major complaint about RoR is that it does not encourage efficiency and innovation. If the telephone company becomes more efficient, and thus cuts expenses, the result is a lower revenue requirement and lower prices. The company does not receive the benefits of efficiency. In the above example, for instance, if the company lowers its expenses substantially—to $70 million—the earned return would be 15%, and the state commission could require the company to lower its service prices. Conversely, others complain that the RoR method encourages "gold plating" by the telephone company. By purchasing the most expensive equipment—even if less costly equipment would suffice—the company could drive up its rate base and, by so doing, its revenue requirement.

These complaints have led policymakers to explore alternative regulatory regimes that provide an incentive for companies to pursue efficiencies and innovations. This incentive is the opportunity for the company to keep the increased earnings that result from reductions in expenses or rate base. Instead of focusing on a company's costs—as RoR does—and then establishing prices, incentive systems focus on service prices. By focusing on prices, incentive systems, at least in theory, replicate what happens in a competitive marketplace. In a competitive marketplace, prices are set by the market and companies decide whether they can earn a profit at the going price.

There are different forms of incentive regulation. Three of the most prevalent are social contract, rate freeze, and price cap regulation.

In *social contract* regulation, the company and the regulator strike an agreement in which the company agrees to perform specific network upgrades or to add specific services in return for greater pricing flexibility (the amount of flexibility varies according to the specific provisions of the agreement) and the end of RoR. The benefit to the consumer in the social contract approach is better or more services; the benefit to the company is the ability to reap the benefits of efficiencies or innovations through higher earnings.

Rate freeze regulation is a variant of social contract regulation. In rate freeze regulation, the company agrees to freeze specific prices for a specified period of time in return for an end to RoR. The services for which prices are usually frozen are those deemed to be essential to the subscriber and that face little or no competition; in most rate freeze plans those services include basic local service, especially for residential subscribers. A company may agree to freeze the price of residential local service for a five-year period, in return for which RoR is ended and the company is given greater pricing flexibility for nonbasic services or competitive services. The benefit to the subscriber is the assurance that basic service prices will not rise for five years; the benefit to the company is greater pricing flexibility for nonbasic services and the chance to earn a higher return, if the company is able to increase efficiencies.

The most prevalent form of incentive regulation is *price cap* regulation (*see section 7.3*). Price cap regulation is a system of service baskets and pricing indices. Price cap regulation regulates price movements, and seeks to do so in a manner that will encourage the company to outperform the rest of the economy. Instead of regulating a company's specific costs, price cap regulation takes into account inflation and productivity in determining how much a company can change its prices. Price movements are regulated through a price cap index (PCI) that represents a cap on price movements. Calculation of the PCI takes inflation into account through the use of the annual changes in the Gross Domestic Product Price Index (GDP-PI), which measures inflation in the whole economy. Through the GDP-PI, average inflation is considered in a company's pricing; however, the company is required to do better than the average because of the "X factor." The X factor is a productivity factor that is subtracted from the price index and serves to keep prices down. As would be expected, the determination of what the X factor should be is a contentious and highly politicized process. The price cap index is calculated as follows:

$$PCI_{current} + GDP\text{-}PI - X +/- Z = PCI_{new}$$

In this formula, the GDP-PI represents inflation; X is the productivity factor; and Z is the impact of "exogenous costs." Exogenous costs are those unexpected costs over which the company has no control; examples of exogenous costs are changes in the tax laws or the impact of a natural disaster like a hurricane. Companies are not expected to bear the full burden, or to reap the full benefits, of changes over which they have no control.

As an example of the price cap calculation, and in the interests of simplicity, let us assume that the price cap index was begun, or initialized, this past year at 100% or 1.0. Let us also assume that the change in the GDP-PI during this past year was an increase of 2%, that the productivity offset has been established by regulators as being 3.3%, and that there are no exogenous costs. The formula would be:

$$1.00 + 0.02 - 0.033 +/- 0.00 = 0.987$$

The new PCI would be .987. This means that prices can only be changed in a way that results in an actual price cap index that is equal to or lower than 0.987. An actual price cap index is calculated for each service basket and compared to the PCI.

The price cap system depends not just on indices, but also on service baskets. A benefit of price caps is that it discourages cross-subsidization between services. A company may be tempted to subsidize its competitive services with profits from its monopoly services by pricing its competitive services low, even below cost, and making up the difference through high prices for its monopoly services. In price cap regulation, monopoly services and competitive services are placed in separate baskets. The price movements of services within a basket are dependent upon one another: if there is an increase in the price of one service, the price of another service must decrease in order to stay within the index. However, price movements of services in different baskets are independent of one another. For example, if residential local service is in one basket and long distance is in another, the company cannot make up for any decreases in long distance rates by increasing local service rates.

As a highly simplified example of how price caps works, assume a basket with two services, X and Y. Assume that this past year the company sold 100 units of X at $1.00 per unit and 200 units of Y at $3.00 per unit. The company is now setting prices for the upcoming year and wishes to increase the price of X to $1.20 and to decrease the price of Y to $2.90. In order to calculate whether these price movements would be allowable under the price cap, the company calculates the actual price index (API) for the service basket. The API measures the impact of proposed price increases on the service basket by multiplying the current API by the weighted average of the proposed price changes. The calculation for the API is as follows:

$$API_{new} = API_{old} * (Price\ Ratios\ of\ Services *$$
$$Revenue\ Weights\ of\ Services)$$

As a first step in this process, the company calculates the revenue weights of X and of Y during this past year. In other words, in the service basket, what percentage of total revenues did each service represent:

	Price	Quantity	Revenue	Revenue Weight
X	$1.00	100	$100	.142857
Y	$3.00	200	$600	.857143
			$700	1.00000

The next step is the calculation of the new API. The new API is calculated by multiplying the price ratio (the proposed price divided by the current price) by the past year's revenue weight and then multiplying the result by last year's API. The result measures the impact of the proposed price changes on the basket of services:

Service	(A) Current Price	(B) Proposed Price	(C) Price Ratio B/A	(D) Past Yr Rev Weight	(E) C*D
X	$1.00	$1.20	1.200	.142857	.171104
Y	$3.00	$2.90	.967	.857143	.828857
Actual Price Index					.999961

Note that the API for the past year was 1.00 or 100%. Multiplying the result of the above calculation (.999961) by 100% yields this year's API. The actual price index calculated for these proposed price changes is .999961; however, the PCI calculated earlier was 0.987. The proposed price changes are not allowable. The company performs the same type of calculation for any other baskets of services, with each actual price index having to fall at or below the PCI in order for price changes to be allowable. Note that the company's revenue requirement is never considered in deciding whether prices are allowable.

A.2   COST ISSUES

There is no one definition of "cost," a point that is nowhere more evident than in the field of telecommunications. Regulators, economists, consumers, and industry executives all argue about what costs involve, what type of costs are at issue, and what type of costs should be used in decision making. Underlying the arguments about costs are issues of timing, allocation, and purpose. Some definitional issues regarding cost are discussed next.

Accounting Cost Versus Economic Cost

The purpose of accounting costs is to capture and report a firm's expenditures according to a specific plan, like a chart of accounts, or according to specified rules like the tax laws. *Accounting costs* are historical and tend to document what a company has done. The purpose of economic costs is to explain how a market works, or should work. *Economic costs* are forward-looking and are often not tied to the costs of any specific firm; instead, economists use economic costs to apply theories to firms and to markets. Accounting costs are historical, embedded costs; economic costs are forward-looking and incremental.

Historical Versus Forward-Looking Costs

Historical costs are those which have actually been expended. A company's books document the historical costs a company has incurred in purchasing investments or paying expenses. *Forward-looking costs* are theoretical and look forward to what the company's costs will be in the future—either in the short or long term. While *historical costs* document how much a company spent in purchasing equipment for a factory line, forward-looking costs seek to estimate what the cost characteristics of the company will be at full production of that factory line.

Embedded Versus Incremental Costs

These include all the costs spent by a firm in producing all of its products. *Incremental costs* are the changes in costs caused by a specific increase or decrease in production. *Embedded costs* look at total costs as already having been expended; *incremental costs* look at changes in cost.

Fixed Versus Variable Costs

Whether a cost is fixed or variable depends on the time frame involved. *Fixed costs* are those that do not vary over a specific range of production. If 3 million widgets can be produced on one assembly line, the cost of the assembly line equipment (assume it is $1 million) is fixed for the production of those widgets. Stated another way, the cost of the assembly line equipment is $1 million whether 3 widgets or 3 million widgets are produced. *Variable costs*, on the other hand, vary with production. Labor costs are usually considered variable costs. If the company wishes to increase production, the company pays for more hours of labor; if the company wishes to decrease production, the company pays for fewer hours of labor. The company pays far less for 3 widgets of production than it does for 3 *million* widgets of production. The issue of fixed versus variable cost is important in pricing considerations. While economists agree that prices should include variable costs, there is disagreement about how a company should recover its fixed costs.

Sunk Costs

These are costs that have been expended but cannot be recovered. The issue of sunk costs is important in telecommunications because telephone companies have argued that regulators have caused them to have obsolete equipment on the books, the cost of which they will not be able to recover. For example, telephone companies argue that, because of long depreciation lives, they have not recovered the cost of the copper wire that is so ubiquitous in their networks. If copper wire becomes obsolete and is replaced with fiber, telephone companies argue that the copper wire will become a sunk cost.

Marginal Cost

This is an incremental cost. It is the change in total cost of production that is caused by a very small change in production. *Marginal cost* is often defined as the change in cost caused by the addition of one additional subscriber or one additional unit of production.

Direct Costs Versus Allocated Costs

Direct costs are those specifically attributed to the production of a service or product. For example if it takes $3 worth of plastic to make one widget, that $3 is a direct cost of production. If the widget were not produced, the $3 cost would not be incurred. *Allocated costs* are costs that are not so directly tied to the production of a service or product. The electricity to power an assembly line is an example of an allocated cost. If the assembly line is working, electricity costs will be expended whether the widget in question is produced or not. What portion of the electricity costs should the widget in question bear if it is produced? Some allocation method is required to assign the electricity costs to the unit of production, in this case the widget.

Shared Costs, Common Costs, and Joint Costs

Shared costs are those that are general to all products and services. The cost of a company's headquarters and the compensation paid the company president are two examples of shared costs. Arguably, all products benefit to some degree from these expenditures. *Common costs* are those costs incurred across several products, or even across all of the products within a firm. The costs of a building are common costs for the products produced in that building. The issue of how to allocate those costs to product prices is open to much debate. *Joint costs* are a specific type of common costs. They are costs that are common to two or more products or services and they vary in proportion to the output of both of the products or services, not in proportion to the output of one of the services or products. An example of joint costs involves the production of leather and beef.[1] Cattle feed is required for the production of both leather and beef, but it is impossible to determine what feed costs are necessary for each product. If production is cut back, less feed is needed, but it is impossible to determine how much of the decrease in feed costs is assignable to leather and how much is assignable to beef production.

Fully Allocated Cost Study

Such a study is based on accounting costs. This type of study seeks to allocate each dollar on the company's books to an appropriate category, product, or service. Economic efficiency is not the goal of a fully allocated cost study. Separations studies (*see section 4.3*) are an example of a fully allocated cost study. Formerly found in Part 67 of the FCC's Rules and Regulations and now found in Part 36 of those Rules and Regulations, separations procedures have as their goal the allocation of a company's investment and expenses between state and interstate jurisdictions. The separations process first allocates a company's investment to specific categories of plant—central office versus cable and wire, for example. The investment is then directly assigned to state or interstate jurisdiction wherever possible; if a circuit is used specifically for an interstate service, it is assigned to interstate. Those investments not directly assignable are allocated based on a relevant usage factor, such as minutes of traffic for switching equipment or messages for operator systems. Once the investment is allocated, directly identifiable expenses are allocated based on their associated investment. If 20 percent of local switching investment is allocated to interstate, 20 percent of local switching maintenance expense is allocated to interstate as well. Expenses not directly identifiable—overhead expenses or general operating expenses—are allocated based on the allocation of the directly identifiable expenses. Once a separations study is completed, all of the regulated assets, expenditures, and revenues of a company are allocated to either the state or interstate jurisdiction.

Another example of a fully allocated cost study is the Cost Allocation Manual (CAM) discussed in section 7.4. The purpose of the CAM is to allocate a company's expenses and investment between regulated and unregulated activities. Wherever possible direct assignment is used in the CAM process, with investments and expenses identifiable as being used for regulated or nonregulated activities being directly assigned, and investment and expenses not so clearly capable of being directly assigned being allocated based on a relevant usage factor.

Long-Run Incremental Cost Study

Unlike fully allocated cost studies, incremental cost studies are concerned not with allocating each dollar on a firm's books, but rather with analyzing a company's economic costs. Incremental cost studies usually look at a company's activities over the long run, assuming production at a company's maximum capacity. Rather than looking at a company's historic costs, incremental cost studies look at the costs of an efficiently engineered network or efficiently engineered services. Instead of looking at the whole company, incremental cost studies focus on one product or service (or group of products and services) and attempt to isolate the economic costs of production of that product or service by identifying the change in production costs caused by that product or service. Because most shared, common, and fixed costs would be incurred by a company even if the products or services in question were not produced, incremental cost studies often exclude shared, fixed, and common costs and focus on the variable costs and the specific common costs caused by the product or service.

While fully allocated cost studies are concerned with a company's full revenue recovery, incremental cost studies are more concerned with the firm's efficient pricing, rather than with the firm's ability to recover every dollar spent in providing a product or service. Incremental cost studies, sometimes called marginal cost studies, have been used by regulators to assure that companies did not engage in predatory pricing. The results of the incremental cost study were regarded as a pricing floor. If companies priced below incremental cost, they were pricing below cost and therefore acting in a predatory manner to drive out competitors.

TELRIC and TSLRIC

These are particular forms of long-run incremental cost studies that are currently being used to support decisions by both the FCC and the state commissions. TELRIC is an acronym for *Total Element Long Run Incremental Cost*. Instead of using historical costs, TELRIC looks at the costs of an optimally designed network that uses the most efficient equipment and network configuration. TELRIC seeks to isolate the change in cost caused by the addition of a network element, such as a local loop, holding all other network elements constant. TELRIC focuses on the incremental costs of the network element; it does not include shared, joint, or common costs unless they would be avoided if the network element were not provided. TELRIC, as administered by the FCC, does allow the imputation of an Annual Carrying Factor (ACF). The ACF makes some allowance for maintenance costs, insurance costs, floor space, depreciation, taxes, and cost of capital.

TELRIC is important because it is the standard used to determine the reasonableness of an incumbent telephone company's prices for unbundled network elements (UNEs). These UNEs are the parts of the network that the incumbent's competitors lease in order to provide their own services. The pricing of UNEs is important in sending the appropriate signals to those wishing to enter the market.

TSLRIC is *Total Service Long Run Incremental* Cost. While TELRIC looks at network elements, TSLRIC looks at services. If TELRIC applies to the local loop, TSLRIC applies to local service. TSLRIC is equal to the firm's total cost of producing all of its products and services (including the service in question) minus the firm's total cost of producing its products and services less the service in question. In its purest form, TSLRIC would include no shared or joint costs and few common costs.

A.3 PRICING ISSUES

Residual Pricing

This was a method often used by state commissions in determining which services to increase in price when a company proved that its current prices were insufficient to generate adequate revenue requirements (see discussion of RoR in section A.1). For example, if a telephone company demonstrated that its current service prices generated a revenue stream that was $2 million less than its allowed revenue requirement, the state commission would have to allow the company to raise its service prices to generate that additional $2 million. The services that the state commission have jurisdiction over include local service, state toll, state access charges, installation charges, and other miscellaneous services like call waiting or call forwarding.

In determining which service prices to increase, and by how much, state commissions tended to regard local service as the last service to be increased in price, and tried to raise as many other service prices as possible before increasing local rates by the "residual" amount of the revenue requirement still to be recovered. For example, if the company served

50,000 lines, recovery of the full $2 million from local service prices would entail an increase of $3.33 per month:

$$\$2 \text{ million}/(50,000 \text{ lines} \times 12 \text{ months}) = \$3.33$$

State commissions would avoid increasing local rates that much at one time, and so would seek to increase other service prices first. If the telephone company generated 20 million minutes of state long distance, the full $2 million could be recovered through a $.10 increase in per minute state toll, a hefty increase. If the company tended to install 1,000 lines per year, and to sell 10,000 call waiting services and 5,000 call forwarding services, the state commission might decide to allow the company to raise rates in the following pattern:

Long distance rates:	$.05 per minute	=	$1 million
Installation charges:	$5.00 per install	=	$5,000
Call waiting:	$2.00 per month	=	$240,000
Call forwarding	$3.00 per month	=	$180,000
Total increases:		$1,425,000	

This would leave a residual of $575,000, which would result in a local rate increase of $.96 {$575,000/(50,000*12 months)}. An increase under a dollar would be much more politically palatable than a $3.33 increase. The use of residual pricing led to the toll disparity (*see section 4.2*) in which state toll prices grew to be significantly higher than interstate toll prices.

Value of Service Pricing

In this method, regulators based prices not on cost, but rather on the perception of the value of the service to the subscriber. The use of rate groups for local service prices was based on value of service pricing. In the rate group method of pricing, subscribers paid for local service rates based on the number of people they could call without paying long distance charges; the fewer callable lines, the lower the local service price. Subscribers in communities of 500 or fewer lines would pay less than subscribers in communities of 500 to 1,000 lines, for example. Rather than basing local service rates on cost, regulators based the price on the perceived value of the service, assuming that the value to the subscriber in the 1,000-line community was greater than the value to the subscriber in the 500-line community. Value of service pricing was totally different in approach from cost-based pricing.

Price and Income Elasticity

Elasticity measures how important a service or product is to consumers. For example, price elasticity of demand measures the impact of a price change on the demand for a product. If demand is affected by price, then the demand for the product is elastic:

- If when price goes up, demand goes down, demand is elastic.
- If when price goes up, demand is constant, demand is inelastic.

Income elasticity of demand measures the impact of income changes on demand. If demand is affected by changes in income, then demand for the product is elastic:

- If when income goes down, expenditure level goes down, demand is elastic.
- If when income goes down, expenditure level is constant, or doesn't decrease in proportion, demand is inelastic.

Elasticity, especially income elasticity of demand, has been used to help determine whether a product is a necessity or not. A product for which the income elasticity of demand is inelastic can be regarded as a necessity because people will buy the product even if their income declines. If the product is also price inelastic, consumers will buy the product regardless of price increases. Consumers of such a product may need regulatory protection to assure they are not taken advantage of.

Ramsey Pricing

This deals with how a company that produces multiple products should recover its costs. According to Ramsey pricing, those products for which there is less elasticity of demand should bear a higher percentage of fixed and common costs. In other words, those customers who are willing to pay more for a product should be charged more. In an efficient market, prices tend toward marginal cost (*see section 1.2*). However, if a company prices all of its products at marginal cost, it will go out of business, because it will not recover the fixed costs and common costs that are not included in marginal costs. In determining how much fixed and common cost to recover from different groups of customers, or through different products, Ramsey pricing suggests that the company keep elasticity in mind, and allocate more fixed and common costs to those products for which customers exhibit a lesser degree of demand elasticity. In other words, companies should recover more overhead costs from those customers who are willing to pay for them.

NOTE

1. We are indebted to the Web site utilityregulation.com for this example, which is included in its essay "Costing Definitions and Concepts." http://www.utilityregulation.com/essays/et12.htm accessed August 24, 2003.

Glossary

These are many of the terms used in the text, with reference to where they appear or are further described. A number of the economic terms are further detailed in Appendix A.

Access: Availability and use of local telephone company facilities for origination and termination of interexchange (long distance) service. *See also* Interconnection.

Access charge: Assessments initiated by the FCC in 1983 to help meet the costs of providing local access telephone service after the breakup of AT&T. There were two main types: the subscriber line charge (SLC) assessed to end users, and the common carrier line charge (CCLC) paid by interexchange carriers seeking interconnection to local exchange companies. Interexchange carriers also paid access charges for use of local company switches and other facilities (*see section 7.3*).

ARPANET: Advanced Research Projects Agency Network, the key predecessor of the modern Internet/World Wide Web. Also called DARPANET for the full name of the Defense Advanced Projects Research Agency, which helped to fund it and NSFnet (*see section 8.5*).

AT&T: AT&T Corporation, formerly the American Telephone and Telegraph Company, or "Bell System," defendant in the divestiture case. Corporate owner of Western Electric and Bell Telephone Laboratories and, until January 1, 1984, the 22 Bell Operating Companies. In 1996, AT&T Technologies, formerly Western Electric, became a separate firm, Lucent Technologies, and included most of Bell Labs. In 2005, AT&T announced its takeover by SBC.

Auction: The use of sophisticated marketplace bidding techniques by the FCC to determine spectrum assignments (*see section 8.3*).

376

Baby Bell: Regional Holding Company or Regional Bell Operating Company after the 1984 breakup.

Bell Communications Research (Bellcore): Cooperative entity controlled by the Regional Bell Operating Companies (1984–96) and established to allow single point of contact for emergency government needs, shared research and development, and shared policy representation in Washington. As detailed in the text, its roles narrowed considerably in the 1990s and it was sold late in 1996. *See also* Central Services Organization.

Bell Operating Company (BOC): A corporate entity owned by AT&T and part of the Bell System until January 1, 1984, and now independent, whose primary function is to provide local telephone services to a specific geographic area. There are 22 of these companies. *See also* Regional Holding Company.

Bellcore: *See* Bell Communications Research.

Bottleneck: Originally, this meant the pre-divestiture ability of AT&T to control access to a local exchange (and thus to limit access by competitors) through its agents, the Bell Operating Companies. The concern is now focused on the four RBOCs which control access to most business and individual customers.

British Telecom (BT): Long a PTT (part of the British Post Office), but privatized in the 1980s.

Broadband: A facility (increasingly an optical fiber cable) having bandwidth (or capacity) sufficient for high speed transmission of multiple voice, video, or data channels simultaneously.

Cable modem: "Black box" device (usually supplied on a monthly lease from the local cable television carrier) that connects one or more computers to a broadband cable connection, allowing for fast and always-on Internet access. *See also* Broadband, DSL.

CCLC: Common carrier line charge. *See* Access charge.

CDMA: Code Division Multiple Access, one means of transmitting digital cellular signals.

Cellular: A type of wireless or radio mobile telephone service utilizing many transmitters each providing service in one cell or part of a larger region (*see section 8.2*). *See also* 2G, 3G.

Central Services Organization (CSO): Entity established by AT&T during divestiture planning to coordinate post-divestiture technical planning for and by the Regional Holding Companies (referred to as the Central Staff Organization in the Plan of Reorganization; *see section 6.4*). In 1985 renamed Bell Communications Research. *See also* Bell Communications Research.

Civil Investigative Demand (CID): Form of subpoena issued by the government to the object of an investigation compelling production of books and records. May lead to further legal action, as in the AT&T 1974 case (*see section 6.1*).

CLEC: Competitive local exchange carrier, that is, one competitive with an ILEC. *See also* ILEC.

Coax: Coaxial cable, or telecommunications cables with two cores, allowing vastly increased capacity for video or other signals (*see section 5.1*).

Common carrier: In communications, an entity available for hire to anyone without discrimination at posted rates to be a conduit for voice, data, or other electronic communications. *See also* Specialized common carrier.

Communications Act of 1934: The basic statute creating and governing operation of the Federal Communications Commission (FCC), whose duties include establishing the regulatory framework and obligations for communications common carriers. Often revised since, with major recent changes in 1984, 1992, and 1996.

Computer Inquiries: FCC proceedings seeking to determine what to regulate in the interface between data processing and telecommunications. There were three of these: Computer I (1966–71), II (1975–82), and III (1985–92; *see section 7.4*).

Consent decree: Agreement negotiated by the parties to an antitrust lawsuit which ends litigation. In return for an informal finding of no wrongdoing or guilt on the defendant's part, defendant usually agrees to a different course of conduct. Examples here are the 1956 (*see section 4.4*) and 1982 decrees (*see sections 6.3–6.4*) ending cases against AT&T.

Convergence Coming together of once-distinct technologies and related industries—in this case the digitally driven merging of the formerly separate (and separately regulated) media and telephone industries (*see section 8.4*).

CPE: Customer premises equipment, such as telephones, modems, and other devices that can be connected to the Public Switched Telephone Network (PSTN).

Cream-skimming: Ability and practice of an OCC (other common carrier) to provide, as AT&T complained in its pleadings, only the most lucrative low-cost long distance telephone service while evading the requirement to support local service costs.

CSO: *See* Central Services Organization, Bell Communications Research.

Customer premises equipment (CPE): Telecommunications equipment located on the premises of a customer, including the telephone set itself.

Digital: A means of encoding information using a binary system made up of zeroes and ones. An adjective referring to information in a discrete, rather than continuous (analog), form.

Divestiture: As a formal legal term, the transfer or disposal of assets by a corporation. In this volume, it refers to the 1984 spinning-off by AT&T of its local operating companies (the BOCs) without compensation.

DoD: The federal cabinet-level Department of Defense.

DoJ: The federal cabinet-level Department of Justice.

Domsat: Domestic communications satellite.

Dot.com: Refers to an Internet-based business (*see section 10.3*).

DSL: Digital subscriber line, or a broadband connection to an individual household or office computer or LAN, supplied by the local telephone carrier (ILEC or CLEC), and providing high speed Internet access and "always-on" capacity. *See also* Broadband, Cable modem.

ENFIA: Exchange Network Facilities for Interstate Access, an AT&T discount tariff created in 1978 in response to MCI's Execunet service (*see section 6.2*).

ESP: Enhanced service provider.

Exchange: Geographic area served by a local telephone company (long distance service then is *interexchange*). *See also* Local access and transport area.

Execunet: MCI discount MTS service first offered in 1974 and subject of considerable legal wrangling (*see section 6.2*).

FCC: Federal Communications Commission, the federal telecommunications regulatory agency established by Congress in 1934. *See also* Communications Act of 1934.

Fiber/Fibre: Refers to optical fiber cables (*see chapter 8*).

Final judgment: Official opinion of a court which ends litigation; sometimes referred to as *final decision*. *See also* Consent decree.

Footprint: Geographical area covered by a satellite signal or, less often, another telecommunications service.

FX: Foreign exchange telephone service (*see section 5.5*).

GSM: General System for Mobile, one of the technical standards for digital cellular service which is the standard across Europe and one of those used in the United States (*see section 8.2*).

IEEE: Institute of Electrical and Electronic Engineers, the primary membership organization for electrical engineers.

ILEC: Incumbent local exchange carrier—either an RBOC local exchange company or an independent local company—operating at the time of the 1996 Telecommunications Act.

Independent telephone company: Any of the approximately 1,400 telephone companies—most quite small—not owned by AT&T (to 1984) or the RBOCs (since 1984). They provide local service to geographic areas (roughly half the country, but only about 20 percent of the population) not served by the Bell Operating Companies.

Information services: Generation of data (including electronic publishing) provided by the interaction between telecommunications and computers.

Interconnection: In this book, usually means combining of competing local and interexchange carriers into a single network. The conditions of interconnection by directly competing firms is a controversial regulatory question (*see, e.g., section 9.5*). *See also* Access.

Interexchange carrier or company (IXC): Firm providing interexchange (or long distance) service. After 1984, this refers specifically to inter-LATA service. Examples include AT&T and Sprint.

Internet: A worldwide system of interconnected digital computer networks that use multiple protocols and evolved from the ARPANET (*see section 8.5*). The World Wide Web is a collection of interactive documents accessible via the Internet.

ISP: Internet (sometimes information) service provider (*see section 8.5*).

IXC: *See* Interexchange carrier.

Jurisdiction: Legal concept that determines the forum in which an issue will be heard. The jurisdictional question in the AT&T breakup of which body should resolve the complaints against AT&T—the Federal Communications Commission or a federal court—was hotly contested, as the two operate under different rules with different enforcement powers (*see section 6.1*).

Last mile: Wire or wireless connection to individual household. *See also* Bottleneck.

LAN: Local area (computer) network.

LATA: *See* Local access and transport area.

LEC: *See* Local exchange carrier.

Line of business (LOB): Under the Modification of Final Judgment, any business other than exchange telephony in which the Regional Holding Companies wished to operate from 1984 to 1996 (*see sections 6.5 and 7.1*).

Local access and transport area (LATA): Geographic area in which local exchange and some toll telephone service was provided after January 1, 1984. Typically, only one Bell Operating Company would operate in a given LATA, but a LATA may contain one or more independent telephone companies (*see section 6.4*).

Local exchange carrier (LEC): Local telephone company (after 1984, one which provided intra-LATA service).

Local loop: Physical connection between a telephone company's local office and a customer's premises.

Lucent Technologies: Once the manufacturing arm of AT&T, spun off by trivestiture in 1996 (*see sections 7.2 and 10.3*).

MCI: Originally Microwave Communications Inc., later simply MCI Inc.—an interexchange carrier. Taken over by WorldCom in 1998; merged and scandal-driven firm renamed MCI in 2003, and announced as a Verizon takeover in 2005.

Modification of Final Judgment (also **Modified Final Judgment; MFJ**): The legal settlement which ended the litigation stage of the AT&T divestiture (*see section 6.4*). The MFJ was termed a "modification" because it superseded the final judgment in the 1956 antitrust case. (Note: both terms are used interchangeably, with the shorter term appearing more often.)

MTS: Message telephone service, or regular toll long distance service (after 1984, inter-LATA service). Can also mean *mobile* telephone service (cellular).

Multiple system operator (MSO): Owner and operator of two or more local cable television systems.

NAP: Network access point.

Narrowband: Transmission facility (wire or radio) designed to carry one type of signal, typically voice. *See also* Broadband.

NARUC: National Association of Regulatory Utility Commissioners, the membership organization (almost a "trade" group) of state public utility commissions. *See also* PUC.

NCR: The computer manufacturing arm of AT&T, taken over in 1991 and spun off in the trivestiture at the end of 1996. The initials used to stand for National Cash Register, the Dayton, Ohio, company's original product focus (*see section 8.4*).

NTIA: National Telecommunications and Information Administration, part of the Department of Commerce, and created in 1978. NTIA develops and expresses mid-term and long-range telecommunication policy for the executive branch, and licenses federal agency use of the spectrum.

Open Network Architecture (ONA): Breaking down bundled network service offerings by ILECs to allow information service providers access only to the specific services and facilities (basic service elements) they need to provide information service (*see section 7.4*).

Open Video Service (OVS): Created by the Telecommunications Act of 1996, this is the FCC term for telephone company-supplied cable television services offered on a common carrier basis. None has been offered (*see section 9.2*).

Other common carrier (OCC): Now-obsolete term for a long distance firm other than AT&T. *See also* Specialized common carrier.

PANS: "Pretty awesome new stuff [or service]" (*slang*). *See also* POTS.

PCS: Personal communication service or system, one name applied to second generation (2G) or digital services.

POTS: "Plain old telephone service" (*slang*). *See also* PANS.

Price cap: A system of incentive rate regulation used by both the FCC and state public utility commissions (*see section 7.3 and appendix A.1*).

PTT: Post, telegraph, and telephone—usually referring to government-owned and -operated postal and telecommunications facilities and their regulation (*see section 7.5*).

Public Switched Telephone [or Telecommunications] Network (PSTN): Backbone national telephone network, consisting of twisted-copper pair lines, microwave links, coaxial cable, fiber optic cable, and domsat links, cooperatively operated by the major telephone carriers, both national and regional.

PUC: Public Utility Commission (though their names vary), which is a state-level regulator of telecommunications (and other things such as power and transport) within the borders of that state. There are PUCs in every state, Puerto Rico, and the District of Columbia (Washington). *See also* NARUC.

Rate-of-Return Regulation: Abbreviated RoR or sometimes RoRR, this is fully described in appendix A.1 (*see also section 4.2*).

Ratemaking: The regulatory process of determining fair and equitable rates for telecommunications service (*see section 3.3*).

RBOC: Regional Bell Operating Company; *see* Regional Holding Company.

Regional Holding Company (RHC): Any of seven entities created by AT&T during divestiture planning to "hold" local operating companies and provide central management and support services. Began independent operation January 1, 1984, and lasted until late 1990s when mergers reduced them to four. Also, and more commonly called Regional Bell Operating Companies (RBOCs) (*see sections 6.5, 7.1, and 9.6*).

Resale Process of bulk purchasing of excess capacity from an existing carrier in order to resell it at a discount to end users. Or, the carrier with the excess capacity may resell that capacity itself.

Roaming: Using a cellphone away from the home calling area.

SCC: *See* Specialized common carrier.

Separations: Determining whether costs and revenues are to be attributed to local or long distance service (*see section 4.2*).

Sherman Act: The basic U.S. antitrust legislation of 1890, which outlawed monopolies. Section 2 provided the legal basis on which the divestiture case rested. *See also* Tunney Act.

SLC: Subscriber line charge. *See also* Access charge.

Solid state: Refers to electrical components where all the elements are fixed in a silicon chip (*see section 5.1*).

Specialized common carrier (SCC): Communications firm established to provide specially designed services, usually for business customers (*see section 5.4*). *See also* Other common carrier.

Stay: Court decision to temporarily or permanently suspend implementation of an administrative agency decision or action.

Tariff: Legal document filed (by a telephone carrier with either the FCC or state PUC) to indicate the conditions and price of specific services.

TCP/IP: A computer communications protocol, that is the protocol for the basis of the Internet.

TDMA: Time Domain Multiple Access, a transmission standard used by some carriers in the United States for 2G digital cellular services (*see section 8.2*).

Telco: Shorthand or slang term for telephone company.

TELPAK: AT&T discount tariff from 1961 to 1981 (*see section 5.5*).

TELRIC: Total element long-run incremental cost, a term developed by the FCC to describe how states should determine financial elements of agreements between incumbent LECs and prospective competitors in its August 1996 Report and Order implementing local competition provisions of the 1996 Telecommunications Act (*see section 9.5 and appendix A*).

Trivestiture: AT&T term to describe its voluntary breakup into three separate companies in 1995–96 (*see section 7.2*).

Tunney Act: Formally known as the Antitrust Procedures and Penalties Act of 1974, a federal law compelling a judge to hold hearings before accepting a consent decree in an antitrust case brought by the government (*see section 6.4*).

2G: Second generation (digital) mobile service.

3G: Third generation (digital with graphics and Internet capability) mobile service.

Unbundled: An economic term meaning to devolve something to its most simple elements. An "unbundled tariff" is one where each element needed to provide that service is priced separately.

UNE: Unbundled network element (*see section 9.5*).

Universal service: Term first used by AT&T's Vail in 1910 to mean one company (AT&T) providing end-to-end service for everyone. Though not appearing in the Communications Act of 1934, the term came to mean making analog voice telephony access available to everyone. As some cannot afford even basic local service, subsidies are provided to encourage everyone to subscribe, making the telephone system more "universal" (*see section 9.4*).

VOD: Video on demand; cable or satellite service allowing user to order up specific material, usually at extra cost (*see section 8.4*).

VoIP: Voice over Internet Protocol, the technical standard used to allow voice telephony by means of Internet links (*see section 8.5*).

Waiver: In the AT&T divestiture case and its aftermath, a petition filed by either a Regional Holding Company or a Bell Operating Company for exemption from one or more of the line-of-business restrictions of the MFJ (*see section 7.1*).

WAN: Wide area network.

Wide Area Telephone Service (WATS): Discount long distance service sold in bulk; may be either "inward WATS" (the party called pays, also known as 1-800 service) or "outward WATS" where the party calling pays.

World Wide Web: Millions of documents (and illustrations) accessible through the Internet. *See also* Internet.

WTO: World Trade Organization, an outgrowth of the General Agreement on Tariffs and Trade (*see section 7.5*).

Chronology

This brief chronology is designed to provide a time-ordered reference to some of the events mentioned in the text. It is by no means complete, but rather points out highlights. For a fuller record, see the sources listed at the end.

Chapter 2: Telegraph to Telephone (to 1893)

1844	(May)	Inception of regular telegraph service in the United States
1847	(Apr)	Post Office leases (and later sells) telegraph system to private interests, spelling the end of government control
1856		Formation of Western Union
1861	(Oct)	First transcontinental telegraph service opens
1866	(Jun)	Expansion of Western Union by merger of several other companies; moves headquarters to New York City from Rochester, NY
1869	(Jan)	Formation of predecessor to Western Electric
1872	(Mar)	Western Electric Manufacturing Co. formed
1876	(Mar)	Bell received key patent on telephone
	(May)	Bell publicly demonstrates his telephone instrument
	(Fall)	Western Union turns down offer of Bell patent for $100,000

1878	(Jan)	First telephone switchboard inaugurated, in New Haven, CT
	(Feb)	New Haven issues first telephone directory (21 listings)
	(May)	Theodore Vail becomes General Manager of Bell Telephone Co.
1879	(Fall)	Inception of use of telephone numbers (Lowell, MA)
1881	(Nov)	Predecessor to AT&T purchases Western Electric, becoming sole supplier of Bell telephones early the next year
1885	(Feb)	AT&T created as New York-based subsidiary of American Bell
1887	(Feb)	Creation of Interstate Commerce Commission
1888	(Mar)	Supreme Court settles various appeals of telephone patents, upholding Bell's claims
1889	(Spring)	First coin-operated pay telephones (Hartford, CT)
1890	(Jul)	Congress passed Sherman Antitrust Act
1892	(Oct)	New York–Chicago long distance service opened (extended to Boston in February 1893)

Chapter 3: Era of Competition (1893–1921)

1893	(Mar)	Basic Bell patents begin to expire—anyone can now make and sell telephone equipment and service
1896	(Aug)	Initial use of dial telephone (Milwaukee, WI)
1899	(Dec)	AT&T becomes the parent company while continuing as the long distance operating company
1906	(Apr)	First classified advertising directory ("yellow pages") appears in Michigan
1907	(Apr)	Theodore Vail returns as president of AT&T for second time (until 1919)
1907		First two state public utility commissions (New York and Wisconsin) formed
1908		Initial AT&T use of "Bell System" term in its national advertising
1910	(Jun)	Mann–Elkins Act makes ICC responsible for interstate telephone rate regulation
	(Dec)	AT&T acquires controlling interest in Western Union
1913	(Dec)	Kingsbury Commitment settles threatened antitrust action and limits AT&T acquisitions of independent

telephone companies and requires divestiture of Western Union

1914 (Oct) Congress passes Clayton Antitrust Act, which permits private suits

1915 (Jan) First transcontinental telephone service opened

1918 (Jul) Post Office takes over control of AT&T for as part of country's reaction to World War I (continues until August 1919)

1919 (Jun) Vail retires as head of AT&T to be succeeded by Harry Thayer

1920 (Feb) Transportation Act requires ICC to develop and maintain adequate telephone service in the country

1921 (Jun) Willis–Graham Act amends Transportation Act to permit consolidation and mergers among telephone carriers, terminating Kingsbury Commitment

Chapter 4: Regulated Monopoly (1921–56)

1922 (Jul) AT&T inaugurates a New York City radio broadcasting station (operated until 1926), which experiments with programs, early advertising, and networking

(Aug) Death of Alexander Graham Bell

1925 (Jan) Creation of Bell Telephone Laboratories

1930 (Dec) In *Smith v. Illinois Bell*, Supreme Court holds long distance fees must contribute to cost of local loop

1934 (Jul) Federal Communications Commission established by the Communications Act

1936 (Mar) FCC begins telephone investigation mandated by Congress

1938 (Apr) FCC issues preliminary report of its telephone investigation with a host of recommendations

1939 (Jun) Final FCC telephone investigation report published as Congressional document

1946 (May) Initial Bell System commercial microwave services begin (in southern California and Massachusetts)

1947 (Apr) First national telephone strike lasts 44 days

(Nov) First national coaxial cable opens between Atlanta and Los Angeles

(Dec) Bell Laboratories scientists develop transistor (first demonstrated publicly in mid-1948)

1949 (Jan) Justice Dept. brings antitrust suit against AT&T

1951	(Sep)	Coast-to-coast television made possible by AT&T-supplied microwave and cable links
1956	(Jan)	Consent Decree ends 1949 antitrust case
	(Sep)	First trans-Atlantic telephone submarine cable (TAT-1) begins service

Chapter 5: Competition Reappears (1956–74)

1956	(Nov)	*Hush-a-Phone* court decision begins deregulation of terminal equipment
1958	(Mar)	House Antitrust Subcommittee begins extensive hearings into 1956 Consent Decree
1959	(Jan)	House Antitrust Subcommittee issues scathing report on 1956 Consent Decree
	(Aug)	*Above 890* decision begins FCC deregulation of telecommunication services
		Invention of integrated circuit ("chip") by Kilby and Noyce
1961	(Jan)	AT&T institutes "Telpak" tariff (terminated in 1981) and WATS (wide area telephone service) service
1963	(Oct)	MCI founded by Jack Goeken
1965		FCC "Seven-Way Cost Study" of AT&T rates
1966	(Nov)	FCC begins its first Computer Inquiry
1968	(Jun)	FCC *Carterfone* decision expands *Hush-a-Phone* precedent
	(Aug)	William McGowan takes over leadership of MCI, reorganizes firm and its plans
	(Sep)	AT&T files tariffs permitting connection of customer equipment through use of a company-provided protective connecting arrangement (PCA)
1969	(Aug)	FCC *MCI* decision authorizes inception of company service
1970	(Jan)	FCC bans co-located telephone-cable cross-ownership
	(Sep)	Inception of Office of Telecommunications Policy in Executive Office of the President
1971	(Mar)	FCC concludes Computer Inquiry I
	(Jun)	FCC authorizes specialized common carriers (SCCs)
1972	(Jan)	MCI begins service on Chicago–St. Louis original line
	(Apr)	John deButts becomes chairman of AT&T

	(Jun)	FCC "open skies" domestic satellite policy issued
1973	(Sep)	DeButts' speech before NARUC convention shows AT&T is ready to fight further FCC deregulatory moves
	(Nov)	Justice Dept. issues Civil Investigative Demand against AT&T

Chapter 6: Breaking Up Bell (1974–84)

	(Mar)	MCI files private antitrust suit against AT&T
1974	(Apr)	*Westar I*, first American domsat, placed into orbit by NASA for Western Union
1974	(Nov)	Justice Dept. brings suit to break up AT&T
1975	(Feb)	AT&T formally replies to government's antitrust case
	(May)	AT&T complains to FCC about MCI Execunet service
	(Jul)	MCI Execunet service tariff rejected by FCC
	(Oct)	FCC institutes equipment certification system in place of telephone PCAs
1976	(Mar)	"Bell Bill" sparks Congressional interest in possible legislation
	(Aug)	FCC begins Computer Inquiry II
1977	(May)	Inception of attempts to "rewrite" Communications Act begin (which last to 1981)
1978	(Jun)	Judge Harold Greene is assigned the case on his first day on the bench
	(Oct)	Jurisdictional issues in AT&T case resolved
1979	(Feb)	Charles L. Brown succeeds John deButts as AT&T chairman
1980	(Apr)	FCC concludes Computer Inquiry II and detariffs CPE
	(Jun)	Initial judgment in MCI case against AT&T provides for $1.8 billion award
1981	(Jan)	*U.S. v. AT&T* trial begins
	(Feb)	William Baxter nominated as head of Antitrust Division of Department of Justice
	(Sep)	Greene rejects AT&T attempt to have case dismissed
1982	(Jan)	Provisional settlement of the case is announced
	(Feb)	Justice Dept. issues competitive impact statement on proposed MFJ
	(Apr)	Public comments on proposed settlement filed with court
	(Jun)	Two days of oral argument on proposed settlement

	(Jul)	With demise of HR 5158, Congressional attempts to rewrite the 1934 Act end
	(Aug)	Formal MFJ agreed to by Judge Greene, AT&T, and Justice Dept.
	(Oct)	AT&T files LATA plan
	(Dec)	AT&T publishes *Plan of Reorganization*
1983	(Jan)	Appeals court overturns MCI record $1.8 billion award from AT&T
	(Feb)	Washington–New York leg of intercity fiber optic trunk goes into service
		Supreme Court issues summary affirmation of MFJ
	(Apr)	Judge Greene approves LATA plan with minor exceptions
	(Aug)	Judge Greene approves AT&T Plan of Reorganization with modifications
	(Nov)	First formal information on new RBOCs issued; their stocks begin to trade

Chapters 7–8: Operating Under the MFJ and Innovating New Services (1984–96)

1984	(Jan)	Divestiture
		First LOB exception request filed (by BellSouth) with Judge Greene
	(Oct)	Congress regulates cable television for the first time; law includes ban on co-located telephone/cable entities
1985	(May)	MCI antitrust.award chopped down to $37.8 million
	(Jun)	FCC initiates subscriber line ("access") charge
	(Aug)	FCC initiates Computer Inquiry III
1986	(May)	Initial FCC decision on Computer III puts forth Open Network Architecture
	(Jun)	James Olson replaces Charles Brown as AT&T's CEO
	(Jul)	Sprint long distance service begun, soon uses first coast-to-coast fiber-optic network
1987		Triennial Review of MFJ events:
	(Jan)	Huber's *The Geodesic Network* report released by Justice Dept.
	(Feb)	Justice recommends lifting LOB restrictions on RBOCs
	(Sep)	Judge Greene rejects Justice Dept. recommendations
1988	(Mar)	Judge Greene issues decision refining information services

	(Dec)	Initial trans-Atlantic fiber optic cable in service (TAT-8)
1989	(Mar)	FCC inaugurates "price cap" regulation of long distance carriers
	(Apr)	First trans-Pacific optical fiber cable (United States to Japan via Hawaii) enters service
1990		FCC extends price cap regime to RBOC interstate services
1991	(Apr)	Judge Greene's information service decision is overturned
	(Jul)	AT&T leases its signature New York headquarters building to Sony
		Judge Greene allows RBOCs to provide content for information services
	(Sep)	AT&T hostile takeover of NCR for $7.5 billion
	(Oct)	Appeals Court overturns Judge Greene on information services
	(Dec)	AT&T stops providing telegraph service
1992	(Jul)	FCC "video dial tone" order issued
	(Oct)	Congress re-regulates cable industry rates
	(Nov)	AT&T announces $4 billion purchase of a third of McCaw Cellular
1993		British Telecom (BT) purchases 20 percent stake in MCI for $5.3 billion
	(Aug)	Federal District Court finds ban on telephone/cable cross-ownership to be violation of First Amendment—first of several similar cases
		AT&T announces takeover of McCaw Cellular in $12.6 billion deal
	(Sep)	Clinton Administration announces National Information Infrastructure (NII) initiative
	(Oct)	Bell Atlantic announced takeover of TCI, largest cable system operator, in $33 billion deal
1994	(Feb)	Bell Atlantic/TCI deal called off over management disagreements, FCC rules
	(Apr)	Judge Greene blocks AT&T/McCaw deal as violating MFJ
	(Jun)	Bell Atlantic and NYNEX announce plans to merge
	(Aug)	Greene approves AT&T/McCaw Cellular Consent Decree
	(Sep)	FCC approves AT&T takeover of McCaw Cellular
1995	(Sep)	AT&T announces trivestiture

1996 (Jan) AT&T announces pending layoffs of up to 40,000 workers

Chapters 9–10: Telecommunications Act and Meltdown (1996–2004)

1996 (Feb) Telecommunications Act of 1996 passed
Network Systems renamed Lucent Technologies

(Mar) First RBOC merger announced: SBC to take over Pacific Telesis

(Apr) Second RBOC merger announced: Bell Atlantic to merge with NYNEX
Judge Greene formally terminates MFJ

(Jun) Bell Atlantic restructures deal—will take over NYNEX

(Aug) FCC issues massive interconnection decision as required by 1996 Act; key provisions are stayed by Federal Appeals Court in October

(Oct) BT announces $22 billion takeover of MCI
As part of AT&T trivestiture, Lucent Technologies becomes independent (including three-quarters of Bell Labs)

(Nov) FCC/PUC Joint Board issues Universal Service report required by 1996 Act
RBOCs sell Bellcore to Science Applications International

(Dec) NCR becomes independent of AT&T

1997 (Mar) New York PSC approves Bell-Atlantic-NYNEX merger, with conditions

(May) FCC issues final universal service report and order
FTC approves Bell Atlantic-NYNEX merger without conditions

(Jun) AT&T and SBC discuss merger but call it off in face of fierce opposition and other disagreements

(Jul) SBC sues to overturn Telecommunication Act's local entry provisions
Court of Appeals, as expected, rejects core of FCC interconnection decision
MCI financial troubles place its takeover by BT in some doubt

(Aug) FCC approves Bell Atlantic/NYNEX merger

1998 MCI–WorldCom merger completed

	(May)	United States and 20 states file antitrust suit against Microsoft
		Under new leadership (M. Armstrong), AT&T embarks on takeover of large cable companies, beginning with TCI
	(Sep)	Iridium LEO satellite system begins partial service
		Microsoft antitrust trial
1999	(Jan)	Supreme Court upholds most provisions of 1996 Telecommunications Act
		Quest takes over US West and becomes Qwest
		Bell Atlantic completes merger with GTE and becomes Verizon
	(Nov)	Federal District Court finds Microsoft is an illegal monopoly
2000	(Mar)	Internet and telecommunication stocks begin to drop on NYSE, NASDAQ, initiating a dramatic industry meltdown
	(Mar)	Iridium LEO satellite service terminates service in bankruptcy
	(Jun)	District Court orders Microsoft broken up after antitrust trial
	(Jul)	Comsat is taken over by Lockheed Martin
		AT&T announces it will divide into four separate firms
		Lucent spins off business systems division to become Avaya Communications
	(Dec)	SBC receives FCC permission to provide long distance service in Kansas and Oklahoma—one of the first such grants
2001	(Jan)	FCC auctions 3G mobile service spectrum for nearly $17 billion
		Government settles Microsoft antitrust case before final decision is reached by the court
	(May)	French Alcatel calls off intended merger with/takeover of Lucent
	(Jul)	Comcast announces takeover of AT&T Broadband, creating largest cable multiple system operator, and trimming AT&T even further
	(Sep)	Terrorist attacks in New York and Washington show fragility of mobile telephone networks and also need for substantial improvement in government emergency communications

Lockheed Martin dismembers remains of Comsat, exiting telecommunications business

Lengthy congressional debate over Tauzin–Dingell bill to remove restrictions on RBOCs to encourage them to rapidly expand broadband networks

2002 (Jan) Global Crossing declares bankruptcy

Lucent spins off microelectronics components business as Agere Systems

Only about 70 facilities-based CLECs still in operation as RBOCs retain more than 90 percent of customers

(Jun) WorldCom announces massive financial fraud and collapses in bankruptcy a month later

2003 (Aug) FCC issues long-awaited Triennial Review order

(Dec) Qwest in Arizona is the last of the RBOCs to receive Section 271 authority; all RBOCs in all states are now cleared to offer long distance

2004 Cingular purchases control of AT&T Wireless, beginning consolidation of mobile markets; deal is finalized late in the year

(Mar) Court of Appeals finds most of FCC Triennial Review order is unlawful

(Jul) AT&T withdraws active role in local telephone residential service; will continue to service its existing customers

(Jul) FCC initiates policy review on its potential role in VoIP

(Dec) Merger of Sprint and Nextel cellular mobile carriers announced

Sources (see References for full citations):

AT&T, *Events in Telecommunications History* (1992)

Federal Communications Commission, *FCC Log* (1984)

Fifty Years of Telecommunications Reports (1985)

"Telephone Calendar," *Telephony* (July 5, 1976), pp. 226–262.

Von Auw, *Heritage and Destiny* (1983), Appendix A "The Bell System in Transition: A Selective Chronology, 1965–1982, pp. 407–421.

Press reports (as cited in chapter notes).

Historical Statistics

The material in the following tables is intended to underpin the previous chapters. The data offered are of a summary nature, and over as long a period as the information is available. While there is a great deal of industry financial data available from the Federal Communications Commission's extensive Web site, we focus here on indicators of telephone, cellphone, and computer availability and use.

Table D.1 These very basic data are offered to provide some context for both the tables that follow and the text discussions of service availability and cost. Based on official U.S. Census Bureau data, most of them from the decennial census database, we see here the growth of U.S. population and the number of households. The U.S. Bureau of Labor Statistics provides the Consumer Price Index (which goes back only to 1913). This indicates the impact of the Depression (note how the CPI declines in 1940 from data of 1930), slow inflation (the 1950s), and far more rapid inflation (the 1980s).

Table D.2 Here is more than a century of data on the expansion of American telephone service. The first two columns provide two different standard measures of the availability of telephone service: the number of telephone instruments available per 100 people (this is for the nation as a whole, and would show strong differences between urban and rural areas, especially before about 1950), and the proportion of households with a telephone. Here again, note the impact of the Depression, where the number of households with a telephone actually declined—the only de-

TABLE D.1
Basic Historical Indicators: 1850–2003

Year	U.S. Population (Millions)	U.S. Households (Millions)	Consumer Price Index (1982–84 = 100)
1850	23.2	N/a	N/a
1880	50.2	N/a	N/a
1890	62.9	12.7	N/a
1900	76.0	16.0	N/a
1910	92.0	N/a	N/a
1920	106.5	24.5	20.0
1930	123.5	30.0	16.7
1940	132.1	35.2	14.0
1950	151.7	43.6	24.1
1960	180.7	52.8	29.6
1970	204.9	62.9	38.8
1980	227.7	80.8	82.4
1990	250.1	93.3	130.7
2000	281.4	103.9 (1999)	172.2
2003	291.0	N/a	184.7

Source: Population and households from U.S. Census Bureau; Consumer Price Index from U.S. Bureau of Labor Statistics.

TABLE D.2
Indicators of U.S. Telephone Service Expansion: 1880–1980

Year	Telephones per 100 People	Households With Telephone	Bell System Miles of Wire (000)	Bell System Employees	Independent Companies Miles of Wire (000)	Independent Companies Employees
1880	.1	N/a	34	3,300	N/a	N/a
1890	.4	N/a	240	8,600	N/a	N/a
1900	7.6	N/a	2,000	37,100	N/a	N/a
1910	8.2	N/a	11,600	121,300	N/a	N/a
1920	12.3	35%	25,400	231,300	4,700	N/a
1930	16.3	41	76,200	324,300	4,900	35,700
1940	16.5	37	91,300	282,200	N/a	N/a
1950	28.1	62	144,300	534,800	9,200	63,000
1960	40.8	78	307,900	595,900	28,600	85,000
1970	58.3	91	602,000	793,200	N/a	142,000
1980	N/a	93	N/a	N/a	N/a	N/a

Sources: For data through 1970, Bureau of the Census *Historical Statistics of the United States: Colonial Times to 1970* (Washington, DC: Government Printing Office, 1976), Part 2, tables R1–12, R17–30, T31–45. Updates from Bureau of the Census, *Statistical Abstract of the United States*, annual. "Bell System" data for 1990 and 2000 combines AT&T and the several Regional Bell Operating Companies as well as Lucent Technologies.

cade where that is true. The final four columns offer measures of the Bell System (through 1980 this is the old unified AT&T; for 1990 and 2000, data shown are primarily for the Regional Bell Operating Companies as AT&T dwindled in importance), and the independent (of AT&T) telephone companies. Through 1980 the "miles of wire" was a solid indicator of system size, though since that time, satellite links and fiber optic cables, neither included here, would show growing proportions of total system size. The number of employees is a good indicator of the growing importance of telephony in American life.

Table D.3 Cellphone ownership and usage is covered here for a much shorter period—essentially the past two decades. As is evident here, cellphones only became important in the late 1990s as their cost declined and more people bought them. Not evident here is the sharp distinction between cellphones in urban and rural areas. As usage costs have declined, use of cellphones has increased, though the scattered evidence only hints at that. As with wireline telephony in earlier years, note how the industry has grown in employees by nearly 100-fold in the period shown here.

TABLE D.3
Mobile Telephone Usage: 1985–2003

Year	Total Subscribers (Millions)	Proportion of Total Population (%)	Cell-Sites	Minutes of Use per Month	Industry Employees (000)
1985	.3	N/a	900	N/a	2.7
1990	5.3	.2	5,600	N/a	21.4
1995	33.8	12.8	22,600	119	68.2
2000	109.5	38.8	104,300	255	184.5
2003	158.7	54.7	163,000	N/a	205.6

Source: FCC Industry Analysis and Technology Division, Wireless Competition Bureau, citing Cellular Telecommunications and Internet Association (CTIA). Accessed April 2005 from http://www.fcc.gov/Bureaus/Common_Carrier/Reports/FCC-State_Link/IAD/trend504.pdf. Cell-site numbers rounded to nearest 100. Proportion of total population figured by authors based on population data provided in Table D.1.

Table D.4 "Computers" are varyingly defined—everything from mainframes to laptop PCs. This table includes all computers—only the final two columns are focused on household PCs and Internet access. The proportion of households with a computer, and the number of computers per 100 people are two ways of defining the growing role of computers in daily life.

TABLE D.4
Indicators of U.S. Computer Industry Growth: 1975–2003

Year	Computers in Use (Millions)	Computers per 100 People	Households	
			Owning a Computer	With Internet Access
1975	.2	.09	N/a	–
1980	3.3	1.4	N/a	–
1985	21.5	9.0	N/a	–
1991	62.0	19.2	N/a	–
1995	96	36.5	24.1% (1994)	N/a
2000	160	58.0	51.0	41.5
2003	N/a	N/a	62.0	54.6

Source: Computer Industry Almanac Inc. Press release: http://www.c-i-a.com/pr1296.htm (accessed April 2005). Households with a computer from FCC Industry Analysis and Technology Division, Wireline Competition Bureau, citing NTIA and Census data. For the original chart, see http://www.fcc.gov/Bureaus/Common_Carrier/Reports (accessed April 2005). 2003 data from U.S. Statistical Abstract 2004–5: http://www.census.gov/prod/2004pubs/04statab/infocomm.pdf (accessed April 2005), table 1150, p. 731.

References

Literature about the U.S. telecommunications industry and its involved history and regulation is rapidly expanding after decades when little serious or scholarly material was available. These pages include the most important resources on which this volume was based and that are useful for further research.

The list appears in two parts: primary sources (chiefly Bell System and U.S. government documents), and secondary resources (books and articles). Most periodical references occur only in the chapter notes, and are not repeated here. Excluded is most biographic and technical history; the focus is on industry, economic, and policy studies. For technical history, see Christopher H. Sterling and George Shiers (2000), *History of Telecommunications Technology: An Annotated Bibliography* (Lanham, MD: Scarecrow Press).

Primary Sources

(U.S. government publications are published in Washington, DC, by the Government Printing Office unless otherwise indicated. Within a given organization, items are listed chronologically.)

American Telephone and Telegraph Co. (All are published by AT&T in New York)
    Annual Report (1880–present).
    Bell System Statistical Manual (1920–81; annual).
    Broadcasting Network Service (1934). Long Lines Department.
    Brief of the Bell System Companies on Commissioner Walker's Proposed Report on the Telephone Investigation (1938).

Depreciation: History and Concepts in the Bell System (1957).

Our Company and How It Operates, various editions.

The Bell System's Role in the Development of Nationwide Network Television (1967).

Defendants' Third Statement of Contentions and Proof in Civil Action No. 74-1698 U.S. v. AT&T (1980). Washington, March 10 (3 vols.). Referenced as "Gold Book" in notes.

_____, Historical Archives and Publications Group (1992). *Events in Telecommunications History.*

_____, Information Department (1914). *Telephone Memoranda: A Few Facts and Opinions Foreign and Domestic.*

Bell, Alexander Graham (1908). *The Bell Telephone: The Deposition of Alexander Graham Bell in the Suit Brought by the United States to Annul the Bell Patents.* Boston: American Bell Telephone Co. (reprinted by Arno Press, 1974).

Bell Telephone Laboratories.

A History of Science and Technology in the Bell System (1975–85). Seven vols.

Bell Laboratories Innovations in Telecommunications: 1925–1977 (1979).

Engineering and Operations in the Bell System (1983).

Iardella, Albert B., ed. (1964). *Western Electric and The Bell System: A Survey of Service.* New York: Western Electric.

Langdon, William Chauncy (1935). *The Early Corporate Development of the Telephone.* New York: AT&T.

U.S. Congress.

Sherman Antitrust Act, 15 USC sec. 1 *et seq.* (July 2, 1890).

Communications Act of 1934, Public Law 416, 73rd Cong. (June 18, 1934).

Telecommunications Act of 1996, Public Law 104-104, 104th Cong. (February 8, 1996).

_____. Congressional Budget Office (1984). *The Changing Telephone Industry: Access Charges, Universal Service, and Local Rates.*

_____. General Accounting Office.

Telephone Communications: Bell Operating Company Entry Into New Lines of Business (1986). GAO/RCED-86-138, April.

Telephone Communications: Bypass of the Local Telephone Companies (1986). GAO/RCED-86-66, August.

Telephone Communications: Issues Affecting Rural Telephone Service (1987). GAO/RCED-87-74, March.

Telephone Communications: Controlling Cross-subsidy Between Regulated and Competitive Services (1987). GAO/RCED-88-34, October.

U.S. Communications Policy: Issues for the 1990s (1991). GAO/IMTEC-91-52B, September.

Telecommunications: Concerns about Competition in the Cellular Telephone Service Industry (1992). GAO/RCED-92-220, July.

Telecommunications: FCC's Oversight Efforts to Control Cross-Subsidization (1993). GAO/RCED-93-34, February.

Information Superhighway: Issues Affecting Development (1994). GAO/RCED-94-285, September.

_____. House of Representatives (1934). *Preliminary Report on Communication Companies.* 73rd Cong., 2nd Sess., House Report No. 1273, three vols. [The "Splawn" Report.]

_____. Committee on Energy and Commerce. Subcommittee on Telecommunications and Finance (committee title has varied).

Options Papers. (1977). 95th Cong., 1st Sess., Committee Print 95-13, May.

Transition in the Long-Distance Telephone Industry: Hearing (1986). 99th Cong., 2nd Sess., February.

Modified Final Judgment: Hearings (1987). 100th Cong., 1st Sess., July and October.

Modified Final Judgment: Hearings (1989). 101st Cong., 1st Sess., May and June, two vols.

Telecommunications Policy Act: Hearings (1990). 101st Cong., 2nd Sess., March–May, two vols.

Competition Policy in the Telecommunications Industry: A Comprehensive Approach: Hearing (1992). 102nd Cong., 1st Sess., August 1991–February, three vols.

National Communications Infrastructure: Hearings (1994). 103rd Cong., 2nd Sess., January–February, three vols.

Communications Act of 1934 as Amended by the Telecommunications Act of 1996: Committee Print (1996). 104th Cong., 2nd Sess., March.

_____. Committee on the Judiciary, Subcommittee on Antitrust.

Consent Decree Program of the Department of Justice; Part II–American Telephone & Telegraph Co.: Hearings (1958). 85th Cong., 2nd Sess., March–May, three vols.

Report of the Antitrust Subcommittee . . . On Consent Decree Program of the Department of Justice (1989). 86th Cong., 1st Sess., January 30.

AT&T Consent Decree: Hearings (1989). 102nd Cong., 1st Sess., August.

Telecommunications: The Role of The Department of Justice: Hearing (1995). 104th Cong., 2nd Sess., May.

Antitrust Issues in Telecommunications Legislation: Hearing (1996). 104th Cong., 2nd Sess., May.

_____. Select Committee on Small Business (1966). *Congress and the Monopoly Problem: History of Congressional Action in the Antitrust Field 1890–1965, Seventy-five Years—Committee Print.* 89th Cong., 2nd Sess.

_____. Library of Congress, Congressional Research Service (1984). *The American Telephone & Telegraph Company Divestiture: Background, Provisions and Restructuring* by Angela A. Gilroy.

_____. Office of Technology Assessment (1990). *Critical Connections: Communications for the Future.*

_____. U.S. Senate. *Study of Communications by an Interdepartmental Committee.* 73rd Cong., 2nd Sess., Senate Committee Print (1934) (reprinted in John M. Kittross, ed., *Administration of American Telecommunications Policy*, Vol. 2, 1980. New York: Arno Press).

_____. Committee on Commerce, Science, and Transportation.

AT&T Proposed Settlement: Hearings (1982). 97th Cong., 1st sess., January–March, two parts.

Long-distance Competition: Hearings (1985). 99th Cong., 2nd Sess., September. *Federal Telecommunications Policy Act of 1986: Hearings* (1986). 100th Cong., 1st Sess., September.

_____. Committee on the Judiciary.

DOJ Oversight: U.S. V. AT&T: Hearings (1981–82). 97th Cong., 1st sess., August, January and March, two vols.

Proposed Modifications to The AT&T Decree: Hearing (1987). 100th Cong., 1st Sess., April.

The AT&T Consent Decree's Manufacturing Restriction: Hearing (1991). 102nd Cong., 1st Sess., May.

S.1822: the Communications Act of 1994: Hearings (1994). 103rd Cong., 2nd Sess.

U.S. Department of Commerce.

_____. Bureau of the Census.

Census of Electrical Industries: Telephones and Telegraphs (1906–39; published every 5 years; title varied).

Annual Survey of Communication Services (1990–present, annual).

_____. National Telecommunications and Information Administration.

Issues in Domestic Telecommunications: Directions For National Policy (1985). Special Publication 85-16.

NTIA Competition Benefits Report (1985). Special Publication 85-17.

NTIA Trade Report: Assessing the Effects of Changing the AT&T Consent Decree (1987).

NTIA Telecom 2000: Charting the Course for a New Century (1988). Special Publication 88-21.

The Bell Company Manufacturing Restriction and the Provision of Information Services (March 1989).

The NTIA Infrastructure Report: Telecommunications in the Age of Information (1991). Special Publication 91-26.

U.S. Department of Justice, Antitrust Division.

Complaint, United States of America v. American Telephone and Telegraph Co., et al. (1974). Civil Action 74-1698, U.S. District Court for the District of Columbia.

Plaintiff's Third Statement of Contentions and Proof in Civil Action No. 74-1698, U.S. v. AT&T. 1980. Referenced as "Green Book" in notes. Two vols.

[Huber, Peter W.] (1987). *The Geodesic Network: 1987 Report on Competition in the Telephone Industry.*

U.S. Federal Communications Commission.

Annual Report (1935–97) (1936–56 issues reprinted by Arno Press, 1971).

Statistics of Communications Common Carriers. 1937–present, annual (title has varied).

Investigation of the Telephone Industry in the United States. 76th Cong., 1st Sess., House Document 340, 1939 (reprinted by Arno Press, 1974).

_____. Common Carrier Bureau.

Preliminary Analysis of the 1982 AT&T Consent Decree. January 10, 1982.

Access Charge Proceeding: Background and Briefing Material, September 1982.

_____. Office of Plans and Policy.

John Haring (1985). *The FCC, the OCCs and the Exploitation of Affection.* OPP Working Paper 17.

Kelley, Daniel (1982). *Deregulation After Divestiture: The Effect of the AT&T Settlement on Competition.* OPP Working Paper 8.

Pepper, Robert M. (1988). *Through the Looking Glass: Integrated Broadband Networks, Regulatory Policies and Institutional Change.* OPP Working Paper 24.

Setzer, Florence O. (1984). *Divestiture and the Separate Subsidiary Requirement.* OPP Working Paper 11.

Werbach, Kevin (1997). *Digital Tornado: The Internet and Telecommunications Policy.* OPP Working Paper 28.

U.S. Information Infrastructure Task Force.

The National Information Infrastructure: Agenda For Action (1993). September 15.

National Information Infrastructure: Progress Report, September 1993–1994 (1994). September.

Global Information Infrastructure: Agenda for Cooperation (1995). February.

Common Ground: Fundamental Principles for the National Information Infrastructure—First Report of the National Information Infrastructure Advisory Council (1995, March).

U.S. Post Office, Postmaster General (1914). *Government Ownership of Electrical Means of Communication.* 63rd Cong., 2nd Sess., Senate Document No. 399.

U.S. Supreme Court:

The Telephone Cases: Cases Adjudged in the Supreme Court at October Term, 1887. New York: Banks & Bros "United States Supreme Court Reports, Vol. 126," 1888.

Munn v. Illinois, 94 US 113 (1877).

Smith v. Illinois, 282 US 133 (1930).

Nebbia v. New York, 291 US 502 (1934).

United States v. Pullman Co., 330 US 806 (1947).

Louisiana Public Service Commission v. FCC, 476 US 355 (1986).

AT&T Corp. v. Iowa Utilities Board, 525 US 366 (1999).

National Cable & Telecommunications Association v. Gulf Power Co., et al., 534 US 327 (2002).

Verizon Communications Inc., et al. v. Federal Communications Commission, et al., 535 US 467 (2002).

Verizon Maryland Inc. v. Public Service Commission of Maryland, et al., Case 00-1531 (2002)

Secondary Sources

Adams, Stephen B. and Orville R. Butler (1999). *Manufacturing the Future: A History of Western Electric.* New York: Cambridge University Press.

Aitken, Hugh G. J. (1985). *The Continuous Wave: Technology and American Radio, 1900–1932.* Princeton, NJ: Princeton University Press.

_____ (1994). "Allocating the Spectrum: The Origins of Radio Regulation." *Technology and Culture*, 35:686–716 (October).

Alleman, James H. and Richard D. Emmerson, eds. (1989). *Perspectives on the Telephone Industry: The Challenge for the Future.* New York: Harper/Ballinger.

Archer, Gleason L. (1938). *History of Radio to 1926.* New York: American Historical Society (reprinted by Arno Press, 1971).

Asmann, Edwin N. (1980). *The Telegraph and the Telephone: Their Development and Role in the Economic History of the United States—The First Century 1844–1944.* Lake Forest, IL: Lake Forest College Print Shop.

Augarten, Stan (1984). *Bit by Bit: An Illustrated History of Computers.* New York: Ticknor and Fields.

Averch, Harvey and Leland L. Johnson (1962). "The Behavior of the Firm Under Regulatory Constraint." *The American Economic Review*, 52: 1052–1069 (December).

Banning, William Peck (1946). *Commercial Broadcasting Pioneer: The WEAF Experiment, 1922–1926.* Cambridge, MA: Harvard University Press.

Barbash, Jack (1952). *Unions and Telephones: The Story of the Communication Workers of America.* New York: Harper.

Baughcum, Alan and Gerald R. Faulhaber, eds. (1984). *Telecommunications Access and Public Policy.* Norwood, NJ: Ablex.

Baumol, William J. and Gregory J. Sidak (1994). *Competition in Local Telephony.* Cambridge, MA: MIT Press.

Bell, Trudy E. (1985). "The Decision to Divest: Incredible or Inevitable?" *IEEE Spectrum* 2: 46–55 (November).

Bernstein, Jeremy (1984). *Three Degrees Above Zero: Bell Labs in the Information Age.* New York: Scribner's.

Bernt, Phyllis and Martin Weiss (1993). *International Telecommunications* (Indianapolis: Sams).

Blondheim, Menahem (1994). *News Over the Wires: The Telegraph and the Flow of Public Information in America, 1944–1897.* Cambridge, MA: Harvard University Press.

Boettinger, H. M. (1983). *The Telephone Book: Bell, Watson, Vail and American Life.* Second Edition. New York: Stearn.

Bolling, George H. (1983). *AT&T: Aftermath of Antitrust.* Washington, DC: National Defense University Press.

Bolter, Walter G., ed. (1984). *The Transition to Competition: Telecommunications Policy for the 1980s.* Englewood Cliffs, NJ: Prentice Hall.

_____ et al., eds. (1990). *Telecommunications Policy for the 1990s and Beyond.* Armonk, NY: Sharpe.

Bonbright, James C., Albert L. Danielsen, and David R. Kamerschen (1988). *Principles of Public Utility Rates*, Second Edition. Arlington, VA: Public Utilities Reports.

Borchardt, Kurt (1970). *Structure and Performance of the U.S. Communications Industry: Government Regulation and Company Planning.* Cambridge, MA: Graduate School of Business Administration, Harvard University.

"Boston to Washington Coaxial Opened" (1947). *Broadcasting* (November 17), p. 100.

Bracken, James K. and Christopher H. Sterling (1995). *Telecommunications Research Resources: An Annotated Guide.* Mahwah, NJ: Lawrence Erlbaum Associates.

Bradley, Stephen P. and Jerry A. Hausman, eds. (1989). *Future Competition in Telecommunications.* Boston: Harvard Business School Press.

Braun, Ernest and Stuart Macdonald (1982). *Revolution in Miniature: The History and Impact of Semiconductor Electronics.* New York: Cambridge University Press.

Bray, John (2002). *Innovation and the Communications Revolution from the Victorian Pioneers to Broadband Internet.* London: IEE.

Brennan, Timothy (1987). "Why Regulated Firms Should Be Kept out of Unregulated Markets: Understanding the Divestiture in *United States* v. *AT&T.*" *The Antitrust Bulletin*, 32: 741–793 (Fall).

_____ (1989). "Divestiture Policy Considerations in an Information Services World." *Telecommunications Policy*, 13: 243–254 (September)

Brenner, Daniel L. (1996). *Law and Regulation of Common Carriers in the Communications Industry*, Second Edition. Boulder, CO: Westview.

Brock, Gerald W. (1981). *The Telecommunications Industry: The Dynamics of Market Structure.* Cambridge, MA: Harvard University Press.

_____ (1994). *Telecommunication Policy for the Information Age: From Monopoly to Competition.* Cambridge, MA: Harvard University Press.

_____ ed. (1995). *Toward a Competitive Telecommunication Industry: Selected Papers From the 1994 Telecommunications Policy Research Conference.* Mahwah, NJ: Lawrence Erlbaum Associates.

_____ and Gregory L. Rosston, eds. (1996). *The Internet and Telecommunications Policy: Selected Papers From the 1995 Telecommunications Policy Research Conference.* Mahwah, NJ: Lawrence Erlbaum Associates.

_____ (2003). *The Second Information Revolution.* Cambridge, MA: Harvard University Press.

Brooks, John (1976). *Telephone: The First Hundred Years.* New York: Harper & Row.

Brooks, Thomas R. (1977). *The Communications Workers of America: The Story of a Union.* New York: Mason/Charter.

Brotman, Stuart N. (1987). *The Telecommunications Deregulation Sourcebook.* Norwood, MA: Artech House.

_____ ed. (1990). *Telephone Company and Cable Television Competition: Key Technical, Economic, Legal, and Policy Issues.* Norwood, MA: Artech House.

Brown, Charles L. (1991). "The Bell System," in Fritz Froehlich et al., eds. *The Froehlich/Kent Encyclopedia of Telecommunications*, Vol. 2. New York: Marcel Dekker.

Bruce, Robert V. (1973). *Alexander Graham Bell and the Conquest of Solitude.* Boston: Little, Brown.

Cantelon, Philip L. (1993). *The History of MCI: The Early Years, 1968–1988.* Dallas, TX: Heritage Press for MCI.

Cason, Herbert N. (1910). *The History of the Telephone.* Chicago: A. C. McClurg.

Ceruzzi, Paul E. (2003). *A History of Modern Computing*, Second Edition. Cambridge: MIT Press, 2003.

Chapius, Robert J. (1982). *100 Years of Telephone Switching (1878–1978).* Amsterdam: North-Holland.

_____ and Amos E. Joel (1990). *Electronics, Computers and Telephone Switching: A Book of Technological History.* Amsterdam: North-Holland.

Charalambides, Leonidas C., et al. (1990). "Balancing the Ex-Bell System Companies." *Telecommunications Policy*, 14: 52–63 (February).

Clarke, Arthur C. (1945). "Extra-Terrestrial Relays: Can Rocket Radio Stations Give World-Wide Radio Coverage?" *Wireless World* (October), pp. 305–308.

Coates, Vary and Bernard Finn (1979). *A Retrospective Technology Assessment: Submarine Telegraphy—The Transatlantic Cable of 1866*. San Francisco: San Francisco Press.

Cohen, Jeffrey E. (1992). *The Politics of Telecommunications Regulation: The States and the Divestiture of AT&T*. Armonk, NY: Sharpe.

Cole, Barry G., ed. (1991). *After the Break-up: Assessing the New Post-AT&T Divestiture Era*. New York: Columbia University Press.

Coll, Steve (1986). *The Deal of the Century: The Breakup of AT&T*. New York: Atheneum.

Coon, Horace (1939). *American Tel & Tel: The Story of a Great Monopoly*. New York: Longmans (reprinted by Books for Libraries, 1971).

Crandall, Robert W. (1991). *After the Break-up: U.S. Telecommunications in a More Competitive Era*. Washington, DC: Brookings.

Danielian, N. R. (1939). *A. T.& T.: The Story of Industrial Conquest*. New York: Vanguard (reprinted by Arno Press, 1974).

Danielsen, Albert and David Kamerschen, eds. (1986). *Telecommunications in the Post-Divestiture Era*. Lexington, MA: Lexington Books.

Dilts, Marion May (1941). *The Telephone in a Changing World*. New York: Longmans, Green.

"Diversification and Regulated Industries—What's Next for the Holding Companies?" (1985). *Comm/Ent*. 7:195.

Dordick, Herbert S. (1990). "The Origins of Universal Service: History as a Determinant of Telecommunications Policy." *Telecommunications Policy*, 14: 223–231 (June).

Drake, William J., ed. (1996). *The New Information Infrastructure: Strategies for U.S. Policy*. New York: Twentieth Century Fund.

Drucker, P. F. (1985). "Beyond the Bell Breakup." *The Public Interest*, 77: 3–27 (Fall).

Dugan, James (1953). *The Great Iron Ship*. New York: Harper.

Electronics, editors of (1981). *An Age of Innovation: The World of Electronics 1930–2000*. New York: McGraw-Hill.

Endlich, Lisa (2004). *Optical Illusions: Lucent and the Crash of Telecom*. New York: Simon & Schuster.

Enis, B. M. and T. Sullivan (1985). "The AT&T Settlement: Legal Summary, Economic Analysis and Marketing Implications." *Journal of Marketing*, 49:127–136 (Winter).

Evans, David S., ed. (1983). *Breaking Up Bell: Essays on Industrial Organization and Regulation*. New York: Elsevier/North-Holland.

Fahie, J. J. (1899). *A History of Wireless Telephony, 1838–1899*. Edinburgh: William Blackwood & Sons (reprinted by Arno Press, 1974).

Faulhaber, Gerald R. (1987). *Telecommunications in Turmoil: Technology and Public Policy*. Cambridge, MA: Ballinger.

Ferris, Charles and Frank W. Lloyd, eds. (1983–present). *Cable Television Law: A Video Communications Practice Guide*, later retitled as *Telecommunications Regulation: Cable, Broadcasting, Satellite and the Internet*. Albany, NY (later Newark, NJ): Matthew Bender (updated twice annually).

———— (1996). *Guidebook to the Telecommunications Act of 1996*. Albany, NY: Matthew Bender.

50 Years of Telecommunications Reports (1985). New York: Business Research Publications.

Fortune, editors of (1983). "Breaking up the Phone Company: Special Report." *Fortune* (June 27).

Frieden, Robert M. (1987). "The Third Computer Inquiry: A Deregulatory Dilemma." *Federal Communications Law Journal*, 38: 383–410.

Friedlander, Amy (1995). *Natural Monopoly and Universal Service: Telephones and Telegraphs in the U.S. Communications Infrastructure, 1837–1940*. Reston, VA: Corporation for National Research Initiatives.

Gabel, Richard. (1967). *Development of Separations Principles in the Telephone Industry*. East Lansing: Michigan State University Press.

Gallagher, Dan (1992). "Was AT&T Guilty? A Critique of US v ATT." *Telecommunications Policy*, 16: 317–326 (May/June).

Gannes, S. (1985) "The Judge Who's Reshaping the Phone Business." *Fortune*, (April 1), p. 134.

Garnet, Robert W. (1985). *The Telephone Enterprise: The Evolution of the Bell System's Horizontal Structure, 1876–1909*. Baltimore: Johns Hopkins University Press.

Gershon, Richard A. (1992). "Telephone–Cable Cross-Ownership." *Telecommunications Policy*, 16: 110–121 (March).

Glaeser, Martin G. (1957). *Public Utilities in American Capitalism*. New York: Macmillan.

Goldstein, Fred R. (2005). *The Great Telecom Meltdown*. Norwood, MA: Artech House.

Goodman, Matthew, et al. (1993). "Telephone Company Entry into Cable Television: A Re-Evaluation." *Telecommunications Policy*, 17: 158–162 (March).

Goulden, Joseph C. (1968). *Monopoly*. New York: Putnam.

Green, Celillianne (1984). "The 1982 Consent Decree: Strengthening the Antitrust Procedures and Penalties Act." *Howard Law Journal*, 27: 1611.

Greene, Harold (1985). "AT&T Divestiture and Consumers." *University of Bridgeport Law Review*, 5: 251.

Griboff, Howard (1992). "New Freedom for AT&T in the Competitive Long Distance Market." *Federal Communications Law Journal*, 44: 435–471.

Hahn, Robert W. and John A. Hird (1991). "The Costs and Benefits of Regulation: Review and Synthesis." *The Yale Journal of Regulation*, 8:261–262 (Winter).

Harlow, Alvin F. (1936). *Old Wires and New Waves: The History of the Telegraph, Telephone, and Wireless*. New York: Appleton-Century (reprinted by Arno Press, 1971).

Harris, Robert G. (1990). "Divestiture and Regulatory Policies: Implications for Research, Development and Innovations in the U.S. Telecommunications Industry." *Telecommunications Policy*, 14: 105–124 (April).

Hausman, Jerry A. (1997). "Valuing the Effects of Regulation on New Services in Telecommunications." Brookings Papers on Economic Activity. *Microeconomics*, 1997: 1–38.

Hecht, Jeff (1999). *City of Light: The Story of Fiber Optics*. New York: Oxford University Press.

Henck, Fred W. and Bernard Strassburg (1988). *A Slippery Slope: The Long Road to the Breakup of AT&T*. Westport, CT: Greenwood Press.

Herring, James M. and Gerald C. Gross (1936). *Telecommunications: Economics and Regulation*. New York: McGraw-Hill (reprinted by Arno Press, 1974).

Higgins, Richard S. and Paul H. Rubin, eds. (1995). *Deregulating Telecommunications: The Baby Bells' Case for Competition*. New York: Wiley.

Hillman, Jordon J. (1984–85). "Telecommunications Deregulation: The Martyrdom of the Regulated Monopolist." *Northwestern Law Review*, 79: 1183 (Winter).

Horwitz, Robert Britt (1986). "For Whom the Bell Tolls: Causes and Consequences of the AT&T Divestiture." *Critical Studies in Mass Communication*, 3: 119–154 (June).

———— (1988). *The Irony of Regulatory Reform: The Deregulation of American Telecommunications*. New York: Oxford University Press.

Howe, Keith M. and Eugene F. Rasmussen (1982). *Public Utility Economics and Finance*. Englewood Cliffs, NJ: Prentice Hall.

Hubbard, Geoffrey. (1965). *Cooke and Wheatstone and the Invention of the Electric Telegraph*. London: Routledge & Kegan Paul.

Huber, Peter (1993). *The Geodesic Network II: 1993 Report on Competition in the Telephone Industry*. Washington: Geodesic Company. (For Huber's 1987 report, see under primary sources, U.S. Department of Justice.)

_____ et al. (1996). *The Telecommunications Act of 1996: Special Report*. Boston: Little, Brown.

_____ et al. (1999). *Federal Telecommunications Law*, Second Edition. Gaithersburg, MD: Aspen Law & Business.

Hugil, Peter J. (1999). *Global Communications Since 1844: Geopolitics and Technology*. Baltimore: Johns Hopkins University Press.

Huurdeman, Anton R. (2003). *The Worldwide History of Telecommunications*. New York: Wiley Interscience.

Hyman, Leonard S., et al. (1987). *The New Telecommunications Industry: Evolution and Organization*. Arlington, VA: Public Utilities Reports, two vols.

Irwin, Manley R. (1971). *The Telecommunications Industry: Integration vs Competition*. New York: Praeger.

_____ (1984). *Telecommunications America: Markets Without Boundaries*. Westport, CT: Greenwood Press/Quorum Books.

Israel, Paul (1992). *From Machine Shop to Industrial Laboratory: Telegraphy and the Changing Context of American Invention, 1830–1920*. Baltimore: Johns Hopkins University Press.

Jeter, Lynne W. (2003). *Disconnected: Deceit and Betrayal at WorldCom*. New York: Wiley.

Johnson, Elizabeth (1986). "Telecommunications Market Structure in the USA: The Effects of Deregulation and Divestiture." *Telecommunications Policy*, 10: 57–67 (March).

Johnson, Leland L. (1982). *Competition and Cross-subsidization in the Telephone Industry*. Santa Monica, CA: Rand Corporation.

_____ (1992). *Telephone Company Entry Into Cable Television: Competition, Regulation and Public Policy*. Santa Monica, CA: Rand Corporation.

_____, and David P. Reed (1992). "Telephone Company Entry into Cable Television: An Evaluation." *Telecommunications Policy*, 16: 122–134 (March).

Kahaner, Larry (1986). *On The Line: The Men of MCI—Who Took on AT&T, Risked Everything, and Won!* New York: Warner Books.

Kellog, Michael K., et al. (1992). *Federal Telecommunications Law*. Boston: Little, Brown; *Supplement*, 1995.

Kennedy, Charles H. (1994). *An Introduction to U.S. Telecommunications Law*. Norwood, MA: Artech.

Kieve, J. L. (1973). *The Electric Telegraph in the U.K.: A Social and Economic History*. New York: Barnes & Noble.

Kleinfield, Sonny (1981). *The Biggest Company on Earth: A Profile of AT&T*. New York: Holt, Rinehart & Winston.

Knauer, Leon T., et al. (1996). *Telecommunications Act Handbook: A Complete Reference for Business*. Rockville, MD: Government Institutes.

Kraus, Constantine and Alfred Duerig (1988). *The Rape of Ma Bell: The Criminal Wrecking of the Best Telephone System in The World*. Secaucus, NJ: Lyle Stuart.

Lavey, Warren G. and Carlton, Dennis W. (1983). "Economic Goals and Remedies of the AT&T Modified Final Judgment." *Georgetown Law Journal*, 71: 1497.

Law & Business, editors of (1982). *The AT&T Settlement: Terms, Effects, Prospects*. New York: Harcourt, Brace Jovanovich.

_____ (1982). *The Breakup of AT&T: Opportunities, Prospects, Challenges*. New York: Harcourt, Brace Jovanovich.

Lesler, David Mitchell (1985). "Sweeping up the Divestiture's Debris: Application of the Successor Employer Doctrine to Ma Bell and Her Relatives." *Federal Communications Law Journal*, 37: 455.

Lipartito, Kenneth (1989). *The Bell System and Regional Business: The Telephone in the South 1877–1920*. Baltimore: Johns Hopkins University Press.

Lubar, Steven (1984). *InfoCulture: The Smithsonian Book of Information Age Inventions*. Boston: Houghton Mifflin.

MacAvoy, Paul W. (1996). *The Failure of Antitrust and Regulation to Establish Competition in Long-Distance Telephone Services*, Cambridge, MA and Washington, DC: MIT Press/AEI Press.

_____ and Kenneth Robinson (1983). "Winning by Losing: The AT&T Settlement and Its Impact on Telecommunications." *Yale Journal on Regulation*, 1:1–42.

_____ (1985). "Losing by Judicial Policymaking: The First Year of the AT&T Divestiture." *Yale Journal on Regulation*, 2: 225–262.

MacMeal, Harry B. (1934). *The Story of Independent Telephony*. [no city given]: Independent Pioneer Telephone Association.

Malik, Om (2003). *Broadbandits: Inside the $750 Billion Telecoms Heist*. New York: Wiley.

Martin, Dick (2005). *Tough Calls: AT&T and the Hard Lessons Learned from the Telecom Wars*. New York: AMACOM.

Mathison, Stuart L. and Philip M. Walker (1970). *Computers and Telecommunications: Issues in Public Policy*. Englewood Cliffs, NJ: Prentice Hall.

McMaster, Susan E. (2002). *The Telecommunications Industry*. Westport, CT: Greenwood Press.

Meisel, John B. (1992). "ONA, Unbundling and Competition in Interstate Access Markets." *Telecommunications Policy*, 16: 194–209 (April).

Meyer, John B., et al. (1980). *The Economics of Competition in the Telecommunications Industry*. Boston: Oelgeschlager, Gunn & Hain.

Mitnick, Barry M. (1980). *The Political Economy of Regulation: Creating, Designing, and Removing Regulatory Forms*. New York: Columbia University Press

Mueller, Milton L. (1989). "The Switchboard Problem: Scale, Signaling and Organization in the Era of Manual Telephone Switching, 1878–1898." *Technology and Culture* (July), pp. 558–559.

_____ (1993). "Universal Service in Telephone History." *Telecommunications Policy*, 17: 352–369 (July).

_____ (1993). *Telephone Companies in Paradise: A Case Study in Telecommunications Deregulation*. New Brunswick, NJ: Transaction.

_____ (1997). *Universal Service: Competition, Interconnection, and Monopoly in the Making of the American Telephone System*. Cambridge, MA: MIT Press.

Newburg, Paula R., ed. (1989). *New Directions in Telecommunications Policy*. Chapel Hill, NC: Duke University Press, two vols.

Noam, Eli M., ed. (1983). *Telecommunications Regulation Today and Tomorrow*. New York: Law & Business, Harcourt Brace Jovanovich.

_____ (2001). *Interconnecting the Network of Networks*. Cambridge, MA: MIT Press.

Noll, A. Michael (1987). "Bell System R&D Activities: The Impact of Divestiture." *Telecommunications Policy*, 11: 161–178 (June).

_____ (1991). "The Future of AT&T Bell Labs and Telecommunications Research." *Telecommunications Policy*, 15: 101–105 (April).

_____ (1997). *Highway of Dreams: A Critical View Along the Information Superhighway*. Mahwah, NJ: Lawrence Erlbaum Associates.

Noll, Roger G. and Susan R. Smart (1991). "Pricing of Telephone Services," in Barry G. Cole, ed. *After the Breakup: Assessing the New Post-AT&T Divestiture Era*. New York: Columbia University Press.

Oettinger, Anthony, et al. (1977). *High and Low Politics: Information Resources for the 80s*. Cambridge, MA: Ballinger.

Page, Arthur W. (1941). *The Bell Telephone System*. New York: Harper.

Paglin, Max D., ed. (1989). *A Legislative History of the Communications Act of 1934*. New York: Oxford University Press.

_____ et al., eds. (1999). *The Communications Act: A Legislative History of the Major Amendments, 1934–1996*. Silver Spring, MD: Pike & Fischer.

Paine, Albert Bigelow (1921). *Theodore N. Vail: A Biography*. New York: Harper.

Peters, Geoffrey M. (1985). "Is the Third Time the Charm? A Comparison of the Government's Major Antitrust Settlements with AT&T This Century." *Seton Hall Law Review*, 15:252.

Phillips, Charles F., Jr. (1993). *The Regulation of Public Utilities: Theory and Practice*, Third Edition. Arlington, VA: Public Utilities Reports.

Phineiro, John (1988). "Research Pathfinder: AT&T Divestiture & the Telecommunications Market." *High Technology Law Journal*, 2: 303–355.

Pleasance, Charles A. (1989). *The Spirit of Independent Telephony*. Johnson City, TN: Independent Telephone Books.

Pool, Ithiel de Sola, ed. (1977). *The Social Impact of the Telephone*. Cambridge, MA: MIT Press.

Reid, T. R. (1984). *The Chip: How Two Americans Invented the Microchip and Launched a Revolution*. New York: Simon & Schuster.

Rogers, Paul (1979). *The NARUC Was There: A History*. Washington, DC: National Association of Regulatory Utilities Commissioners.

Rosenberg, Edwin A., et al. (1994). *Regional Telephone Holding Companies: Structures, Affiliate Transactions, and Regulatory Options*. Columbus, OH: National Regulatory Research Institute.

Schlesinger, Leonard A., et al. (1987). *Chronicles of Corporate Change: Management Lessons From AT&T and Its Offspring*. Lexington, MA: Lexington Books.

Shapiro, Peter D. (1974). *Public Policy as a Determinant of Market Structure: The Case of the Specialized Communications Market*. Cambridge, MA: Harvard University Program on Information Technologies and Public Policy.

Sharkey, William W. (1982). *The Theory of Natural Monopoly*. New York: Cambridge University Press.

Shepherd, William G. (1997). *The Economics of Industrial Organization: Analysis, Markets, Policies*, Fourth Edition. Upper Saddle River, NJ: Prentice Hall.

Shooshan, Harry M. III, ed. (1984). *Disconnecting Bell: The Impact of the AT&T Divestiture*. New York: Pergamon.

Simon, Samuel and Michael J. Whelan (1985). *After Divestiture: What the AT&T Settlement Means for Business and Residential Telephone Service*. White Plains, NY: Knowledge Industry Publications.

Smith, George David (1985). *The Anatomy of a Business Strategy: Bell Western Electric and the Origins of the American Telephone Industry*. Baltimore: Johns Hopkins University Press.

Southworth, George C. (1962). "Survey and History of the Progress of the Microwave Art." *Proceedings of the Institute of Radio Engineers*, 50:1206 (May).

Standard & Poor. *Industry Surveys: Telecommunications, Basic Analysis*. New York: Standard & Poor, various annual editions.

Stehman, J. Warren (1925). *Financial History of the American Telephone and Telegraph Company*. Boston: Houghton Mifflin (reprinted by A. M. Kelley, 1967).

Sterling, Christopher H. (1996). "Changing American Telecommunications Law: Assessing the 1996 Amendments." *Telecommunications and Space Journal*, 3: 141–165. Paris: SERDI/Mahwah, NJ: Lawrence Erlbaum Associates.

_____ (2006). "Pioneering Risk: Learning from the U.S. Teletext/Videotex Failure." *IEEE Annals of the History of Computing*, in press.

_____, Jill F. Kasle, and Katherine Glakas, eds. (1986). *Decision to Divest: Major Documents in U.S. v. AT&T, 1974–1984*. Washington, DC: Communications Press, three vols.

_____ and Jill F. Kasle, eds. (1988). *Decision to Divest: The First Review, 1985–1987*. Washington, DC: Communications Press.

_____ and John Michael Kittross (2002). *Stay Tuned: A History of American Broadcasting*, Third Edition. Mahwah, NJ: Lawrence Erlbaum Associates.

Stigler, George J. (1971). "The Theory of Economic Regulation." *The Bell Journal of Regulation and Management Science*, 2:3–21 (Spring).

Stirba, Dick, ed. (1986). *Computer III—A New Era: A Guide to the FCC's Landmark Deregulatory Ruling*. Alexandria, VA: Telecom.

Stone, Alan (1989). *Wrong Number: The Breakup of AT&T*. New York: Basic Books.

_____ (1991). *Public Service Liberalism: Telecommunications and Transitions in Public Policy*. Princeton, NJ: Princeton University Press.

Telecommunications Law Reform (1980). Washington, DC: American Enterprise Institute.

Temin, Peter and Louis Galambos (1987). *The End of the Bell System: A Study in Prices and Politics*. New York: Cambridge University Press.

Teske, Paul Eric (1990). *After Divestiture: The Political Economy of State Telecommunications Regulation*. Albany: SUNY Press, 1990.

_____ ed. (1995). *American Regulatory Federalism and Telecommunications Infrastructure*. Hillsdale, NJ: Lawrence Erlbaum Associates.

Thompson, Robert Luther (1947). *Wiring a Continent: The History of the Telegraph Industry in the United States, 1832–1866*. Princeton, NJ: Princeton University Press.

Thorne, John, et al. (1995). *Federal Broadband Law*. Boston: Little, Brown.

Train, Kenneth E. (1991). *Optimal Regulation: The Economic Theory of Natural Monopoly*. Cambridge, MA: MIT Press.

Trebing, Harry M. (1981). "Equity, Efficiency, and the Viability of Public Utility Regulation," in Warner Sichel and Thomas G. Geis, eds. *Applications of Economic Principles in Public Utility Industries*. Ann Arbor: University of Michigan Press.

Tunstall, Jeremy (1986). *Communications Deregulation: The Unleashing of America's Communications Industry*. New York: Basil Blackwell.

Tunstall, W. Brooke (1985). *Disconnecting Parties—Managing The Bell System Break-up: An Inside View*. New York: McGraw-Hill.

Turmoil in Telecommunications: 8 Months That Shook the Industry (1983). Washington, DC: Television Digest.

"Vertical Integrations: Should the AT&T Doctrine Be Extended?" (1983). *Antitrust Law Journal*, 52: 239.

Vogelman, Joseph H. (1962). "Microwave Communications." *Proceedings of the Institute of Radio Engineers*, 50:907 (May).

von Auw, Alvin (1983). *Heritage and Destiny: Reflections on the Bell System in Transition*. New York: Praeger.

Wasserman, Neil H. (1985). *From Invention to Innovation: Long-Distance Telephone Transmission at the Turn of the Century*. Baltimore: Johns Hopkins University Press.

Waterson, Michael (1988). *Regulation of the Firm and Natural Monopoly*. New York: Basil Blackwell.

Weinhaus, Carol L. and Anthony G. Oettinger (1988). *Behind the Telephone Debates*. Norwood, NJ: Ablex.

Whalen, David J. (2002). *The Origins of Satellite Communications, 1945–1965*. Washington, DC: Smithsonian Institution Press.

Wilson, Geoffrey (1976). *The Old Telegraphs*. London: Phillimore.

Wilson, Kevin G. (2000). *Deregulating Telecommunications: U.S. and Canadian Telecommunications, 1840–1997.* Lanham, MD: Roman & Littlefield.

Winston, Brian. (1986). *Misunderstanding Media.* Cambridge, MA: Harvard University Press. (Revised as *Media, Technology and Society: A History from the Telegraph to the Internet.* London: Routledge, 1998).

Winston, Clifford (1993). "Economic Deregulation: Days of Reckoning for Microeconomists." *Journal of Economics Literature,* 31: 1263–1289 (September).

Zielinski, Charles A. and Gilbert E. Hardy (1983). *State Regulation of the Bell System After the Modified Final Judgment: Some Questions and Suggested Answers.* Columbus, OH: The National Regulatory Research Institute (Occasional Paper No. 7).

Author Index

A

Adams, S. B., 63
Aitken, H. G. J., 84
Amos, E. J., 83
Archer, G. L., 84, 114
Atlas, R. D., 356
Augarten, S., 141
Averch, H., 33

B

Banning, W. P., 114
Baumol, W. J., 61
Bell, A. G., 50
Berman, D. K., 354
Bernstein, J., 35
Bernt, P., 226
Blizinsky, M., 224
Blondheim, M., 61
Boettinger, H. M., 114
Bonbright, J. C., 34, 35, 115, 177
Braun, E., 114, 115
Bray, J., 113, 142
Brock, G. W., 59, 60, 62, 84, 85, 116,
    142, 143, 144, 224, 225, 226, 357
Brooks, J., 62, 83, 84, 117
Brown, C. L., 61
Butler, O. R., 63

C

Cantelon, P. L., 143, 144, 172
Casey, 256, 303
Ceruzzi, P. E., 142
Chapuis, R. J., 114, 142
Charny, B., 255
Clarke, A. C., 121
Coates, V., 61
Cole, B. G., 177
Coll, S., 144, 173
Cowhey, P., 227
Coy, P., 357
Crockett, R. O., 355

D

Danielsen, A. L., 34, 35, 115, 177
Darby, L. F., 355, 357
Dreazen, Y. J., 354, 355
Dugan, J., 61
Dunning, J., 114

E

Eisner, J., 223
Elstrom, P., 310, 356, 357, 358

411

F

Fagan, M. D., 83
Fahie, J. J., 83
Feder, B. J., 357, 358
Ferris, C., 256, 303
Fink, D. J., 59
Finn, B., 61
Fredebeul-Krein, M., 227
Freytag, A., 227
Friedlander, A., 62, 83, 84

G

Gabel, R., 85, 115, 116
Garcia-Murillo, M. A., 358
Garnet, R. W., 84
Glaeser, M. G., 34
Glakas, K., 117, 173, 174, 175, 176
Gordon, 256
Goulden, J. C., 83, 84, 85, 83, 84, 85
Green, H., 355
Greene, H., 174, 175, 180, 181, 182,
    184, 185, 211, 221, 350
Guyon, J., 256

H

Haddad, C., 356
Hahn, R. W., 34
Hammond, B., 305
Hansell, S., 256
Hausman, J. A., 34
Henck, F. W., 116, 142, 143
Higgins, R. S., 220
Hird, J. A., 34
Horwitz, R. B., 34, 35, 60, 61, 84, 85,
    114, 116
Howe, K. M., 33, 34, 35
Hubbard, G., 50, 59
Huber, P. W., 34, 35, 182, 183, 222,
    225, 226
Hudson, K., 356
Hugil, P. J., 84

I, J

Israel, P., 60
Jaffe, G., 355

Joel, A. E., 114
Johnson, L. L., 33

K

Kadlec, D., 312
Kahaner, L., 143
Kamerschen, D. R., 34, 35, 115, 177
Kasle, J. F., 117, 173, 174, 175, 176,
    221, 222, 223, 303
Kellog, M. K., 34, 35, 225
Kieve, J. L., 59
Kittross, J. M., 114, 116
Klimenko, M., 227

L

Labaton, S., 256, 357, 358
Leahy, M., 357
Lee, J., 357
Lloyd, F. W., 256, 303
Lohr, S., 356
Lowry, 357, 358
Lubar, 142

M

MacAvoy, P. W., 223
Macdonald, S., 114, 115
MacInnes, I., 358
Markoff, J., 355
McConnell, B., 256
McGeehan, P., 355
Mehta, S. N., 354, 355, 356
Mills, M., 256
Mitnick, B. M., 34
Morgenson, G., 356
Mosk, M., 354
Mueller, M. L., 83, 84, 85, 116, 224

N

Neustadt, R. M., 257
Noam, E. M., 227
Noguchi, Y., 355, 356, 357
Noll, A. M., 223, 358
Noll, R. G., 177

Norris, F., 358

O, P

Oettinger, A. G., 116, 224
Paine, A. B., 84
Palmeri, C., 356
Parsons, S. G., 306
Passer, H. C., 59
Pearlstein, S., 354, 355, 356
Phillips, C. F., Jr., 33, 34, 35, 115

R

Rangos, K., 223
Rasmussen, E. F., 33, 34, 35
Reid, T. R., 142
Rey, R. F., 255
Rider, J. D., 59
Rohde, L., 358
Romero, S., 354, 355, 356
Rose, F., 355
Rosenbush, S., 355, 356, 357, 358
Rubin, P. H., 220

S

Schement, J. R., 224
Schiesel, S., 256, 356, 357, 358
Seichenwald, K., 356
Shapiro, P. D., 144
Sharkey, W. W., 115
Shepherd, W. G., 33
Shields, T., 312
Shiers, 398
Sidak, G. J., 61
Sigel, E., 256
Singhal, A., 227
Sloan, A., 312
Smart, S. R., 177
Smith, G. D., 61, 62

Smith, J., 358
Sorkin, A. R., 358
Southworth, G. C., 114
Stehman, J. W., 84, 85
Sterling, C. H., 114, 117, 173, 174, 175, 176, 221, 222, 223, 256, 257, 302, 303, 398
Stigler, G. J., 34
Stone, A., 116
Strassburg, B., 116, 142, 143

T

Temin, P., 143, 144, 173, 175
Thompson, R. L., 60, 61, 62
Thorne, J., 34, 35, 225, 256
Timmons, H., 355
Train, K. E., 115
Trebing, H. M., 34, 35

V, W

Vogelman, J. H., 114
von Auw, A., 144, 393
Walsh, M. W., 356
Wasserman, N. H., 83
Waterson, M., 115
Weinhaus, C. L., 116, 224
Weiss, M., 226
Whalen, D. J., 142
Wilson, G., 59, 225
Winston, B., 142
Winston, C., 34
Witte, G., 358
Woolliscroft, D. J., 59
Wymbs, C., 227

Y, Z

Yang, C., 355
Zielinski, C. A.,
Zolnierek, J., 223

Subject Index

A

Antitrust Suit (1913), 79
Antitrust Suit (1949), 108–112
AT&T, 5, 17, 19, 57–59, 67–82, 86,
    88–91, 106–113, 118, 122–141,
    145–169, 171, 178–193, 196–199,
    201–204, 206–209, 219, 228–229,
    233–234, 240–241, 244–246,
    252–254, 259–261, 263–264,
    266–268, 286, 288–290, 294, 296,
    301, 316–317, 320, 324–328,
    330–331, 334–335, 339–341,
    343–344, 346–348, 350, 352–353
    ENFIA tariff, 153–154
    Kingsbury commitment, 80–81
    NARUC speech, 137–138, 211
    seven-way cost study, 130–131
    T1 carrier system, 228
    Vail, 72
AT&T Corp v. Iowa Utilities Board, 296
Automatic switching, 89–90

B

Bandwidth, 3, 8–10
Bankruptcy, 327–328
Bell Labs, 188, 232, 349

Bell system, 1, 19, 21, 88–89, 100,
    102–103, 107–108, 109, 112, 123,
    126–127, 130, 145, 147, 149–151,
    158–162, 165–168, 185, 191,
    194–195, 199, 212, 241, 243, 266,
    287–288, see also RBOCs
Bell-Atlantic/TCI merger, 243–244, 289
    lawsuits, 147–167
    PCAs, 126–127
Bell telephone company, 52–58, 64–65, 68,
    70–72, 74–76, 78, 81, 337, 345
Broadband, 318–319

C

Cable television (CATV), 241–243
Caller ID, 230
Carterfone, 125–126
CCLC, 274–275
Cellular systems, 232–234
    roaming agreements, 233
Circuit switching, 10
Clayton Antitrust Act, 155, 343
Coalition for Affordable Local and Long
    Distance Service (CALLS), 275
Coaxial cable, 87, 123–124
Communication Act of 1934, The, 21,
    30–32, 103, 105–106, 145–146,
    197, 202, 207, 234–235, 242, 252,
    258–260, 263–264

Communication satellites, 121–123
Competition, 15–18, 64–65, 73, 264–268, 281
  interconnection, 64, 78
  Sherman Act of 1890, 77
  Willis-Graham Act, 64, 82–83, 93
Competitive local exchange carriers (CLECs), 6, 19, 264–265, 267, 269, 272, 275–276, 279–282, 285–287, 294, 296, 298–300, 302, 331–333, 335, 340, 352–353
  and ILECs, 279–285, 294–301, 352–353
Computers, 120–121
Consent decree (1956), *see* Final Judgment
Consumer Communications Reform Act of 1976 (CCRA), 145
Cost proxy model, 276–277
Crossbar switches, 90
Customer premise equipment (CPE), 5, 12, 100–101, 103, 206–207, 212

D

Datran, 138–139
D.C. Commission, 2
Department of Justice (DOJ), 210–211, 287, 290–291, 293
Deregulation, 212–214, 340–342
Digital transmission, 7–9, 228–231
  codec device, 8
  regenerative repeaters, 228
Dole Bill (1986), 260–261
Domsats, 135
Dot.coms, 329–330

E

Economics, 12–18, 47–48, 93
  cost-benefit analysis, 18
  diminishing returns, 14
  marginal cost, 14, 16, 18
  perfect competition, 15–18
  pressure, 320–322
  supply and demand, 12–20
Electrical industry, 37–40, 43–45, 47
  government, 44–45
  patents, 38, 47, 51, 53, 55–56, 70–71

telegraph, 38–41
Eligible Telecommunications Carrier (ETC), 272–273, 275, 278

F

Federal Communications Commission (FCC), 2, 6, 10, 18–19, 26, 30–33, 88–89, 99, 101–103, 106, 124–135, 138–140, 145–146, 149, 152–154, 168–172, 178, 183–184, 189, 194, 196–197, 199, 201–212, 214–218, 232–245, 251–254, 258–261, 263–270, 272–273, 275–287, 291, 294–298, 300–302, 317, 340–344
  computer inquiries, 204–209
  local competition order (1996), 281
  spectrum lotteries, 235–237
  telephone investigation, 106–109
  TELRIC, 283–284, 295
Fiber links, 314–316
Fiber optic cable, 3–4, 8
Final Judgment (1956), 112–113, 158, *see also* Modified final judgment
Foreign Exchange (FX) service, 139–140, 147
Fraud, 325–327
Frequency division multiplexing (FDM), 8–10, 46, 87

G

General Agreement on Tariffs and Trade (GATT), 219
General Agreement on Trade in Services (GATS), 220
Global One, 219
GSM technology, 233–234, 336

H

Harmonic telegraph, 50–51
High-capacity channels, 8–10
Human speech, 2–3
Hush-a-Phone, 124–125
Hybrid, 3

I

ILECs, 279–285, 294–301, 337, 352–353
Interexchange carriers (IXCs), 4–6, 19,
        100–101, 139, 190, 192, 194–195,
        197, 199, 200, 211
    AT&T, 5, 17, 19, 57–59, 67–82, 86,
        88–91, 106–113, 118, 122–141,
        178–193, 196–199, 201–204,
        206–209, 219
Interference, *see* Noise
International Standards Organization
        (ISO), 11
International Telecommunication Union
        (ITU), 2, 11
Internet, 246–249, *see also* Modems
    ARPANET, 248–249
Internet Service Providers (ISPs), 11–12
Interstate Commerce Commission (ICC),
        30–31, 78–79, 101

J

Jurisdictional separations, 99–103
    Smith v. Illinois Bell, 101, 103

L

LATA, 159–160, 168–170, 189–191, 195,
        209, 264, 275, 286–288
Layoffs, 323–325
Loading coils, 67
Local area networks (LANs), 248
Local exchange carriers (LECs), 4–7, 21,
        202–204
    central office (CO), 5–7
    tandem, 5
    Verizon, 5
Local loop, 6–7, 100–101, 103, 194
Local switch, 194
Long-distance discount service (LDDS),
        293
Lucent technologies, 187–188, 330–331,
        349

M

Mann-Elkins Act, 78–79
MCI, 131–136, 139–141, 147–149,
        151–153, 155–157, 163, 189–190,

        192, 218, 229, 259, 267, 286,
        288, 292–294, 301, 327, 332, 334,
        345, 348, 352–353
Mergers, 243–244, 288–294
Microwave frequencies
    above 890, 127–129
Microwave technology, 87–89
Modems, 3, 10, 249–253, 318
Modified Final Judgment (MFJ), 158–159,
        178–183, 185, 188–191, 196, 201,
        204, 210–211, 260, 264, 268, 300,
        352, 354
Morse, 38–44
Multiplexing, 8–10, 46, 86–87
Munn v. Illinois, 64

N

NARUC, 137–138, 211
National Exchange Carrier Association
        (NECA), 194, 200–201
Natural monopoly, 93–95
NECA, 273
Noise, 4
NYNEX, 164–167

O

Other common carriers (OCCs), 154, 190,
        192, 196
OECD, 218
Open network architecture (ONA),
        207–209
Overcapacity, 314

P

Packet switching, 10–11
Post Telephone and Telegraph system
        (PTT), 213–217
Private branch exchange (PBX), 196
Public Utility Commissions (PUCs), 2, 22

R

Radio Act of 1912, 104
Rate-of-Return regulation (RoR), 95–100,
        201

RBOCs, 161–167, 178–185, 188,
190–194, 197, 199–202, 204,
207–211, 218, 229–230, 242–243,
248, 260, 264–268, 279, 286–288,
290–292, 295–296, 299–302, 318,
325–326, 332–337, 339–342, 346,
348, 350, 353
Regulation, 23–33, 77–83, 104–107, 204,
279–282
Communication Act of 1934, 21,
30–32, 103, 105–106, 197, 202,
207
concepts, 360–372
depreciation, 169–170
deregulation, 212–214, 340–342
German alliance v. Lewis, 25
in wartime, 81
Mann-Elkins Act, 78–79
monopoly, 93–95
Munn v. Illinois, 24–25
Nebbia v. New York, 25–26
pricing, 372–375
RoR, 95–100, 201
theories, 27–30

S

SCC, 148, 152–153
Separations, see Jurisdictional separations
Sherman Act, 149, 343
Signaling, 229–230
Signal-to-noise ratio, 4, 7
SLC, 274–275
Solid state electronics, 119–120
Specialized common carriers, 134–135,
140
Spectrum, 235–240
3G spectrum, 239–240, 342–343
Splawn report, 104
Sprint, 229, 316
Stock decline, 322–323
Supply and demand, 12–20
market structure, 13–18
Switching technology, 10–11, 65–67,
89–90, 100–101, 103, 194, 229,
234

T

Technology, 2–10, 37
electrical industry, 37–40, 43–45, 47
Telecommunication Act of 1996, 5,
31–32, 204, 208–209, 243, 258,
263–265, 267–302, 339, 344, 346
Telecommunications
economics, 12–18, 47–48, 93
meltdown, 313–314
policy, 20–32
technology, 2–10
Telegraph, 36, 38–41, 50–51
Bell's device, 50–51
impact, 48–49
licensing, 40–41
submarines, 49–50
Western Union, 42–43, 45–49, 55–56,
64, 74, 77, 87, 106, 133–134, 136
wireless, 68–69
Telephone industry, 56–57
Telephone numbers, 54
Telephone systems, 4
Telephony, 2, 36, 50
long distance, 68
voice, 50–51
TELPAK, 129–130
Time division multiplexing (TDM), 8–10
Transducer, 2
Transistors, 91–92
Transmission medium, 2–4, 86–87
attenuation, 3
digital transmission, 7–9
Transport facilities, 194
Twisted pair copper, 3–4, 7, 87

U, V

Unbundled Network Elements (UNEs),
280–282, 285–286, 298–299
Verizon, 266, 287–288, 316, 334, 336
Voice over Internet Protocol (VoIP),
11–12, 253–255, 300, 338,
353–354

W

Walker report, 107–108

Western Electric, 57, 76–77, 107–108,
       111–113, 147, 149–151, 186
   United States v. Western Electric,
       109–110
Western Union, 42–43, 45–49, 55–56,
       64, 74, 77, 87, 106, 133–134, 136
WiFi, 316–318, 331

Willis-Graham Act, 64, 82–83, 93
Wireless networks, 316–318
Wireless telephony, 231–232
Worldcom, 292–294, 301, 313, 319, 321,
       325, 341, 346, 349
World Trade Organization (WTO),
       219–220